GENERAL ZOOLOGY
LABORATORY MANUAL

Seventh Edition

Stephen A. Miller

College of the Ozarks

McGraw-Hill
Irwin

The McGraw-Hill Companies

McGraw-Hill
Irwin

GENERAL ZOOLOGY LABORATORY MANUAL, SEVENTH EDITION

Published by McGraw-Hill, a business unit of The McGraw-Hill Companies, Inc., 1221 Avenue
of the Americas, New York, NY 10020. Copyright © 2013 by The McGraw-Hill Companies, Inc.
All rights reserved. Printed in the United States of America. Previous editions © 2002, 1999, and
1994. No part of this publication may be reproduced or distributed in any form or by any means,
or stored in a database or retrieval system, without the prior written consent of The McGraw-Hill
Companies, Inc., including, but not limited to, in any network or other electronic storage or transmission, or broadcast for distance learning.

Some ancillaries, including electronic and print components, may not be available to customers
outside the United States.

This book is printed on acid-free paper.

1 2 3 4 5 6 7 8 9 0 QDB/QDB 0 9 8 7 6 5 4 3 2

ISBN 978–0–07–747929–9
MHID 0–07–747929–7

Vice President and Editor-in-Chief: *Marty Lange*
Publisher: *Michael Hackett*
Sponsoring Editor: *Rebecca Olson*
Director of Development: *Liz Sievers*
Marketing Manager: *Patrick Reidy*
Senior Project Manager: *Lisa A. Bruflodt*
Buyer: *Nicole Baumgartner*
Compositor: *S4Carlisle Publishing Services*
Typeface: *10/12 Goudy*
Printer: *Quad/Graphics*

All credits appearing on page or at the end of the book are considered to be
an extension of the copyright page.

Contents

UNIT I

Introductory Concepts and Skills

Exercise 1

The Microscope and Scientific Inquiry   3

Exercise 2

Cells and Tissues   21

Exercise 3

Aspects of Cell Function   41

Exercise 4

Genetics   61

Exercise 5

Adaptations of Stream Invertebrates— A Scavenger Hunt   77

UNIT II

An Evolutionary Approach to the Animal Phyla

Exercise 6

The Classification of Organisms   99

Exercise 7

Animal-like Protists   111

Exercise 8

Porifera   123

Exercise 9

Cnidaria   131

Exercise 10

Platyhelminthes   143

Exercise 11

Mollusca   155

UNIT III

An Evolutionary Approach to Vertebrate Structure and Function

Laboratory Times

Preface

This *General Zoology Laboratory Manual* is intended for students taking their first course in zoology. Exercises are provided that will help students: (1) understand the general principles that unite zoology with other fields of biology, (2) appreciate the diversity found in the animal kingdom and understand the evolutionary relationships that explain this diversity, (3) become familiar with the structure and function of vertebrate organ systems and appreciate some of the evolutionary changes that took place in the development of those organ systems, and (4) develop problem-solving skills.

The seventh edition of *General Zoology Laboratory Manual* has received significant revisions. It has been carefully edited for clarity and conciseness, and changes have been made to make it more interactive. It has also received major content revisions, especially in regard to pedagogy and animal taxonomy.

Content revisions will be obvious in all three units of the laboratory manual. The exercises in unit I have been updated. The coverage of histology in exercise 2 has been revised to make it more manageable. There is less detail presented on epithelial and connective tissue subtypes. Muscular and nervous tissues are introduced in more detail here than in the previous edition, rather than delaying their study until unit III. The cellular respiration investigation in exercise 3 has been revised to make it easier to complete and assess. Most of the content revisions in this edition occur in unit II. A phylogenetic tree has been added to exercise opener pages to remind students of the phylogenetic position of the phylum being studied. The reordering of exercises in unit II reflects this phylogeny. The "Evolutionary Perspective" has been updated for each phylum, and an updated interactive cladogram is included in most of the worksheets in unit II. Taxonomy has been updated within the protozoa (exercise 7), the Annelida (exercise 12), the pseudocoelomates (exercise 13), the Arthropoda (exercise 14), the Echinodermata (exercise 15) and the Chordata (exercise 16). The content of unit III has undergone minor revisions. Some of the physiology present in the sixth edition has been eliminated because of concerns regarding the use of live vertebrates in the laboratory. Other physiological

experiences have been retained and improved. The art program has been improved throughout the laboratory manual with some pieces being revised and others being replaced. One major improvement is the moving of the photographic plates portraying dissections from the pages between units II and III into the exercises where the photographs are used. Students and instructors will find this new placement a significant improvement to the user-friendliness of this edition.

Pedagogy of the seventh edition retains many of the popular features of the sixth edition including the "Prelaboratory Quiz," boldfaced type for important terms, and icons that highlight where sets of instructions begin (▼) and end (▲). The worksheets that can be removed from the manual and handed in for grading have been retained and improved. Many of the student responses to lab manual queries have been moved from the text of the manual to worksheets at the end of each exercise. The intent of this change is to make it easier for instructors to assess student learning through the completed worksheets. Other pedagogical elements are new to the seventh edition. "Learning Objectives" have been dropped in favor of numbering major headings within the text and including "Learning Outcomes" for each numbered heading. At the end of the observations covered by a set of learning objectives, students are directed to a worksheet that has review questions and responses that correspond to the material just studied. These questions take two forms. A set of multiple choice questions are entitled "Learning Outcomes Review." A second set of questions are entitled "Analytical Thinking." These questions are longer-answer questions that engage students in higher order thinking. These worksheets can be used as deemed appropriate by each instructor, but they are designed to encourage students to assess their understanding as they proceed through the exercise. Some exercises, for example exercises 1 (The Microscope and Scientific Inquiry), 3, (Aspects of Cell Function), and 17 (Vertebrate Musculoskeletal Systems), have two or three worksheets. These additional worksheets provide tables where students record data or observations and respond to further queries presented during the exercise. All worksheets are in a form that can be removed and submitted for grading.

ORGANIZATION

Unit I contains exercises that focus on general biological principles, from cell structure and function to ecological and evolutionary principles. Exercise 1 includes an introduction to microscopes and the scientific method. Problem-solving activities have been incorporated into exercises 1–5.

Unit II is a survey of the animal phyla. Exercise 6 is an introduction to taxonomy and cladistics. A brief "evolutionary perspective" begins other exercises in this section and describes the position of the phylum in question within the animal kingdom. Each exercise includes coverage of class representatives, and some activities that illustrate processes unique to the particular animal group. Most of these exercises conclude with a discussion of evolutionary relationships within the phylum studied and an activity involving cladistic analysis of these relationships.

Unit III is a systematic approach to the study of vertebrate structure, function, and evolution. The evolutionary orientation of unit III helps the student understand the changes that occurred in the evolution of vertebrate organ systems. Many activities ask the student to think critically about the adaptive significance of what is being observed. The shark is used in demonstration dissections, and the rat is the primary specimen for student dissection. Other vertebrate representatives are used to illustrate adaptations in other vertebrate classes.

This manual includes more material than can be covered in a one-semester course, which allows some flexibility in course planning based on the preferences of individual instructors. For example, some instructors have a very brief survey of the animal phyla in their labs and focus on units I and III. These instructors select a few activities in each of the exercises in unit II to survey the animal phyla in a few laboratory periods. Other instructors may omit most of unit I if the students had general biological principles in another course. Instructors who spend the bulk of the semester in a survey of the animal phyla can survey the animal phyla in unit II. This survey can be followed by picking and choosing portions of unit III, focusing perhaps on mammalian structure and function and the rat dissections.

An Instructor's Resource Guide is available for all adopters of the seventh edition. It has been placed on the McGraw-Hill website and is available at www.mhhe.com/zoology (click on the cover of Miller and Harley's *Zoology* text). This guide lists all materials required for preparing each laboratory exercise. It also gives formulas for solutions and media, and includes suggestions for incorporating living material into the laboratory. Answers to the "Prelaboratory Quiz," "Learning Outcomes Review" questions, and "Analytical Thinking" questions are also included in the resource guide.

ACKNOWLEDGMENTS

I would like to thank those individuals who helped to make this laboratory manual possible. Students in my general zoology laboratories have inspired my teaching and have been kind critics of the content of this manual. They have been patient as guinea pigs. They were the first to experience many of the laboratory activities included in this manual. They also helped weed out some activities that will never appear in print. Many comments from users of past editions were most helpful in preparation of each new edition. The comments and suggestions of the following reviewers were most helpful:

Barbara J. Abraham   *Hampton University*
Sarah Cooper   *Beaver College*
Donna Charley-Johnson,   *University of Wisconsin–Oshkosh*
Tom Dale   *Kirtland Community College*
Inez Devlin-Kelly   *Bakersfield College*
Peter K. Ducey   *State University of New York at Cortland*
DuWayne Englert   *Southern Illinois University*
Dennis Englin   *The Masters College*
Nels H. Granholm   *South Dakota State University*
W. E. Hamilton   *Penn State University*
Michael Hartmann   *Haywood Community College*
Kenneth D. Hoover   *Jacksonville University*
Barbara Hunnicutt   *Seminole Community College*
Karen E. McCracken   *Defiance College*
Thomas C. Moon   *California University of Pennsylvania*
John H. Roese   *Lake Superior State University*
Richard G. Rose   *West Valley College*
Fred H. Schindler   *Indian Hills Community College*

The writing of this laboratory manual would have been impossible without the help of the McGraw-Hill editors and their staff. The advice of Rebecca Olsen, Sponsoring Editor, has proven extremely valuable. The Developmental Editor for the seventh edition of *General Zoology Laboratory Manual* was Wendy Langerud. Her ability to keep the manuscript moving through all phases of development is greatly appreciated. Also, thanks to Lisa Bruflodt, Project Manager. I am also grateful for the support of my wife, Carol, during this revision and for her understanding when this project encroached upon family time. Carol, you are the light of my life and, with or without zoology, life would be incomplete without you.

Stephen A. Miller, Ph.D.

Getting Started

The following section contains terms and concepts that will be used throughout this laboratory manual. Although most of these terms will be redefined within the exercises that follow, it is helpful to have some knowledge of them before beginning. The metric system is used throughout this laboratory manual. You must be familiar with metric units and know how to carry out conversions between larger and smaller units of measurement. The information under the following Terms of Direction, Symmetry, and Body Planes and Sections heading is used frequently in units II and III for locating the parts of an animal and indicating how an animal or a structure is to be cut in a dissection. This information is located at the beginning of the laboratory manual for convenient, later reference.

The Metric System

Scientific measurements are usually carried out using the metric system. Metric units are used exclusively in this laboratory manual. The main advantage of the metric system is the convenience that results from its decimal nature. With the exception of units of time, units for measuring a given quantity are derived from a basic unit by multiplying or dividing by multiples of 10. The fundamental units of the metric system are

liter, a unit of capacity;
meter, a unit of length;
gram, a unit of mass;
second, a unit of time.

Units larger or smaller than these are formed by adding prefixes such as kilo (1,000), centi (1/100), and milli (1/1,000). Thus, kilometer = 1,000 meters, centimeter = 1/100 of a meter, and millimeter = 1/1,000 of a meter.

Other units used in measuring microscopic distances are the following:

The micrometer (μm) = 1/1,000,000 of a meter or 1×10^{-6} m.
The nanometer (μm) = 1/1,000,000,000 of a meter or 1×10^{-9} m.
The angstrom (Å) = 1/10,000,000,000 of a meter or 1×10^{-10} m.

Some commonly used units of conversion between the metric and English systems are as follows:

English	Metric
1 mile	1.6 kilometers
1 inch	2.54 centimeters
0.39 inch	1 centimeter
1 ounce	28 grams
0.035 ounce	1 gram
1 fluid ounce	30 milliliters
0.033 fluid ounce	1 milliliter
1 pound	0.45 kilogram
2.2 pounds	1 kilogram

Microscope and Seat Assignment

Seat number _____

Microscope number _____

Microscope calibrations using an ocular micrometer (from exercise 1):

Low power (10× objective). 1 ocular unit = _____ μm.
High power (40× objective). 1 ocular unit = _____ μm.

Estimation of the diameter of the field of view (from exercise 1):

Low power (10× objective). Diameter = approximately _____ μm.
High power (40× objective). Diameter = approximately _____ μm.

Terms of Direction, Symmetry, and Body Planes and Sections

Terms of Direction

Terms of direction are used for locating body parts relative to a point of reference.

Aboral   The end of a radially symmetrical animal opposite the mouth. Opposite—oral.

Anterior   The head end. Usually the end of a bilateral animal that meets its environment. Opposite—posterior.

Caudal   Toward the tail. Opposite—cephalic.

Cephalic   Toward the head. Opposite—caudal.

Distal   Away from the point of attachment of a structure on the body. (e.g., The toes are distal to the knee.) Opposite—proximal.

Dorsal   The back of an animal. Usually the upper surface. For animals that walk upright, dorsal and posterior are synonymous. Opposite—ventral.

Inferior   Below a point of reference. Opposite—superior.

Lateral   Away from the midsagittal plane of the body. Opposite—medial.

Medial (median)   On or near the plane of the body that divides a bilaterally symmetrical animal into mirror images. Opposite—lateral.

Oral   The end of a radially symmetrical animal containing the mouth. Opposite—aboral.

Posterior   The tail end. Opposite—anterior.

Proximal   Toward the point of attachment of a structure on the body. (e.g., The hip is proximal to the knee.) Opposite—distal.

Superior   Above a point of reference. Opposite—inferior.

Ventral   The belly of an animal, usually the lower surface. For animals that walk upright, ventral and anterior are synonymous. Opposite—dorsal.

Symmetry

Symmetry describes how the parts of an organism are arranged around an axis or a point (fig. .001). The following three categories are the most common:

Asymmetry   The arrangement of body parts without a central axis or point (e.g., the sponges).

Bilateral symmetry   The arrangement of body parts such that a single plane passing dorsoventrally through the longitudinal axis divides the animal into right and left mirror images (e.g., vertebrates).

Radial symmetry   The arrangement of body parts such that any plane passing through the oral–aboral axis divides the animal into mirror images (e.g., the cnidarians). Radial symmetry can be modified by the arrangement of some structures in pairs, or in other combinations, around the central axis (e.g., biradial symmetry in the ctenophorans and pentaradial symmetry in the echinoderms).

Body Planes and Sections

References to body planes are used to indicate the position of some structure relative to an imaginary plane passing through the body. References to body, organ, and tissue sections indicate how something is cut to observe internal structures.

Cross section   A section cut perpendicular to the longitudinal axis of a structure.

Frontal   A plane or section perpendicular to both sagittal and transverse planes. Divides an animal into dorsal and ventral regions.

Longitudinal section   A section cut parallel to the longitudinal axis of a structure.

Median   A plane or section passing through the longitudinal axis of a bilateral animal. Divides the animal into halves that are mirror images of each other.

Sagittal   A plane or section passing dorsoventrally through the body of a bilateral animal parallel to the longitudinal axis.

Transverse   A plane or section perpendicular to both sagittal and frontal planes. Divides an animal into anterior and posterior regions.

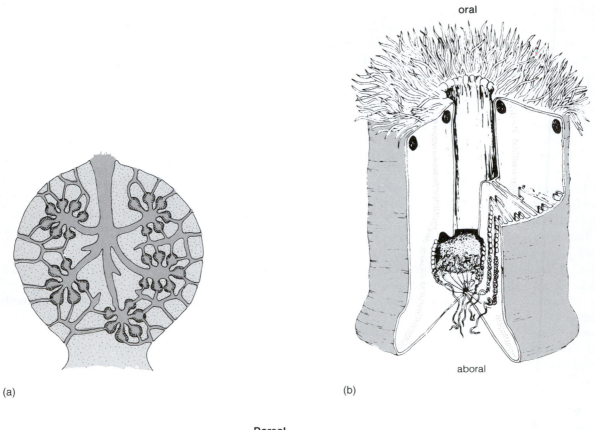

(a)

oral

aboral

(b)

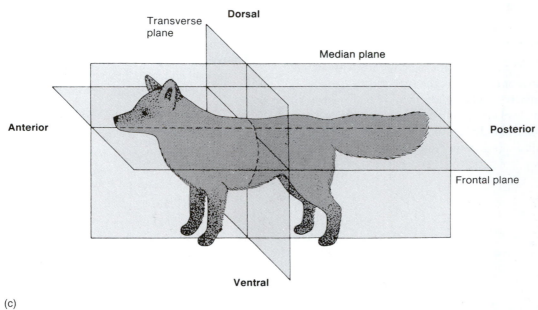

Transverse plane

Dorsal

Median plane

Anterior

Posterior

Frontal plane

Ventral

(c)

Figure .001   Symmetry, body planes, and terms of direction: (*a*) asymmetry, leucon sponge; (*b*) radial symmetry, a sea anemone; (*c*) bilateral symmetry, a vertebrate.

UNIT

I

Introductory Concepts and Skills

Certain basic concepts unite the diverse subdisciplines within zoology. All zoologists must have a working knowledge of these important concepts. For example, outside of evolution, modern biology loses much of its meaning and excitement. For this reason, units II and III emphasize evolutionary relationships. Similarly, the modern zoologist cannot function without some background in cell and tissue structure and function, genetics, and ecology. Unit I stresses these unifying concepts. All of these topics set the stage for material that is to come in units II and III. As you proceed through this course, you will find that what you learned in unit I is built upon and emphasized in different ways.

Exercise 1

The Microscope and Scientific Inquiry

Prelaboratory Quiz

Study this week's laboratory exercise and then complete the following quiz to assess your preparation for the laboratory.

The Microscope

1. During this week's laboratory, you will be learning to use compound and dissecting microscopes. Which of these two microscopes would you use to examine the whole-body structure of a honeybee?
   a. compound microscope
   b. dissecting microscope

2. The ability of a microscope to discern small objects that are very close together or to discern detail in an object is called
   a. total magnification.
   b. parfocal.
   c. resolution.
   d. discernment.

3. Which of the following parts of a compound microscope is adjusted to regulate the amount of light coming through the specimen?
   a. objective lens
   b. diaphragm
   c. condenser lens
   d. tube

4. Which of the following parts of a compound microscope is nearest the observer's eye when using the microscope?
   a. objective lens
   b. revolving nosepiece
   c. base
   d. ocular lens

5. The low-power objective of a compound microscope has a magnification of
   a. 4×.
   b. 10×.
   c. 40×.
   d. 100×.

6. Which of the following objective lenses should be used when beginning to focus a compound microscope?
   a. 4×
   b. 10×
   c. 40×
   d. 4× or 10×, whichever is the lowest power objective present on the microscope

7. A(n)_____ is a small, nonunit scale in the eyepiece of a microscope that can be calibrated and used for measuring the size of microscopic objects.
   a. ocular micrometer
   b. stage micrometer
   c. objective micrometer
   d. focal micrometer

8. In this week's laboratory you will be making a temporary preparation that is useful in viewing fresh material. This temporary preparation is called a
   a. dry mount.
   b. wet mount.
   c. hanging mount.
   d. bubble mount.

Scientific Method

9. Define "science."

   _____

   _____

   _____

10. A generalized statement that has predictive value and that is supported by years of testing is called a(n)
    a. hypothesis.
    b. scientific theory.
    c. educated guess.
    d. postulate.

11. Parameters that could influence the results of an experiment and that must be held constant during the experiment are called
    a. dependent variables.
    b. controlled variables.
    c. independent variables.
    d. contingent variables.

12. The parameters that are monitored during an experiment while gathering data are called
    a. dependent variables.
    b. controlled variables.
    c. independent variables.
    d. contingent variables.

13. A test of a hypothesis should include all of the following EXCEPT
    a. replicates.
    b. a single controlled variable.
    c. a single independent variable.
    d. a control group.

14. "If/then statements" are used to
    a. reformulate a hypothesis that was proven wrong.
    b. predict the outcome of an experiment based on the hypothesis and the experimental methods.
    c. design a test of the hypothesis.
    d. formulate an initial hypothesis.

15. Replication of a scientific experiment is important because

    _____

    _____

    _____

    _____

1.1 BASIC MICROSCOPY

LEARNING OUTCOMES

1. Describe the usefulness of the compound microscope in zoology.
2. Name the parts of the compound microscope and their functions and describe the care of the compound microscope.
3. Demonstrate the ability to use the compound microscope in estimating the size of microscopic objects, examining objects prepared as wet mounts, and determining the three-dimensional attributes of objects being viewed.
4. Describe the usefulness of the dissection microscope in zoology.
5. Demonstrate the ability to use the dissection microscope in examining specimens of varying sizes and colors.

The microscope is a tool that all biologists must be able to use correctly. The microscope enables biologists to extend their vision to a resolution of fractions of a micrometer. This degree of resolution requires sophisticated electron microscopes, not generally used by undergraduate students. Lower magnifications, however, are just as useful for most aspects of animal biology. Magnifications ranging from 5× to 1,000× will be commonly used in the studies that follow. All good-quality microscopes are expensive. Proper care and use of the microscope is, therefore, essential. The purpose of this exercise is to familiarize the student with the use and care of the two microscopes most frequently used in this course: the compound microscope and the dissecting microscope.

THE COMPOUND MICROSCOPE

The compound light microscope is used to examine details of the cellular and tissue structure of animals. The usefulness of a light microscope is limited not by its ability to magnify objects, but by its ability to discern small objects that are very close together. This is called **resolution** and is related to the wavelength of visible light used to illuminate an object. Shorter wavelengths of light (toward the blue end of the spectrum of visible light) are scattered less than longer wavelengths as they pass through material being viewed, and thus distort the image less. Using short wavelengths of visible light, compound microscopes can resolve two points that are about $0.2\ \mu$m apart. Two points closer together than this appear as one point regardless of the magnification used. Resolution refers to the sharpness of the image. Electron microscopes are used when greater resolution is required.

As you work through this exercise, keep in mind that, although basically similar, microscopes vary depending upon make and quality. Expect that your microscope will differ in some details from the description that follows. Your instructor will point out these differences.

The compound microscope that is used in most zoology laboratories contains at least two lenses. These lenses bend (refract) light rays into one focal point—the eye(s) (fig. 1.1a). The **objective lens system** is nearest the object; the **ocular lens system** is nearest the eye. The total magnification of an object is the product of the two lens systems.

Ocular lens system	10×	10×
Objective lens system	10×	40×
Total magnification	100×	400×

▼ A microscope must be handled very carefully at all times. Always carry it upright with one hand under the base and the other hand around the arm. Obtain a microscope as directed by your instructor and place it on the bench in front of you. Unwind just enough of the cord to reach the outlet. Do not let extra cord hang over the edge of the bench. Plug in the microscope. Find and learn the functions of the microscope parts described in figure 1.1b and table 1.1. Compare your microscope to the variations shown in figure 1.2. ▲

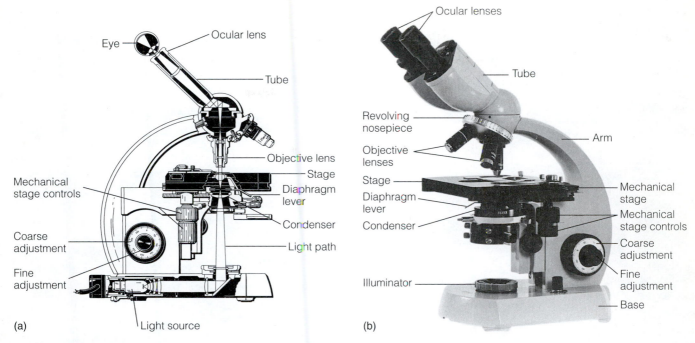

Figure 1.1 The compound microscope: (*a*) focusing of light rays; (*b*) parts of a microscope.

TABLE 1.1   The Compound Microscope Parts and Their Function

Ocular Lens	Eyepiece containing a lens normally with a 10× magnification (usually stamped on the rim of the eyepiece). Often contains a pointer to aid in locating objects within the field in view.
Tube	Section between ocular lens and objective lenses. Contains system of lenses and mirrors.
Revolving Nosepiece	Holds objective lenses.
Objective Lenses	Lenses closest to the specimen. Usually stamped as follows: Scanning—4× Low power—10× High power—40× Oil immersion—100×*
Arm	Support running from base to tube.
Stage	Platform on which slide is positioned. Note the manner in which the slide is held in place.
Condenser	Lens just under the stage and the specimen. Concentrates light under the specimen.
Diaphragm	Controls the amount of light passing through the specimen and controls contrast and resolution of the image. Opens and closes by sliding a lever or turning a disc.
Base	Platform on which microscope rests. Contains either built-in light source or a mirror.
Coarse Adjustment	Larger knob on both sides of the arm. Objective lenses or stage move perceptibly when knobs are turned.
Fine Adjustment	Smaller knobs on both sides of the arm. Movement of the lenses or stage is nearly imperceptible unless looking through ocular.

*This lens, if present, should be used only with instructions from your instructor. When in use, the lens is immersed in a drop of oil that prevents the refraction of light between the microscope slide and lens. Refraction of light is a problem because of the small working distance (distance between the focal point of the objective lens and the specimen). Some immersion oil may be carcinogenic and should be handled with care.

Care of the Microscope

Following is a list of precautions to observe when using the microscope. Become thoroughly familiar with them. They apply to both compound and dissecting microscopes.

1. Report any damage or malfunction immediately.
2. Carry the microscope carefully with one hand under the base and the other around the arm.
3. Do not remove the lenses from the microscope. If they are dirty inside, advise the instructor.

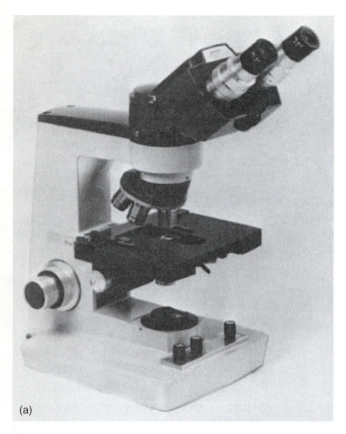 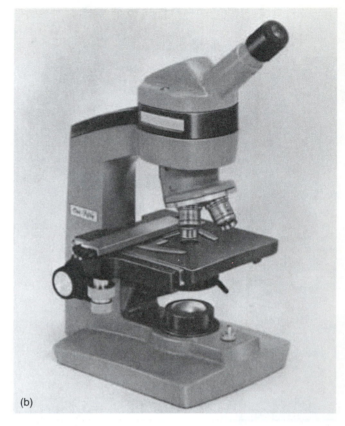

Figure 1.2 Variations in the compound microscope: (*a*) binocular; (*b*) turret nosepiece monocular with mechanical stage.

4. Clean the lenses with lens paper only. If gentle rubbing with lens paper does not adequately clean the lens, notify your instructor.
5. If anything is spilled on the microscope, clean it off immediately; if it is anything other than water, inform the instructor.
6. Unwind just enough cord to reach the outlet.
7. Before putting the microscope away, check for the following:
    a. Are the lenses clean?
    b. Is the lowest-power lens in position?
    c. Is the stage turned down to its lower limit or the nosepiece raised to its upper limit?
    d. Is the stage bare?
    e. Is the dustcover in place?

Use of the Compound Microscope

▼

1. Obtain a slide (letter slide or similar preparation) from your instructor. Clean and examine it visually to locate the position of the object on the slide (holding it up to a light may help).
2. Position the slide on the stage of the microscope so the object to be viewed is centered over the condenser.
3. Check to be sure the light is on. **Always begin with the lowest-power lens in position** and the

iris diaphragm set at a medium opening. If your microscope has a voltage control knob, adjust this knob to a medium setting.
4. Looking from the side of the microscope, lower the objective (or raise the stage) until the objective reaches its lower (or upper in the case of the stage) limit, or until the objective is just about to touch the slide.
5. Now look through the microscope and focus by turning the coarse adjustment so that the slide and the lens move away from each other. Stop when the object comes into focus. Note: always focus initially using the lowest power. If a voltage control knob is present, adjust the light to a comfortable intensity. Adjust the iris diaphragm to achieve best contrast and resolution. If the iris diaphragm is closed too much, contrast will be high but resolution will be poor. Remember that thick specimens require more light than thin specimens.
6. Move the slide first to the left, then the right. Move the slide toward you, then away from you. What do you notice? _____
Compare the orientation of the object on the stage with the orientation of the image. What do you notice? _____
7. If your microscope is binocular, you should not see a double field of view. Adjusting the eyepieces to

fit the distance between your eyes should result in a single field of view. In addition, eyepieces can be adjusted for focal differences in each eye as follows. (If your microscope is monocular skip to step 8.) Notice that at least one eyepiece can be focused independently of the second eyepiece by turning a ring on the eyepiece. (Note the ring on the left eyepiece in figure 1.2a.) Look through the microscope, closing the eye corresponding to the adjustable eyepiece. Now focus on an object using the focusing knob, the open eye, and the nonadjustable eyepiece. Once that eyepiece is focused, reverse the eye being used, opening the eye corresponding to the adjustable eyepiece and closing the other eye. Focus the adjustable eyepiece using only the ring on the eyepiece. Now open both eyes. The object should be in perfect focus. Focusing rings are graduated so that you can record the ring position and return it to that position the next time you use your microscope.

8. A microscope is said to be **parfocal** if it can be switched from a lower to a higher power and remain in focus. If you began with a 4× objective, center the object and switch to the 10× objective, and focus with the coarse focus knob. If you are using a letter slide, the image will probably be filling the entire field of view. Center one portion of the letter (e.g., the "tail" of the "e") and switch from the 10× objective to the 40× objective. **Never focus with the coarse adjustment when the high-power objective is in position.** In switching from low power to high power, what do you notice about the brightness of the field of view? _____
Why is it important to center the object in the field of view before switching to high power? _____
_____
_____ ▲

Estimating the Size of Microscopic Objects

It is useful to know the size of objects viewed in the microscope. A common technique to estimate size is to measure the diameter of the field of view and then to estimate the proportion of the field that the object occupies. From those figures, an approximate size can be obtained. The following exercise gives instructions for measuring the diameter of the lower-power field of view with a millimeter ruler. The same technique can be used with the 4× scanning lens.

▼ Place a millimeter ruler on the stage of your microscope and focus on the scale. Position the edge of the ruler across the diameter of the field of view with the 0 end of the millimeter scale at one side of the field of view. What is the diameter of the field of view of your microscope in millimeters? _____
This value usually ranges between 1.5 and 1.8 mm. If an object takes up one half of a 1.6-mm field of view, what would be its approximate size? _____ ▲

The diameter of the high-power field cannot be measured directly because of its small size. It can be calculated as follows:

$$\frac{\text{Magnification of low-power objective}}{\text{Magnification of high-power objective}}$$

$$\times \text{Diameter of low-power field}$$

▼ What is the diameter of the high-power field of view of your microscope? _____
This value usually ranges between 0.35 and 0.42 mm (350 and 420 μm). ▲

The Ocular Micrometer

More precise measurements can be made using an **ocular micrometer.** An ocular micrometer is a thin circle of glass or plastic etched with a nonunit scale usually ranging from 0 to 100. When placed in the ocular lens of the compound microscope, the scale image is superimposed over the image of the specimen. A stage micrometer is a glass slide etched with a scale of known units, usually divided into 0.01-mm and 0.1-mm graduations.

▼ The ocular micrometer is calibrated using the stage micrometer by aligning the images at the left edge of the scales (fig. 1.3). Calculate the distance between a given number of divisions on the ocular micrometer and substitute those values into the following.

$$1 \text{ ocular micrometer space} = \frac{\begin{array}{l}\text{Distance between selected}\\\text{graduations on the ocular}\\\text{micrometer (from the}\\\text{stage micrometer)}\end{array}}{\begin{array}{l}\text{Number of spaces}\\\text{measured on the oular}\\\text{micrometer}\end{array}}$$

$$= \text{_____ mm}$$
$$= \text{_____ micrometers}$$
$$(1 \ \mu m = 1/1,000 \text{ mm})$$

This should be done for each objective lens and recorded under "Microscope and Seat Assignment" in the "Getting Started" section at the beginning of this manual. Assuming that one uses the same microscope in subsequent labs, these values can be used for measuring microscopic objects. Individual microscopes (even those of the same make and model) will vary with regard to these values and must be calibrated individually. ▲

Making a Wet Mount

The **wet mount** is a temporary preparation that is useful in viewing fresh material. A specimen is placed in a drop of water on a slide, then covered with a thin piece of glass or plastic called a coverslip (fig. 1.4). This technique keeps fresh material moist during observation.

▼ After adding the specimen to a drop of water on a clean slide, place one edge of a coverslip on the slide next to the

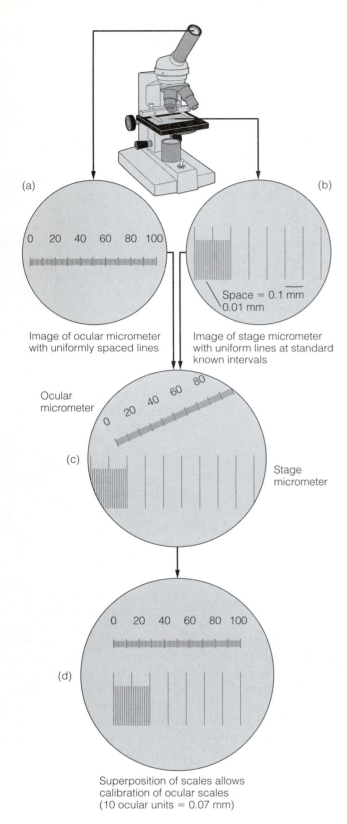

(a)

(b)

Image of ocular micrometer with uniformly spaced lines

Image of stage micrometer with uniform lines at standard known intervals

Space = 0.1 mm
0.01 mm

Ocular micrometer

(c)

Stage micrometer

(d)

Superposition of scales allows calibration of ocular scales (10 ocular units = 0.07 mm)

Figure 1.3 Calibrating an ocular micrometer.

drop. Incline the coverslip at a 45° angle and lower it slowly until the coverslip touches the water drop and rests on the slide. The water now forms a thin sheet under the "floating" coverslip. This procedure should eliminate most air bubbles that might obscure the specimen.

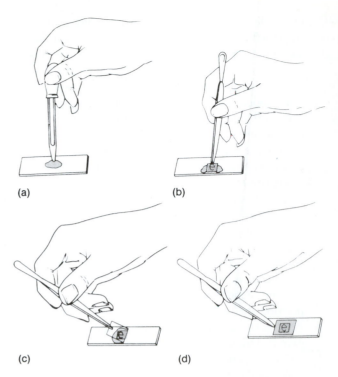

(a)

(b)

(c)

(d)

Figure 1.4 Making a wet mount.

Make a wet mount of a small length of hair. Estimate its diameter under low and high power.

Low power _____

High power _____

Which figure is more accurate?_____
Why? _____
_____

Look for any air bubbles. What do they look like under the microscope? _____▲

Other Exercises

▼

1. Practice making wet mounts and estimating sizes using drops of water from a sample of pond water.
2. Make a wet mount of two crossed hairs or threads of different colors. Using high power, focus on the regions where the hairs or threads cross. When focusing down with the fine adjustment, notice that the hair or thread on top will come into focus first. Can you tell which is on top? This exercise demonstrates that focusing up and down through a relatively thick specimen can give one an appreciation of the three-dimensional structure of a specimen.▲

THE DISSECTING MICROSCOPE

The dissecting microscope is useful when magnifications between 5× and 50× are desired. It is useful for viewing entire specimens of small animals or body parts of larger animals.

It has the added advantage of giving a three-dimensional view of the object.

Use of the Dissecting Microscope

All precautions and rules that apply to the compound microscope also apply to the dissecting microscope. Some dissecting microscopes have built-in light sources; others do not. Obtain a dissecting microscope and light source. Note that the microscope is binocular and has a single focusing knob and a lens system that allows you to change magnifications by rotating a dial or revolving a nosepiece (fig. 1.5).

▼ Your instructor has provided some materials for viewing under the dissection microscope. Adjust the eyepieces to fit the width of your eyes. Adjust the two eyepieces for differences in your eyes as described in step 7 under "Use of the Compound Microscope." Examine your specimen under low and high magnifications. Note the three-dimensional image. Compare the orientation of the image to the orientation of the specimen on the stage. Move the specimen to the right and left. What do you notice?

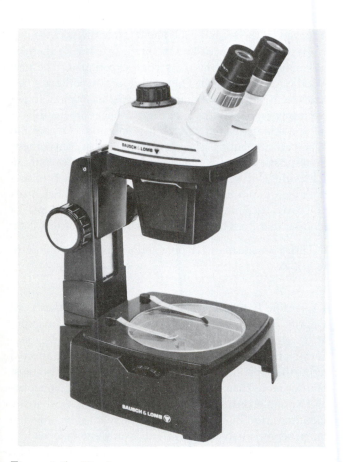

Figure 1.5   The dissecting microscope.

The stage of the microscope is usually equipped with a plate that can be reversed to provide either a white or black background. Use the background that will provide better contrast. Try switching back and forth.

Living and preserved specimens are best viewed while completely immersed in water. This prevents a distorted image due to the refraction of light off moist surfaces. ▲

1.1 LEARNING OUTCOMES REVIEW—WORKSHEET 1.2

1.2 AN INTRODUCTION TO SCIENTIFIC INQUIRY

LEARNING OUTCOMES

1. Discuss why science is a process, not just a body of facts.
2. Describe the kinds of problems that can be investigated using scientific methods.
3. Design and carry out a scientific investigation using an experimental approach to answering a question.
4. Analyze and present data, and draw conclusions, from a scientific investigation.

Science is an objective process used to investigate natural occurrences. It is knowledge of the natural world gained by observation. The goal of science is often said to be the description of theories. A theory in the scientific context is different from the everyday use of the term to mean a guess. Theories are not developed as a result of any one experiment or, as commonly portrayed, a hypothesis that has been supported. Instead, **scientific theories** are generalized statements that have predictive value. These statements are supported by many experiments, each testing one small aspect of a larger concept. Each conclusion is reviewed by many scientists. This process of **peer review** is fundamental to science because it ensures that methods used in any experiment were valid, the results were accurately interpreted, and the conclusions warranted. Peer review is the reason that we can trust that scientific theories are sound. In other words, a theory is a concept that has usually undergone years of testing and has been supported again and again. It is a concept that is as certain as anything can be in the realm of science. The uncertainty implied in the word "theory" reflects the fact that scientists should never imply that they know everything there is to know about any concept. New technologies, ideas, and experiments may bring new information to light that modifies a scientist's understanding of the concept. Therefore, details of a theory may change; however, the essence of a theory is unlikely to change. A theory is thought to accurately reflect how things happened in the past and how they will occur in the future.

In order for our concept of science to be valid, scientists must make three assumptions that are impossible to prove but seem to be true based upon experience. (1) Nature is understandable. Scientists are not perfect in their observations, but the work of science involves testing and retesting by individuals working independently or together. This retesting eventually illuminates errors and helps to improve our understanding of natural events. (2) Nature is real. What we observe with our senses in the laboratory or in field studies is reality. (3) Cause-and-effect relationships are consistent in nature. This concept is sometimes called uniformitarianism. It means that the processes that shaped our natural world in the past are the same as those occurring now and the same as those that will occur in the future. Stated another way, an experiment that we performed yesterday and repeated today should have produced similar results both days. The results should also be similar if the experiment is performed a year from now. It is this latter concept that allows scientists and our students to make reliable predictions regarding the outcome of an experiment.

Science is not without its limits. It is important that students realize that science is a process that can be used to investigate natural events. In order to investigate anything scientifically, scientists use their senses to weigh, measure, or otherwise observe an occurrence. Scientists often use microscopes and electronic devices in their work; however, these instruments are extensions of scientists' senses. Science cannot be used to investigate the supernatural. Many problems arise when individuals try to use scientific methods to study matters of the spirit. Similarly, science is ethically neutral. Science is not good or bad, moral or immoral. Scientists have no special authority when it comes to making ethical decisions. We can, however, all benefit from the information presented by scientists when we make ethical decisions.

SCIENTIFIC METHODS

Scientists use a variety of methods when investigating natural events. Even though experimentation and the scientific method are fundamental scientific approaches to solving problems, many questions are answered by careful, detailed observation. Observational science and historical science (reconstructing past events) are fundamental to the work of a biologist studying the mating behavior of a bird, a paleontologist investigating events that occurred millions of years ago, or a forensic scientist reconstructing a crime.

A Scientific Investigation

The following exercise stresses both careful observation and the use of experimental science. The exercise is organized in a fashion that is often described as "the scientific method." Elements of this exercise, such as asking questions, careful observation, and hypothesis formulation, are a part of any scientific investigation. Other elements of the exercise, such as defining variables and designing a control group, are unique to an experimental approach to an investigation. As you work through this exercise, realize that there is no single "method" in science, just a commitment to logic and objectivity in investigating natural occurrences.

Questions Scientists usually begin with a question or a set of questions based on observations, previous research, reading, or discussions with other scientists. They are questions that can be answered using our senses or extensions of our senses.

▼ Turn to worksheet 1.1 now and compose responses to items 1 and 2. ▲

Hypothesis Scientists begin the process of answering questions by posing plausible explanations. A **hypothesis** is a reasonable explanation for a question. It is more than a guess because it is usually supported by previous work of the scientist or others. A hypothesis is sometimes called an "educated guess."

A hypothesis is stated in the form of a declarative statement and must be falsifiable. Falsifiability is critical, because it is possible to prove a hypothesis false with a single experiment. On the other hand, one cannot prove a hypothesis to be true. A hypothesis may be supported by the results of a particular experiment, but new evidence or another experiment with slightly different conditions might prove it to be false. For example, suppose an experimenter formulated the hypothesis that nutrient A enhanced the regeneration of *Dugesia* (planaria). She carried out an experiment under conditions specified by her experimental protocol and found results consistent with her hypothesis. Did she prove her hypothesis to be true? What if she repeats the experiment under slightly different conditions of light, temperature, and moisture? It is possible that she could prove the hypothesis false. Her hypothesis was supported by the first experiment but may be disproved by the second experiment. Now the scientist can return to the hypothesis and modify it based upon the new information provided by the second experiment. The new hypothesis is then tested. The support for a hypothesis grows increasingly strong with repeated experimentation (often over many years), but the hypothesis is never proven true.

If a scientist is not careful, it is easy to feel ownership of, and attachment to, a particular hypothesis. A hypothesis should not be viewed as something that one wants to confirm. A hypothesis that is proven false provides as much (sometimes more) information as one that is supported. Scientists must avoid becoming personally attached to a hypothesis—a mind-set that leads to blinding oneself to contradictory data.

▼ Turn to worksheet 1.1 now and compose a response to item 3. ▲

A Scientific Experiment—The Test of the Hypothesis All scientific experiments have sets of variables tested, manipulated, or controlled. The **dependent variable** is the parameter that the scientist is measuring. It is what will be affected during the experiment. In many experiments, scientists must monitor more than one dependent variable. The **independent variable** is the parameter being tested. It is what the scientist varies. In any experiment, there must be only one independent variable. When there is only one independent variable, any changes in the dependent variable(s) must be the result of the scientist's manipulation of the independent variable. All other variables that might influence the outcome of an experiment are called **controlled variables.** Controlled variables are held constant in an experiment so that they do not affect the outcome. For example, a student plans an experiment to test the hypothesis that *Dugesia* are negatively geotactic; that is, *Dugesia* move away from the direction of the pull of gravity; In this experiment, gravitation is the independent variable, and the response of *Dugesia* (movement relative to the direction of the pull of gravity) is the dependent variable. Controlled variables would include light intensity, water temperature, and others.

After the hypothesis is formulated and the variables are identified, an experimental procedure is outlined. An experimental procedure is often based upon the work of others, consultation with peers, and a scientist's own creativity. The level of treatment is based on previous experience, background reading, or intuition followed by trial and error experimentation. For example, the procedure used in the *Dugesia* experiment could be determined by background reading or by observation of the animal in its natural habitat. Experimental conditions that closely mimic the animal's natural environment usually produce the most meaningful results.

While outlining an experimental procedure, two final considerations are critical. One consideration is replication. A **replicate** is the duplication of an experiment. When a scientist performs a well-thought-out experiment more than once, the results from all replicates should be consistent. Consistency of results is one source of reassurance that an experiment was well conceived and carried out. (Recall the uniformitarianism concept.) A second consideration is the inclusion of a control group. A **control group** is an experimental setup in which the independent variable has been eliminated or set to some standard value. In the *Dugesia* experiment, a control group might consist of the experiment being carried out in a container oriented in another direction relative to the pull of gravity.

Finally, after an experiment is designed, a scientist should be able to make **predictions** regarding the outcome of the experiment. These predictions are in the form of "if/then statements." If/then statements reinforce the concept of cause-and-effect relationships in nature. Predictions provide a useful reference point for drawing conclusions from an experiment. The "if" portion of the statement is the condition assumed by the hypothesis. The "then" portion of the statement is the outcome expected given the hypothesis and the conditions of the experiment. An if/then statement that predicted the outcome of the *Dugesia* experiment could be worded as follows: "If *Dugesia* respond negatively to the pull of gravity, then they will crawl to the highest point in a dish containing water and a rock tower."

PHOTOTAXIS OF *DUGESIA*

Dugesia (planaria) is a flatworm (phylum Platyhelminthes) that lives on the bottom of freshwater lakes, ponds, and rivers. It feeds on decaying plant and animal materials. Like all animals, *Dugesia* must respond to changes in its environment.

A response to an environmental condition is called a *taxis*. For example, a response to light is called phototaxis. If an animal responds by orienting toward or moving toward light, it is positively phototactic. If an animal responds by orienting or moving away from light, it is negatively phototactic.

▼ The following activities should be completed individually or in groups prior to coming to laboratory. The laboratory time should be used for carrying out your experiment. Examine figure 10.1 to familiarize yourself with the structure of *Dugesia*. Note the general shape of the body. Notice the eyes and the nervous system (see fig. 10.1d). Consult your textbook for additional information on *Dugesia's* ecology and behavior. Record these background observations in worksheet 1.1, item 4.

Hypothesis

Based on the observations that you have made on the structure, habitat, and feeding habits of *Dugesia*, formulate a hypothesis regarding phototaxis of *Dugesia*. Use worksheet 1.1, item 5 to help develop and record your hypothesis.

Materials and Methods

Your instructor will have equipped the laboratory with pond water, dishes, beakers, white and black paper, tape, aluminum foil, light sources, and other equipment that you may find useful in setting up your experiment. Use worksheet 1.1 item 6 to outline your materials and methods. You will be asked to sketch an empty data table and graph as you formulate your materials and methods. You should know the form the data will take before you begin your experiment.

Results

When the laboratory begins, you will carry out your experiment and fill in a final copy of your data table. This table may be in a computer spreadsheet, or it may be constructed using the graph paper provided at the end of this exercise. Perform any calculations required and transfer the data for presentation. You may use the graph paper provided or the data may be entered into a computer spreadsheet.

Discussion

Look at your data. Do they support or refute your hypothesis? If they refute your hypothesis, suggest an alternative hypothesis that could be tested in another experiment. If your hypothesis is supported, describe how the results of your experiment help explain where *Dugesia* live in ponds and streams. Record your conclusions in the "Discussion" section of worksheet 1.1 and turn in your results to your laboratory instructor.

1.2 LEARNING OUTCOMES REVIEW—WORKSHEET 1.2

WORKSHEET 1.1   Phototaxis of *Dugesia*

1. Write two questions that could be answered with a scientific investigation.

2. Write two questions that could not be answered with a scientific investigation.

3. Write a reasonable hypothesis for one of the two questions you composed in item 1 above.

4. Record your background observations on *Dugesia* regarding the presence and location of eyes, organization of the nervous system, habitat, food, and other observations. Note the source of information used in each of these observations (laboratory manual, textbook, other literature sources, personal observations).

5. The hypothesis
    a. Do you think *Dugesia* will display phototaxis? Why or why not?

    b. If it does display phototaxis, do you think it will be positive or negative? Why?

    c. Write your hypothesis. As you compose the hypothesis remember that a hypothesis is in the form of a declarative statement. Is yours? A hypothesis must be falsifiable. Is yours?

6. Materials and methods
    a. What is the single independent variable that you will test in your experiment? Be as specific as possible. Are you concerned with the direction of light, intensity of light, color of light, or some other aspect of the quality or quantity of light?

    b. Design a simple experimental setup using the materials your instructor has made available. Describe or draw an illustration showing that setup. Explain where animals will be placed at the beginning of the experiment.

    c. What dependent variables will you be monitoring during the experiment? These could include the orientation of the *Dugesia* in the dish relative to the light source, the position of the *Dugesia* relative to some fixed point or line in the dish, or some other measuring criterion. Is the time frame for your measurements important? In other words, will you make your observations one time at the end of the experiment, or periodically during the experiment? Remember that this exercise must be designed to be carried out within the time allotted for the laboratory and that you may want to perform one or more replicates of your experiment.

d. Following is a list of variables that will need to be controlled. Next to each variable, describe how each can be controlled. Add other controlled variables as you think of them.
   i. overhead lighting
   ii. window lighting
   iii. heat from a lamp
   iv. other controlled variables

e. What should be the nature of your control group?

f. How many *Dugesia* should be in your control and experimental groups? (Your instructor will probably impose some limits on this aspect of your experiment.)

g. How many replicates of this experiment will you carry out?

h. How will the data from the experiment be recorded? Will you be recording data in timed increments? Sketch a blank table in the "Results" section of this worksheet below. This sketch is a preliminary to the final table you will use to record data. Your instructor may ask you to use a computer spreadsheet to record your data, but you need to have a table planned before going to the computer.

i. The results of an experiment must be communicated to others. The results of an experiment consist of raw data from replicates of an experiment. These data must be combined, averages or percentages calculated, or otherwise presented for interpretation. What is the most effective form for the presentation of your results and conclusions to others in your class? Sketch the table or graph in the "Results" section of this worksheet. Be sure to write the column headings in your table and/or label the axes of your graph. This sketch is a preliminary to the final table or graph you will use to present your data. Your instructor may ask you to use a computer spreadsheet to present your data.

7. Results
   a. Sketch of table used to record data during the experiment.

   b. Sketch of the table or graph used to present your data.

8. Discussion

WORKSHEET 1.2   The Microscope and Scientific Inquiry

1.1   The Compound Microscope—Learning Outcomes Review

1. Which of the following came earliest in the procedure you used to focus your compound microscope after the microscope illuminator was turned on, slide had been cleaned and the object centered over the condenser lens, and the iris diaphragm set at a medium opening?
   a. I lowered the lowest-power objective lens (or raised the stage).
   b. I looked into the ocular lens.
   c. I raised the highest-power objective lens (or lowered the stage).
   d. I used the fine adjustment to bring the object into focus under the 10× objective.

2. As compared to the orientation of the object on the stage, an image in a compound microscope
   a. appears larger, but otherwise normal.
   b. appears larger and upside-down.
   c. appears larger and backwards.
   d. Both b and c are correct.

3. It is important to center an object in the field of view before changing to a higher-power objective because
   a. lower-power objectives have larger fields of view and an object in focus at a lower-power may be out of the higher-power field of view.
   b. the brightness of the field of view increases with higher powers and the object might be obscured by the higher light intensity.
   c. the object may no longer be in focus at the higher magnification.
   d. All of the above are reasons for centering.
   e. a. and c. only are reasons for centering.

4. An object takes up approximately ¼ of the 10× field of view. It is approximately _____ long.
   a. 100 μM
   b. 400 μM
   c. 750 μM
   d. 1 mm

5. The diameter of a 10× field of view is 2 mm. The diameter of the 40× high-power field of view is
   a. 1 mm.
   b. 750 μM.
   c. 500 μM.
   d. 200 μM.

6. Red and black threads are placed on a slide using a wet-mount technique. The crossed threads are examined under low and high power. Under high power, a student observed that when the stage of her compound microscope was raised with the fine focus the black thread came into focus first, then the red thread came into focus. Which of the following statements is true?
   a. The red thread is on top of the black thread.
   b. The black thread is on top of the red thread.
   c. It is impossible to tell which thread is on top, as a compound microscope provides only a two-dimensional image.

7. All of the following are TRUE of a dissection microscope except one. Select the exception.
   a. Dissection microscopes provide three-dimensional images.
   b. Dissection microscopes have total magnifications that usually vary between 50× and 200×.
   c. Images in dissection microscopes are not backwards or inverted.
   d. Dissection microscopes have binocular, not monocular, heads.

8. Dissection microscopes do not have an iris diaphragm. Rather than using an iris diaphragm lever, contrast can be improved by
   a. changing magnification.
   b. increasing or decreasing light intensity.
   c. changing the background by flipping the plate on which the specimen rests.
   d. Contrast cannot be changed with a dissecting microscope.

1.2 An Introduction to Scientific Inquiry—Learning Outcomes Review

9. A hypothesis is
   a. a reasonable explanation for a question.
   b. framed as a declarative statement.
   c. testable if it is falsifiable.
   d. impossible to prove correct.
   e. All of the above are true.

10. In experimental science, a(n) _____ is the parameter being tested in an experiment. There can be only one of these in an experiment for the experiment to be informative.
    a. controlled variable
    b. independent variable
    c. dependent variable
    d. control group

11. In experimental science, a(n) _____ is the parameter being measured in an experiment. There can be more than one of these in an experiment.
    a. controlled variable
    b. independent variable
    c. dependent variable
    d. control group

12. A scientific theory is
    a. a hypothesis proven correct.
    b. an educated guess that will be tested by experimentation and observation.
    c. a generalized statement that can be used to predict outcomes; a theory is based on the testing and retesting of related hypotheses.
    d. scientists' best explanation for how something happens; the essence of a theory is well-established and will likely stand the test of time.
    e. All of the above are true.
    f. Only c and d are true.

1.1 The Compound Microscope—Analytical Thinking

1. Based on your knowledge of the use of compound microscopes, explain why the procedure that you learned required the following:

   a. putting the microscope away with the stage focus turned down to its lower limit and the lowest-power lens in position;

   b. looking from the side of the microscope and lowering the objective lens to its lower limit at the beginning of the focusing procedure;

   c. setting the iris diaphragm at a medium opening at the beginning of the focusing procedure.

2. Explain why the diameter of the field of view must be recalculated, or the ocular micrometer, must be recalibrated, for each objective lens.

1.2 An Introduction to Scientific Inquiry—Analytical Thinking

3. Explain why it is misleading to speak of "the scientific method" as if there is a single method used by all scientists.

4. Science is not without its limits. What are the limitations of science?

5. A controlled variable and a control group are not the same. Explain.

Exercise 2

Cells and Tissues

Prelaboratory Quiz

Study this week's laboratory exercise and then complete the following quiz to assess your preparation for the laboratory.

Cell Structure

1. Using a light microscope, one can clearly understand
   a. details of the organization of cellular organelles.
   b. the size, shape, and diversity of cells.
   c. how cells are organized to form tissues.
   d. All of the above are correct.
   e. Both b and c are correct.

2. The cellular organelle responsible for the synthesis of proteins in a cell is the
   a. mitochondrion.
   b. centriole.
   c. smooth endoplasmic reticulum.
   d. ribosome.

3. Ribosomes are synthesized at the
   a. centriole.
   b. mitochondrion.
   c. Golgi apparatus.
   d. nucleolus.

4. A _____ is a cylindrical arrangement of proteins that produces movement within cells and supports cell processes.
   a. microtubule
   b. microfilament
   c. flagellum
   d. rough endoplasmic reticulum

5. Vesicular structures that contain enzymes capable of digesting biologically important macromolecules are called
   a. vacuoles.
   b. lysosomes.
   c. secretory vesicles.
   d. ribosomes.

Histology

6. What is histology?

7–10. Name the four kinds of animal tissues and give one example of each type of tissue.

Name	Example
7. _____	_____
8. _____	_____
9. _____	_____
10. _____	_____

11. An epithelium that consists of a single layer of cells resting on the basement membrane is called a
    a. squamous epithelium.
    b. stratified epithelium.
    c. stratified squamous epithelium.
    d. simple epithelium.

12. A connective tissue that functions in binding skin to muscle is
    a. dense connective tissue.
    b. loose connective tissue.
    c. adipose tissue.
    d. cartilage.

13. All of the following are components of connective tissues EXCEPT
    a. cells.
    b. protein fibers.
    c. a basement membrane.
    d. a matrix.

14. A connective tissue specialized for storing fat is
    a. adipose tissue.
    b. dense connective tissue.
    c. lymph.
    d. cartilage.

2.1 CELL STRUCTURE

LEARNING OUTCOMES

1. Describe why understanding cell structure and function is important in zoology.
2. Discuss why animal cells vary in size, shape, and the kinds of organelles present.
3. Identify cellular organelles from electron micrographs and describe the functions of these organelles.

To understand cellular function is to understand much of life. While working through this course, you will find that you must repeatedly return to the level of the cell to understand animal function. For example, whereas one can study the contractile properties of a muscle by observing an entire muscle, it is impossible to understand how a muscle contracts without studying the muscle cell. Similarly, it is impossible to understand the function of the nervous system without an appreciation of the nerve cell, or to understand the function of the endocrine system without familiarity with models of the plasma membrane. Just as animal function is tied to cell function, animal structure is dependent on the organization of cells into tissues. Today it is more important than ever for the biologist to understand cell structure and function.

ANIMAL CELLS—LIGHT MICROSCOPE

Detailed study of cell structure requires the use of the electron microscope. Few details of subcellular structure can be studied under the light microscope. The purpose of this study is not to examine details of structure, but to gain an appreciation of some of the diversity found in animal cells. As you work through the exercise, record your observations in table 2.1 in worksheet 2.1. Draw and label the cells observed using pages 39 and 40.

▼ Using the flat end of a clean toothpick, a tongue depressor or a popsicle stick, gently scrape the inside of your cheek. Swirl the scrapings in a drop of water on a slide. Cover with a coverslip, place the slide on the stage of the microscope, and examine using reduced light and low power. Look for small, flattened, and transparent epithelial cells. Stain the preparation with methylene blue by placing a small drop of stain at one edge of the coverslip. Draw the stain under the coverslip by touching the edge of a paper towel to the opposite edge of the coverslip. As water is drawn from under the coverslip, it will be replaced by stain. Examine the preparation under low power, then high power.

The bulk of the cell's interior is **cytoplasm.** The cytoplasm contains organelles that carry out specific functions. These structures usually cannot be seen clearly under the light microscope. The outer boundary of the cell is defined by the **plasma membrane.** Note the **nucleus,** the genetic control center of the cell. If your microscope has an ocular micrometer, measure the diameter of the cell using the calibration factor you calculated in exercise 1. If your microscope does not have an ocular micrometer, estimate the cell's diameter using the method described in exercise 1. Record your observations in table 2.1 in worksheet 2.1.

Examine a spinal cord smear and look for star-shaped nerve cells. These cells show a distinct **nucleolus** within a relatively clear nucleus. The nucleolus is an area of active RNA synthesis. The processes that radiate outward from the cell body of this nerve cell receive electrical impulses from other cells or transmit electrical impulses out of the spinal cord to muscles or glands. One of these processes may be very long, perhaps extending down the leg to a distant muscle. Thus, you are only seeing a small portion of the entire cell. Estimate the size of these cells and record your observations in table 2.1.

Examine a blood smear, first under low power, then under high power. Biconcave red blood cells will be stained pink. Do you see any evidence of a nuclei in any of the red blood cells? Red blood cells are formed in the bone marrow and spleen. They lose their nuclei prior to being released into the bloodstream, where they carry oxygen in combination with the cell's hemoglobin molecules. Estimate their size. Using low power, look for cells containing darkly stained nuclei and relatively clear cytoplasm. Examine one of these white blood cells under high power. White blood cells function in body defenses. Estimate the size of the white blood cell you are observing. Record your observations in table 2.1.

Examine a preparation of *Amphiuma* liver. Note the closely packed cells. Focus on one or two cells under high power. Do you see nucleoli in these cells? If so, how many nucleoli do you see? The very small stained dots in the cytoplasm of these cells are mitochondria. **Mitochondria** are responsible for energy conversions in the cell. Estimate the size of a cell and record your observations in table 2.1 in worksheet 2.1.

Examine a preparation of small intestine epithelium stained with Golgi stain. The cells you need to observe line the cavity of the intestine and are located along one edge of this slide preparation. Look for tall cells lining one edge of the tissue. You should see darkly stained particles near the surface of these cells. This side of the tissue is facing the lumenal surface of the intestine (toward the cavity of the intestine), and the darkly stained particles are secretory products that have been released at the plasma membrane. Find a nucleus of one of these intestinal cells. What is its shape? Next to the nucleus, you should see layered membranes that are stained the same dark color as the secretory products observed earlier. These membranes comprise the Golgi apparatus. The **Golgi apparatus** is responsible for packaging the protein secretions of these cells prior to their release at the plasma membrane. Estimate the longest dimension of this cell and record your observations in table 2.1 in worksheet 2.1. ▲

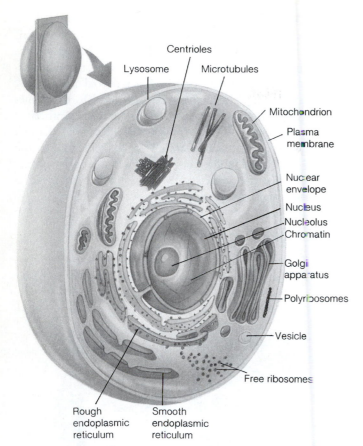

Figure 2.1 Diagrammatic representation of the ultrastructure of an animal cell.

ANIMAL CELLS—ELECTRON MICROSCOPE

The structure of a cell as seen with the electron microscope is called the cell's ultrastructure. The previous activity was designed to give you an appreciation for a small part of the diversity found in animal cells. It was also designed to give you an appreciation for the minuteness of cell organelles. Electron micrographs are needed to study details of cell structure. In the following activity, you will study cell ultrastructure based on the diagrammatic representation in figure 2.1 and the electron micrographs and drawings of figure 2.2–2.12.

The Plasma Membrane

The cell has many membranes. The plasma membrane separates the contents of the cell from its surroundings, but it has many other functions as well. Proteins in the plasma membrane create channels for substances moving passively through the plasma membrane. Proteins also serve as carriers for substances that are actively transported across the membrane. In addition, proteins serve as receptor sites for hormones, antibodies, and other biologically important compounds. The plasma membrane provides a location for enzymatic reactions and is involved with cell adhesion. A current model of the plasma membrane that explains those functions is the **fluid mosaic model** (fig. 2.2). This model describes the plasma membrane as a bimolecular lipid layer. Hydrophilic ends of lipid molecules are oriented to

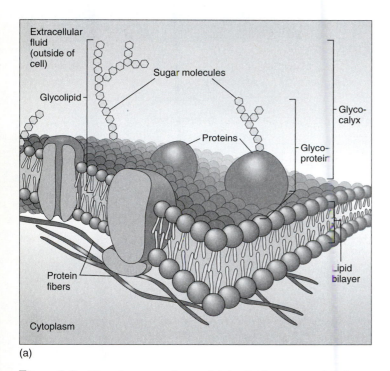

(a)

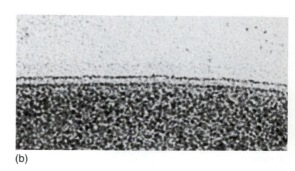

(b)

Figure 2.2 The plasma membrane: (a) the fluid mosaic model, showing the lipid bilayer and intrinsic proteins; (b) transmission electron micrograph of a portion of a plasma membrane. Note the organization into a double layer (×250,000).

the outside of the membrane, while hydrophobic ends are oriented to the inside of the membrane. Some proteins are loosely associated with the surface of the membrane (extrinsic proteins); other proteins are embedded in the membrane (intrinsic proteins). When carbohydrates unite with lipids, they form glycolipids, and when they unite with proteins they form glycoproteins. Surface carbohydrates, lipids, and proteins make up the **glycocalyx,** which is necessary for cell-to-cell recognition. The membrane can be visualized as a very fluid structure that can invaginate to form and pinch off vesicles or unite with vesicles to release materials to the outside of the cell.

Endoplasmic Reticulum and Ribosomes

The **endoplasmic reticulum** (ER) is a complex network of membranes that runs through much of the animal cell (fig. 2.3). This membrane system is the site of many of the synthesis reactions, and it has important transport functions within the cell. Very small particles, called **ribosomes,** are frequently associated with the ER. In this state, the endoplasmic reticulum is called **rough ER.** When devoid of ribosomes, the endoplasmic reticulum is called **smooth ER.** Ribosomes and ER are functionally interrelated. Proteins are synthesized at the ribosome and then transported throughout the cell within the ER. Ribosomes are also found "floating" in the cytoplasm, probably interconnected by fine protein strands. Smooth ER is the site of synthesis of steroids, detoxification of organic molecules, and the storage of ions such as Ca^{++}.

The Golgi Apparatus and Vesicles

The Golgi apparatus is a membranous structure that packages products of cellular metabolism (fig. 2.4). It

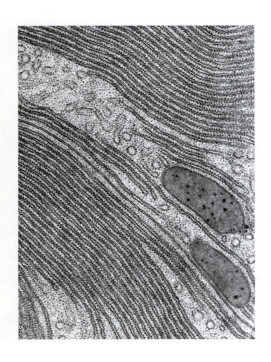

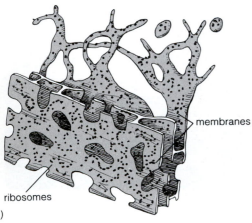

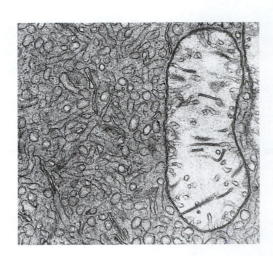

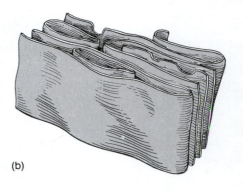

(b)

(a)

Figure 2.3 Endoplasmic reticulum (ER): (*a*) rough ER with associated ribosomes (×60,000); (*b*) smooth ER lacks ribosomes (×64,000). A mitochondrion is shown at the right side of the electron micrograph.

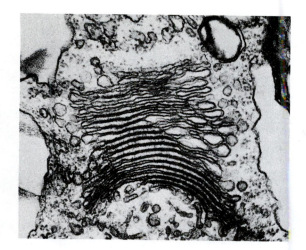

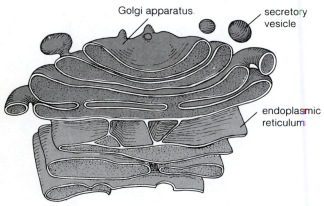

Golgi apparatus

secretory vesicle

endoplasmic reticulum

Figure 2.4 The Golgi apparatus. Note the vesicles pinching free of the cisternae (×37,250).

comprises stacks of **cisternae** that are often associated with the endoplasmic reticulum. In figure 2.4, note the vesicles that have been released to the cytoplasm. Golgi apparatuses are particularly numerous in cells of the pancreas, in cells lining the cavity of the small intestine, in cells of endocrine glands, and in other secretory structures.

The Nucleus and Nucleolus

The nucleus is the genetic control center of the cell and the most obvious structure in most cells. The nucleus contains DNA and protein, usually in a highly dispersed state called **chromatin.** During cell division, chromatin condenses into chromosomes. The nucleus is surrounded by a double membrane called the **nuclear envelope.** The nuclear envelope has relatively large pores and is continuous with endoplasmic reticulum (figs. 2.5 and 2.6). Nuclear pores allow RNA to move between the nucleus and the cytoplasm. Inside the nucleus one frequently finds a nucleolus. This is where the RNA of ribosomes is synthesized.

Mitochondria

Most energy conversions in the cell take place within mitochondria (fig. 2.7). In these reactions, energy in carbohydrates, fats, and proteins is converted into a form usable by the cell called adenosine triphosphate (ATP). Each mitochondrion is enclosed by two membranes. The inner

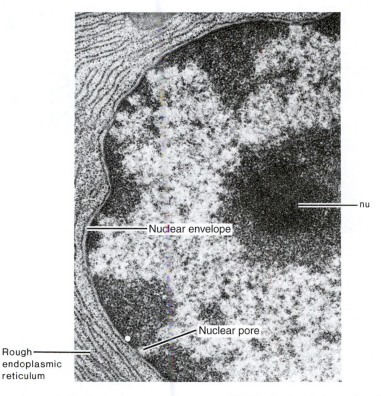

nu

Nuclear envelope

Nuclear pore

Rough endoplasmic reticulum

Figure 2.5 The nucleus. Note the nucleolus (nu), chromatin (containing DNA), the double-layered nuclear envelope, and nuclear membrane pores. Note the rough endoplasmic reticulum outside the nucleus (×31,500).

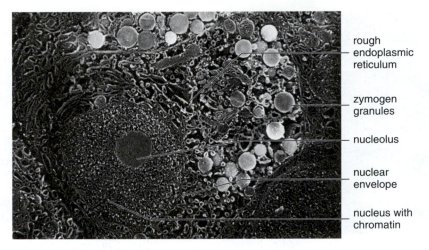

Figure 2.6 A high resolution scanning electron micrograph of a pancreatic cell. This cell produces and secretes precursors of digestive enzymes (zymogen granules). Note the densely folded network of rough endoplasmic reticulum that fills the interior of the cell.

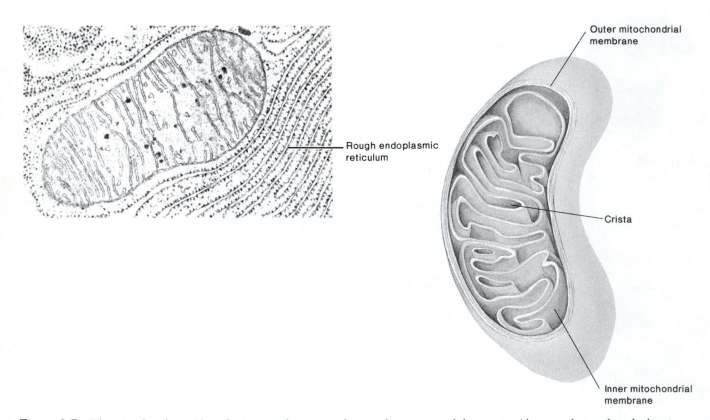

Figure 2.7 The mitochondrion. Note the inner and outer membranes, the cristae, and the matrix. Also note the rough endoplasmic reticulum outside of the mitochondrion (×36,300).

membrane is tightly folded to produce shelflike cristae. Between cristae is a gelatinous matrix containing DNA, ribosomes, and enzymes. Mitochondrial DNA and ribosomes are similar to bacterial DNA and ribosomes. This has led biologists to believe that mitochondria are self-replicating units and that they may have evolved from symbiotic, bacteria-like cells. Enzymes located in the matrix and on the cristae catalyze cellular energy conversions.

Lysosomes

Lysosomes are vesicular structures that contain enzymes capable of digesting all biologically important macromolecules (fig. 2.8). They are responsible for digesting particles ingested by the cell as well as old, nonfunctional organelles. Lysosomes are formed at the Golgi apparatus when enzymes, produced at the ribosomes and transported to the Golgi apparatus by the endoplasmic reticulum, are enclosed within a vesicle.

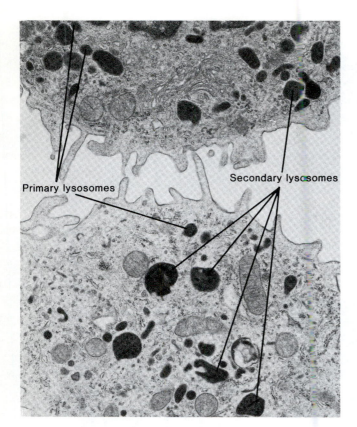

Primary lysosomes   Secondary lysosomes

Figure 2.8 Lysosomes in white blood cells (macrophages) of the lungs. After a lysosome (primary lysosome) fuses with a vacuole containing foreign or waste material, it is often called a secondary lysosome ($\times 10,400$).

The Cytoskeleton: Microtubules and Microfilaments

Microtubules and microfilaments are the basis of cellular locomotion and structure. **Microfilaments** are protein fibers that are arranged in linear arrays or networks. Microfilaments are organized similar to actin, one of the proteins present in muscle. Actin, along with other proteins, is involved with the shortening of muscle cells during contraction. Similar microfilaments are involved with amoeboid locomotion and other cytoplasmic movements, movement of vesicles and other structures within the cell, and the maintenance of cell shape. The array of microtubules and microfilaments within a cell is called the cytoskeleton.

Microtubules are cylindrical arrangements of proteins that, like microfilaments, are the basis of cellular locomotion and structure (fig. 2.9). Microtubules support elongate cell processes (axons of nerve cells), produce movement within cells (movement of chromosomes in cell division), and provide movement for entire cells or substances outside of cells (cilia and flagella).

Cilia and Flagella

Cilia and **flagella** are hairlike organelles that project from the surface of some cells. Cilia and flagella cause cells to move through their media or cause fluids or particles to

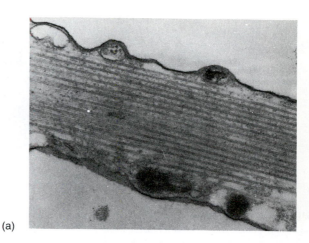

(a)

(b)

Figure 2.9 Ultrastructure of an axopodium (slender cell extension used in feeding and locomotion) of the protozoan *Echinosphaerium nucleofilum* showing microtubules in (a) longitudinal ($\times 30,000$), and (b) cross sections ($\times 4,000$).

move over the cell surface. Cilia and flagella (excluding bacterial flagella) are similar in structure. The primary difference is in length, flagella being longer than cilia. Just inside the plasma membrane of a cilium or flagellum are nine double microtubules. These microtubules originate at the base of the cilium or flagellum at a structure called the **basal body.** These microtubules run the length of the cilium or flagellum and surround a core of two central microtubules (fig. 2.10). This is often referred to as a 9 + 2 organization. Movement of the cilium or flagellum is believed to occur as adjacent double tubules slide past one another. The interaction between microtubules is believed to result from small "arms" on one member of a pair of microtubules interacting with an adjacent pair of microtubules. The basal body (or kinetosome) organizes the formation of the microtubules. It is a short cylinder in which microtubules are arranged in nine triplets around a core that lacks central tubules (9 + 0).

The Microtubule-Organizing Center and Centrioles

The **microtubule-organizing center** occurs outside the nucleus of animal cells. It gives rise to two **centrioles** set at

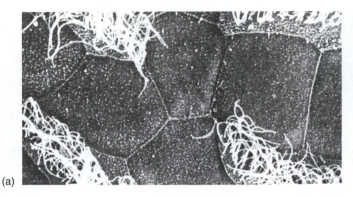
(a)

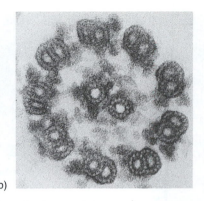
(b)

Figure 2.10 The structure of cilia: (a) scanning electron micrograph of cilia from the gills of an amphibian embryo (*Ambystoma mexicanum*) (×600); (b) Cross section through a flagellum (TEM ×350,000). Notice the 9 + 2 arrangement of microtubules. Each tubule pair has a set of arms that interacts with adjacent microtubules during flagellar movement.

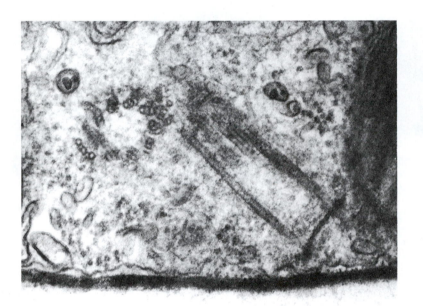

Figure 2.11 Cross section and longitudinal section of a pair of centrioles. Note the 9 + 0 pattern (×100,000).

right angles to each other. The microtubule-organizing center plays a critical role in forming the tubule system (mitotic spindle) that is involved with chromosome movements during cell division. The microtubule-organizing center also produces the array of microtubules and microfilaments that make up the cytoskeleton of a cell. Centrioles are similar in structure to basal bodies (fig. 2.11).

Identification of Subcellular Structures

▼ Identify and describe the function of the subcellular structures shown in figure 2.12. Record your answers in table 2.2 in worksheet 2.1. ▲

2.1 LEARNING OUTCOMES REVIEW—WORKSHEET 2.2

2.2 HISTOLOGY: THE STUDY OF ANIMAL TISSUES

LEARNING OUTCOMES

1. Recognize simple epithelium from lung tissue and describe that epithelium's function.
2. Recognize stratified epithelium from the skin and describe that epithelium's function.
3. Recognize dense connective tissue, loose connective tissue, and adipose tissue. Describe the function of these connective tissues.
4. Recognize skeletal muscle tissue and describe the function of muscular tissue.
5. Recognize nervous tissue from spinal cord smears and describe the function of nervous tissue.

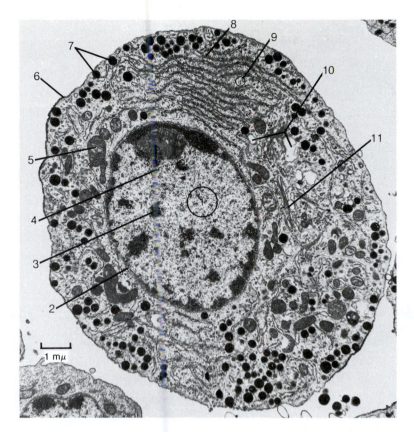

Figure 2.12   Transmission electron micrograph of an animal cell (photomicrograph ×8,000). Identify the organelles indicated and enter their names and functions in table 2.2 in worksheet 2.1.

Tissues are groups of cells of similar structure and origin that function together. The study of the structure of tissues is called **histology.** There are four kinds of tissues: **epithelial, connective, muscular,** and **nervous.** In the following section, you will study epithelial and connective tissues as they occur in vertebrates. (Although invertebrate tissues are organized in a similar manner, there are striking differences that cannot be covered here.)

▼ When examining each tissue, read the accompanying description first and then examine the slide as directed. Keep in mind that you will be looking at sections through organs (assemblages of tissues of various kinds) and that you want to look in specific regions for the desired tissue. Also, thin sections for mounting specimens on slides are often made at angles such that the structure of a cell may not look as you expect. You must look at more than one cell, or possibly more than one slide, to appreciate the actual structure of the tissue. Compare what you see to the appropriate figure. ▲

EPITHELIAL TISSUE

Epithelial tissue consists of a sheet of cells that covers the surface of the body or of one of its cavities. One side of a layer of epithelial tissue is, therefore, exposed to air or body fluids and the other side rests against a noncellular layer produced by the epithelial tissue and the underlying connective tissue called the **basement membrane.** In most preparations, the basement membrane cannot be directly observed, but its position is indicated by the lower limit of the inner cell layer. The chief functions of epithelium are protection, secretion, absorption, and providing surfaces for diffusion. Epithelial tissues are classified according to shape and layering. You will examine two epithelial tissues.

Simple Epithelium

Simple epithelium consists of a single layer of cells, all of which reach the basement membrane. Simple epithelium is found in areas of little wear and tear or where diffusion or absorption occurs.

▼ Simple squamous epithelium consists of thin, flat cells. Nuclei often appear to bulge from the middle of the cell and are flattened in the same plane as the cell is flattened (fig. 2.13a). In the lungs, simple squamous cells line the alveoli (tiny air sacs) and make up the thin layer across which gases diffuse. Examine a prepared slide of lung tissue. Under low power, note the loose, airy consistency. The sectioned larger ducts and vessels represent branches of the pulmonary circulation and respiratory passageways. The alveoli are the smallest, blind-ending sacs. Focus on a group of these under

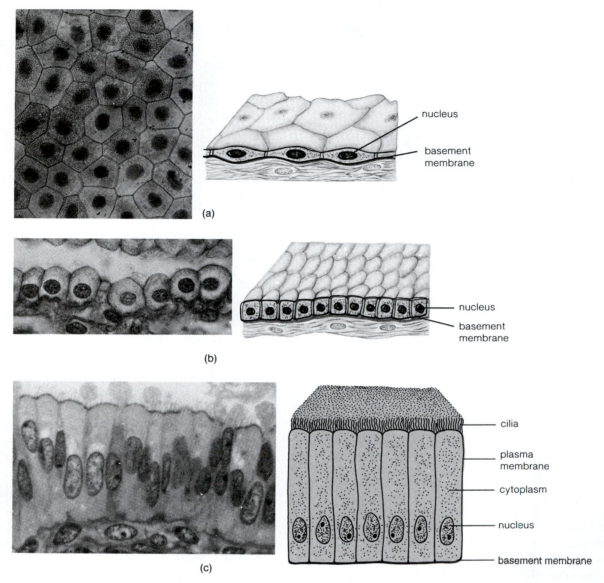

Figure 2.13 Epithelial tissues: (a) simple squamous epithelium (×750); (b) simple cuboidal epithelium from kidney tubules (photomicrograph ×250); (c) simple columnar epithelium from fallopian tubes. Note the cilia responsible for moving the ovum or developing embryo to the uterus (photomicrograph ×1,200).

high power. Note the thin layer of cells bordering the alveoli. You will not be able to see cell boundaries; however, individual cells are represented by the presence of darkly stained nuclei. ▲

Other simple epithelial tissues include the lining of kidney tubules and the lining of fallopian tubes. Kidney tubule epithelium is comprised of cuboidal cells that are approximately as wide as they are high and deep. The nuclei usually appear round (fig. 2.13b). Exchanges of ions, water, glucose, and other molecules occur across these tubules during urine formation. The lining of fallopian tubes is comprised of columnar cells that are taller than they are wide. Nuclei are flattened laterally. The surfaces of these columnar cells may be modified with cilia for the transport of ova and embryos (fig. 2.13c). In the small intestine, the surfaces

of simple columnar cells are modified with microvilli that increase the surface area for secretion and absorption. Recall your observations of intestinal simple columnar epithelium when observing the Golgi apparatus in the cell structure exercise.

Stratified Epithelium

Stratified epithelium is found where there are two or more layers of cells (i.e., the outermost layer of cells does not reach the basement membrane). Only the bottommost layer of cells is reproducing. As new cells are produced, older cells are pushed toward the surface.

Stratified epithelium occurs in areas where wear and tear are common. It serves as a protective barrier between

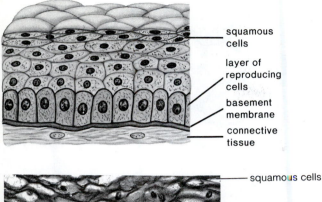

squamous cells

layer of reproducing cells

basement membrane

connective tissue

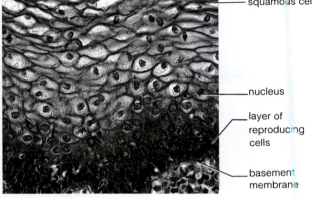

squamous cells

nucleus

layer of reproducing cells

basement membrane

Figure 2.14 Stratified squamous epithelium (photomicrograph ×100).

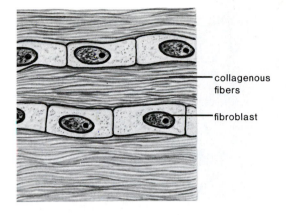

collagenous fibers

fibroblast

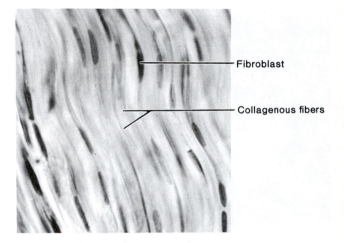

Fibroblast

Collagenous fibers

Figure 2.15 Dense (fibrous) connective tissue (photomicrograph ×400).

the outside and the tissues below. Little transport occurs across stratified epithelium. Stratified squamous epithelium occurs in the skin, vagina, and cornea.

▼ Examine a section of human skin. Under low power, find the outer cell layers. Note the rows of darkly stained cells. The innermost row of cells sits on the basement membrane. This wavy boundary separates dermal and epidermal layers of the skin (fig. 2.14).

Switch to high power. Compare the shape of the inner and outer cell layers. Note the transition between the cuboidal cells near the basement membrane and the squamous cells at the outside. In this tissue, the outer cells are largely dead or dying; they are so far from the blood supply deeper in the skin that they can no longer be nourished by diffusion. In addition, those outer cells contain a tough protein (keratin) that makes this layer relatively impermeable to water, bacteria, and other external agents. Can you appreciate why this tissue is largely protective? ▲

CONNECTIVE TISSUE

Connective tissues bind, anchor, and support the body and its organs. They are made of cells that secrete, and are embedded in, a matrix (ground substance). The cells usually also secrete one or more kinds of protein fibers. The nature of the matrix (fluid, semisolid, or solid) and the kinds of fibers determine the properties of the connective tissue. You will first study two general types of connective tissues and

then one connective tissue that is specialized for fat storage. Other kinds of connective tissues will be studied in unit III.

Dense connective tissue has a gel-like matrix. Its fibers are primarily made of tough, resistant collagen. It is strong and resistant to stretching. Dense connective tissue is found in tendons, ligaments, and in the middle regions of the skin (leather).

▼ Examine a slide of longitudinally sectioned dense (white fibrous) connective tissue (fig. 2.15). (These slides are usually prepared from a piece of tendon.) Note the parallel strands of collagenous fibers separated by rows of tendon cells (nuclei darkly stained). The bulk of this tissue is made of bundles of collagenous fibers. The matrix and cells are sandwiched between the fibers. ▲

Loose connective tissue also has a gel-like matrix. It has multiple fiber types that are oriented randomly in the tissue. The random orientation of fibers makes this tissue resistant to pull from any direction. It is very common and fastens skin to muscle, muscle to muscle, and blood vessels and nerves to other body parts. It is found in the subcutaneous layer of the skin.

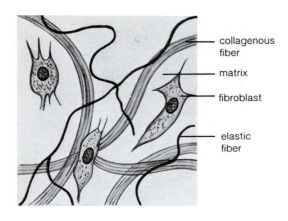

collagenous fiber

matrix

fibroblast

elastic fiber

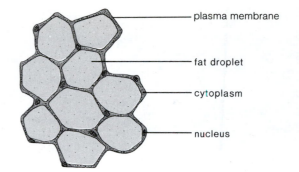

plasma membrane

fat droplet

cytoplasm

nucleus

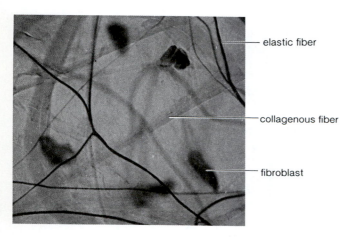

elastic fiber

collagenous fiber

fibroblast

Figure 2.16 Loose connective tissue (photomicrograph ×250).

Figure 2.17 Adipose tissue (photomicrograph ×250).

▼ Examine a slide of loose (areolar) connective tissue under low and high power (fig. 2.16). This slide will most likely be a section from the subcutaneous layer of the skin, which fastens the skin to muscle. Note the irregularly arranged fibers and scattered cells. All cells and fibers are embedded in a clear matrix. ▲

Adipose tissue is one of a number of kinds of connective tissues that have specialized structures and functions. **Adipose tissue** is specialized for fat storage (fig. 2.17). Adipose accumulates in the subcutaneous layer of the skin, around the heart, and around the kidneys. As adipose cells mature, fat accumulates in the cells and pushes cytoplasm to the periphery.

▼ Examine a slide showing adipose tissue. Use a slide especially prepared to show adipose tissue or use the skin section previously studied. If the latter, look deep in the section within the subcutaneous layer. This area contains loose connective tissue as well as adipose tissue. Observe under low and high power. Note the large, clear, fat storage area. The fat has been dissolved by the clearing agent used in preparing the slide. Identify the nucleus and cytoplasm of the fat cells. ▲

Other specialized connective tissues include cartilage, bone, blood, and lymph. Some of these will be studied in unit III.

MUSCULAR TISSUE

There are three types of muscular tissue: skeletal, cardiac, and smooth. These three types of muscular tissue will be studied in more detail in exercise 17. You will examine skeletal muscle here as an introduction to muscular tissue. Muscular tissue is contractile. That is, it has the ability to shorten and accomplish movement. Skeletal muscle tissue is associated with the vertebrate skeleton in combination with other tissues to form muscles (e.g., the biceps). In the microscope, skeletal muscle tissue has a striated appearance that results from a very regular and precise arrangement of subcellular microfilaments (myofilaments). A more detailed explanation of this structure will be presented in exercise 17.

▼ Examine a slide of skeletal muscle using low power. Note the elongate cells (fig. 2.18). The dark structures within the cells are nuclei. Skeletal muscle cells are multinucleate. The long multinucleate cells allow a muscle to shorten along its entire length during contraction. Switch to high power. Notice the striations, or banding. The darker bands are called A bands and the lighter bands are called I bands. ▲

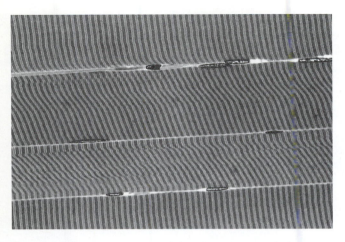

Figure 2.18 Muscular tissue. This photomicrograph shows skeletal muscle. The elongate cells have multiple nuclei. The banding pattern results from the very regular arrangement of contractile proteins that allow muscular tissue to shorten (photomicrograph × 1,000).

NERVOUS TISSUE

The basic unit of the nervous system is the nerve cell or neuron. Other kinds of cells support the function of neurons. The histology of the nervous system involves the study of the interconnections between different kinds of neurons and their supporting cells. In the cell structure portion of this exercise, you examined neurons from the spinal cord. We will not reexamine that preparation at this time. Additional histological observations of nervous tissue will be made in exercise 18.

2.2 LEARNING OUTCOMES REVIEW—WORKSHEET 2.2

WORKSHEET 2.1 Animal Cells

			Cellular Structures	
TABLE 2.1	**Observations of Selected Animal Cells As Seen through the Compound Microscope (Recall that 1 mm = 1,000 μm).**			

Cell Type	Size (μm)	Cell Shape	Cellular Structures Visible	Special Considerations
Cheek epithelial cell				Question 1 below
Nerve cell				Question 2 below
Red blood cell				Question 3 below
White blood cell				Question 4 below
Amphiuma liver cell				Question 5 below
Intestinal epithelial cell				Question 6 below

1. Compare the appearance of the cheek cells before and after staining. What structure(s) was selectively stained? Why did the instructions require that you use reduced light to view these cells?

2. In what way is the anatomy of these cells well suited for their function?

3. These red blood cells were lighter pink in the middle of the cell and darker pink around the periphery of the cells. What does this observation suggest about the shape of the cells? (Hint. Does light pass more easily through a thick or thin object?)

4. Why do you think there are so many fewer white blood cells than red blood cells in this blood smear?

5. Mitochondria are also abundant in certain kinds of fat deposits in mammals adapted to cold environments, in mammals that hibernate, and in newborn mammals. Based on what you know of the function of mitochondria, what do you think could explain their presence in these fat deposits?

6. Remembering that these cells are from the small intestine, what do you think is the function of the protein secretions of these cells?

TABLE 2.2   The Identification and Description of Subcellular Structures in Figure 2.12

Number	Structure Name	Function
1		
2		
3		
4		
5		
6		
7		
8		
9		
10		
11		

7. The nucleus, ribosomes, endoplasmic reticulum, and Golgi apparatus have functions that are interrelated. Explain this interrelationship.

8. Compare and contrast the usefulness of compound and electron microscopes in the study of cell structure.

WORKSHEET 2.2 Cells and Tissues

2.1 Cell Structure—Learning Outcomes Review

1. This organelle is in the form of stacks of disk-like membranes and is especially prominent in secretory cells.
   a. ribosome
   b. lysosome
   c. centriole
   d. Golgi apparatus

2. This cellular structure is the preassembly point for ribosomes. There is often more than one of these in each cell.
   a. nucleus
   b. Golgi apparatus
   c. nucleolus
   d. ribosome

3. One of the preparations you observed did have more than one of the structures described in question 2. Which preparation was it?
   a. spinal cord nerve cell
   b. red blood cell
   c. cheek epithelial cell
   d. *Amphiuma* liver cell

4. All of the following cellular structures are comprised of microtubules EXCEPT one. Select the exception.
   a. endoplasmic reticulum
   b. centrioles
   c. cilia
   d. flagella

5. This is a double-membrane enclosed organelle. It has its own DNA that is separate from nuclear DNA.
   a. ribosome
   b. Golgi apparatus
   c. mitochondrion
   d. basal body

6. These cellular structures help maintain cell shape, contributing to the cytoskeleton, and may help produce cellular movement.
   a. microtubules
   b. microfilaments
   c. endoplasmic reticulum
   d. cristae
   e. Both a and b contribute to the cytoskeleton.

7. When you examined one of the tissue preparations in the histology exercise, you observed microfilaments specialized for a particular function. Which of the following is the tissue in question?
   a. stratified epithelium
   b. simple epithelium
   c. loose connective tissue
   d. muscular tissue

2.2 Histology: The Study of Animal Tissues—Learning Outcomes Review

8. Which of the following tissues would you expect to find making up the inner lining of a blood vessel?
   a. epithelial
   b. connective
   c. muscular
   d. nervous

9. Simple epithelium is likely to occur in areas where
   a. protection is required.
   b. cell movement occurs.
   c. absorption or secretion occurs.
   d. communication between cells is required.

10. One would expect to find a basement membrane in
    a. adipose tissue.
    b. muscular tissue.
    c. simple epithelium.
    d. loose connective tissue.
    e. Both c and d would have basement membranes.

11. General connective tissues, like dense and loose connective tissue, would all have the following components EXCEPT one. Select the one item that would not be present.
    a. a matrix that may be fluid, semisolid, or solid
    b. cells that maintain the tissue
    c. protein fibers that help determine the characteristics of the connective tissue
    d. microvilli on the lumenal surface of cells to promote absorption or secretion

12. Some connective tissues are specialized for particular functions and may lack one or more of the common elements of general connective tissues listed in the previous question. An example of a specialized connective tissue is
    a. a tendon.
    b. a ligament.
    c. blood.
    d. the outer layer of the skin.

2.1   Cell Structure—Analytical Thinking

1. Explain why zoologists often study cells and, at the very least, need to have a working knowledge of cell structure and function?

2. In this exercise you observed that animal cells vary in size and shape. Use examples from your observations of cells to explain how cell size and shape may be related to the function of the cell. Think about the nerve cells, the red blood cells, and the intestinal epithelial cells that you observed.

3. Cells are usually very small. Why do you think tissues are comprised of many small cells rather than fewer large cells? (There are exceptions to this generalization. As you think about your answer, consider the very long skeletal muscle cells that you observed. Why are these cells so very long and what is unusual about the organelles you observed in these skeletal muscle cells?)

2.2 Histology: The Study of Animal Tissues—Analytical Thinking

4. Why do you think blood is considered a connective tissue?

5. The cartilage that supports your ear and the bone that makes up the femur of your leg are both connective tissues. These tissues have different properties. The cartilage of your ear is quite flexible and elastic and the bone of your femur is very hard and strong. What component of these tissues do you think is mostly responsible for the different properties: matrix, fibers, or cells? Explain.

6. The outer layer of the skin (an organ) is comprised of stratified epithelium and below that is dense connective tissue. Just below the dense connective tissue, one finds loose connective tissue and adipose tissue. What is the function of the skin, and what is the contribution to each of these tissues to that function? You may consult your textbook for information of the functions of vertebrate skin.

Exercise 3

Aspects of Cell Function

Prelaboratory Quiz

Study this week's laboratory exercise and then complete the following quiz to assess your preparation for the laboratory.

Transport Into and Out of Cells

1. In this week's laboratory, you will be studying cellular transport processes. Which of the following processes involve(s) passive transport?
   a. diffusion only
   b. osmosis only
   c. bulk transport
   d. diffusion and osmosis

2. Which of the following is necessary in order for materials to move by diffusion?
   a. energy expenditure by a cell
   b. a selectively permeable membrane
   c. a concentration gradient
   d. a membrane carrier

3. In this week's laboratory, you will be observing osmosis across the membrane(s) of
   a. a dialysis bag.
   b. cheek cells.
   c. white blood cells.
   d. red blood cells.

4. If a cell is placed in a hyposmotic solution, it will
   a. swell.
   b. shrivel.
   c. remain unchanged.

5. When large molecules are transported across the plasma membrane of a cell by invagination of the plasma membrane, forming a vacuole around the molecule, the process is called
   a. passive transport.
   b. active transport.
   c. bulk transport.

Aerobic Cellular Respiration

6. In the experiment on aerobic cellular respiration, you will be measuring the rate of cellular respiration by monitoring the rate at which
   a. oxygen is produced.
   b. oxygen is used.
   c. carbon dioxide is used.
   d. carbon dioxide is produced.

7. The purpose of soda lime in the cellular respiration apparatus is to
   a. absorb carbon dioxide.
   b. absorb oxygen.
   c. absorb water.
   d. provide oxygen.

8. The interpretation of the results of a scientific study is presented in this portion of a scientific paper.
   a. abstract
   b. introduction
   c. materials and methods section
   d. discussion section

9. A(n) _____ is a portion of a scientific paper that presents a concise summary of information presented in the paper.
   a. abstract
   b. introduction
   c. materials and methods section
   d. discussion section

10. Tables, graphs, and other data presentations are usually found in the _____ section of a scientific paper.
    a. abstract
    b. introduction
    c. materials and methods
    d. results
    e. discussion

Cell Division

11. The division of the cytoplasm during cell division is
    a. mitosis.
    b. prophase.
    c. anaphase.
    d. telophase.
    e. cytokinesis.

12. When chromosomes are lined up along the equator of a dividing cell, the cell is in _____ of mitosis.
    a. prophase
    b. metaphase
    c. anaphase
    d. telophase

13. Chromatids reach the poles of a cell and begin to "unwind" into chromatin, the nuclear envelope and nucleolus reappear, and spindle fibers disappear during _____ of mitosis.
    a. prophase
    b. metaphase
    c. anaphase
    d. telophase

14. In comparison to the cells that enter into the process of mitotic cell division, the daughter cells
    a. have exactly the same genetic composition.
    b. have one-half the number of chromosomes.
    c. have chromosomes consisting of two chromatids.
    d. have double the number of chromosomes.

15. All of the following are valid comparisons of mitotic cell division and mitotic cell division except one. Select the exception.
    a. Meiotic cell division results in the production of gametes, and mitotic cell division results in the production of other body cells.
    b. Mitotic cell division results in the formation of four daughter cells, and meiotic cell division results in the formation of two daughter cells.
    c. Meiotic cell division results in the formation of haploid cells, and mitotic cell division results in the formation of diploid cells.
    d. Mitotic cell division occurs in all tissues of an animal, and meiotic cell division occurs only in gonads.

3.1 TRANSPORT INTO AND OUT OF CELLS

LEARNING OUTCOMES

1. Explain why cells must exchange materials with their environment.
2. Describe how diffusion occurs, give an example of diffusion occurring in an animal, and describe how diffusion was illustrated in the laboratory.
3. Describe how osmosis occurs, give an example of osmosis occurring in an animal, and describe how osmosis was illustrated in the laboratory.
4. Describe how active transport and bulk transport occur and give examples of active transport and bulk transport occurring in an animal.

A cell is not independent of its environment. It is continuously exchanging gases, nutrients, and wastes between intracellular and extracellular compartments. What can be exchanged and how exchange occurs is largely determined by the properties of the plasma membrane and the substance involved. Exchange occurs by **passive transport, active transport,** and **bulk transport.** Passive transport involves the movement of substances without the expenditure of energy by the cell. In active transport, protein carriers in the plasma membrane use energy in transporting substances across the plasma membrane. Bulk transport occurs when large molecules must be transported; it involves invagination or evagination of broad areas of the plasma membrane.

PASSIVE TRANSPORT— DIFFUSION AND OSMOSIS

Diffusion is the movement of a substance from an area of high concentration to an area of low concentration and is the result of the constant random motion of the molecules involved. As molecules randomly vibrate, they collide with one another and rebound. The result is a net movement of molecules toward areas of lower concentration. The difference in concentration of a substance between two points of reference is a **concentration gradient.** Diffusion is important in virtually all animal systems. It is the means by which gases are exchanged between cells and their environment, many nutrients are taken into cells, waste products are lost from cells, and ions are exchanged in muscle, nerve, and kidney functions.

Brownian Movement

▼ Use a teasing needle to get a small amount of powdered carmine dye into a drop of water on a microscope slide. Cover with a coverslip. Focus on some of the smallest particles using high power. Notice that the dye particles are vibrating slightly. This vibration is Brownian movement. It is the result of water molecules bumping into dye particles. Is there any net movement, or does the particle just move around the same average point? _____ ▲

Diffusion

▼ Moisten the tip of a teasing needle with water. Dip the tip of the needle in a small beaker of methylene blue crystals. Stab the crystals of methylene blue into the center of a plate of agar. Observe the plate of agar during the rest of the laboratory. Use a millimeter ruler to measure the diameter of the circle of dye at time 0 and at 15-minute intervals. Record your observations in table 3.1 in worksheet 3.1. Answer questions 1 and 2 in worksheet 3.1. ▲

Osmosis

Osmosis is the diffusion of water through a selectively permeable membrane. In living systems, it is the plasma membrane that serves as the selectively permeable membrane. A substance's permeability in the plasma membrane is determined by that substance's solubility in lipid, size, and/or charge. Substances soluble in lipid diffuse readily through the lipid portion of the plasma membrane (e.g., oxygen). Water can diffuse rapidly through water channels of the plasma membrane and to a lesser extent through the lipid bilayer. The diffusion of a polar water molecule through the nonpolar lipid bilayer of the plasma membrane apparently occurs because the water molecule is so small that it can actually dissolve in and diffuse through the lipid bilayer.

In osmosis, the selectively permeable membrane separates two solutions containing different concentrations of a nondiffusible solute. Water diffuses along its concentration gradient until concentrations are equal or osmotic pressure balances water's diffusion gradient. **Osmotic pressure** is a measure of the pressure that would be needed to prevent the net diffusion of water from one solution to another. Envision a fish in freshwater. The fish has body fluids that include ions and other dissolved substances that make the fluids approximately 0.9% solute and 99.1% water. The freshwater around the fish is 100% water. Water will diffuse across the fish's permeable gill and mouth surfaces into the fish tissues by osmosis. A marine fish also has body fluids that are approximately 0.9% solute and 99.1% water. The sea water around the fish has a combination of ions that make the environment 3.5% solute and 96.5% water. Water will diffuse out of the fish's tissues and into sea water. These two fish have exactly the opposite osmotic challenges, in one case getting rid of excess water and conserving ions and in the other case conserving water and getting rid of excess salt.

▼ Osmosis is important in kidney function, blood pressure regulation, exchanges at capillary beds, and many other water-related processes in animals. In this exercise, you will observe osmosis across an artificial selectively permeable membrane.

Your instructor will have sections of dialysis tubing cut in approximately 10-cm lengths. A solution of 20% sucrose, beakers, twist-ties, and a balance will also be available. You will form the dialysis tubing into three bags and fill two of the bags with sucrose and one with water. The bags will be placed into beakers as shown in figure 3.1. Designate the beakers as A, B, and C. Add enough of each solution to each beaker so that the bags will be covered when they are placed in the beakers. Dialysis membrane comes from suppliers as a dried, cellophane-like material. When soaked in water, it becomes flexible and is actually a double-layered tube. Soak your three lengths of membrane in water and rub the membrane between your fingers to open up the tube. Close one end of the tube by making a 1-cm lengthwise fold at the end of the tube, twisting the fold tightly, and closing with a twist-tie. Be careful not to puncture the tubing with the wire twist-tie. Repeat with the two other sections of dialysis membrane. Designate the bags as A, B, and C. Add sucrose solution to bags A and C, and add water to bag B. Do not overfill the bags. They should not be turgid when closed. Having too little water or sucrose in the bag is better than having too much. Close the open end of the bag as you did earlier with the opposite end. Rinse the bags with fresh water. Gently blot the bags with a paper towel, and use a clean weigh boat and weigh the bags on the balance provided. Record the mass to the nearest 0.1 gram in table 3.2 in worksheet 3.1. Place the bags in the corresponding beakers all at the same time. Remove the bags after 10 minutes. Blot them gently, and reweigh the bags. Record the mass in table 3.2. Repeat weighing bags and recording masses over the course of one hour.

Terms are used to describe conditions similar to those in the beakers you have prepared. If the solution outside the bag has a lower concentration of nondiffusible solute (sucrose) than inside the bag, the solution is **hyposmotic** to the bag. If the solution outside the bag has a higher

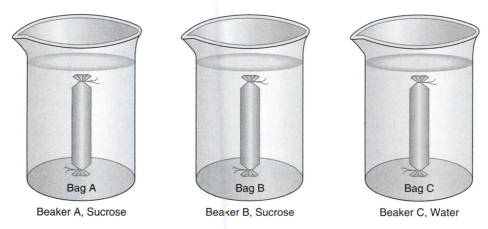

Figure 3.1 Osmosis across an artificial selectively permeable membrane. Beakers A and B contain 10% sucrose, Beaker C contains water. Bags A and C contain 10% sucrose. Bag B contains water.

concentration of nondiffusible solute (sucrose), it is **hyperosmotic** to the bag. If the concentrations are equal, the solution and the bag are **isosmotic** to one another. Graph the change in mass vs. time in the graph paper provided. Answer questions 3–5 in worksheet 3.1. ▲

ACTIVE TRANSPORT

Substances that need to be accumulated within cells against a concentration gradient must move by active transport. Active transport involves the expenditure of energy and the use of protein carriers in the plasma membrane. As you proceed through this course, you will become aware of the importance of this process in nerve cells, muscle cells, and the endocrine and excretory systems.

The laboratory exercise on the excretory system (exercise 22) has a demonstration of active transport of chlorophenol red across the goldfish kidney tubule. Your instructor may want you to turn to that section at this time.

BULK TRANSPORT

Bulk transport occurs when large molecules must be transported across the plasma membrane without being broken down. These molecules are too large to diffuse through membrane pores or to be carried by membrane carriers. For example, nursing mothers provide their infants with maternal antibodies (large proteins) that must be transported across the gut epithelium of the infant without being broken down. Breaking antibodies down would destroy their function. This transport is accomplished by the plasma membrane invaginating to form a vacuole around the antibody, thus incorporating the antibody into the cell. Bulk transport depends on a very fluid plasma membrane and occurs in a variety of ways. None of the mechanisms will be demonstrated, but two common forms it can take are described in the following two paragraphs.

Endocytosis is a form of bulk transport in which materials are taken into a cell. It occurs by **pinocytosis** (cell drinking) or **phagocytosis** (cell eating). Pinocytosis is the nonspecific uptake of small droplets of extracellular fluid and the substances dissolved in the fluid. Invagination of the plasma membrane and vacuole formation incorporate the fluid into the cell. In phagocytosis, cell processes extend around a solid, engulfing it as an amoeba engulfs a food particle or a white blood cell engulfs a foreign substance. Exercise 7 (Animal-like Protists) has a demonstration of endocytosis by *Paramecium*.

Exocytosis occurs when large molecules are moved out of the cell. A vacuole fuses with the plasma membrane and an opening in the plasma membrane develops to release the contents of the vacuole to the outside. The vacuolar membrane is then simply incorporated into the plasma membrane.

3.1 LEARNING OUTCOMES REVIEW—WORKSHEET 3.2

3.2 AEROBIC CELLULAR RESPIRATION: AN EXERCISE IN SCIENTIFIC METHOD

LEARNING OUTCOMES

1. Describe the significance of cellular respiration in animals.
2. Describe the relationship between organismal gas exchange and aerobic cellular respiration.
3. Apply scientific method by designing and carrying out an experiment relating to aerobic cellular respiration.
4. Analyze data and prepare it for presentation in a scientific report.

Aerobic cellular respiration is the process by which most animals make energy available for cellular functions. It consists of a series of reactions, occurring in the cytoplasm and mitochondria, that can be represented by the following equation:

$$C_6H_{12}O_6 + 6O_2 \xrightarrow[\text{mitochondria}]{\text{enzymes}} 6CO_2 + 6H_2O + ATP$$

In these reactions, energy in the form of glucose is converted into a form used by the cell, adenosine triphosphate (ATP). It is assumed that students have covered these reactions in lecture sections.

In detecting aerobic cellular respiration, one can look at the quantity of oxygen used by an organism. When an animal breathes air into its lungs, oxygen is transported by the blood to the cells where it is used in cellular respiration. The rate at which oxygen is used by cells largely determines the rate at which oxygen is delivered to the tissues. This regulation is a complex process involving an interaction of the oxygen levels in body fluids with respiratory centers in the nervous system. The oxygen needs of an animal are directly correlated to the metabolic demands of its cells. Therefore, it is possible to evaluate cellular respiratory rates by measuring the rates of oxygen consumption of an animal. Metabolic demands of cells vary depending upon a variety of internal and external conditions (temperature, time of year, body size, hormonal states, etc.).

In this exercise, you will be designing an experiment to test the influence of one variable on the rate of oxygen consumption in an experimental animal. It will be followed by a formal written report in which you will present and discuss your results. (Exercise 1 has a discussion of scientific method.)

THE RESPIROMETER

A respirometer is an apparatus that is used for measuring oxygen consumption in an experimental animal (fig. 3.2). This exercise assumes that you will be using laboratory mice in your experiment, although other experimental animals can be

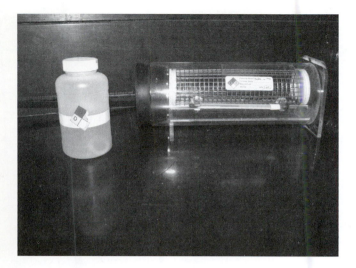

Figure 3.2 A respirometer. A mouse placed in the wire cage is suspended above soda lime. A soap bubble is introduced into the pipette after the respirometer is sealed. As the mouse uses oxygen, and as the carbon dioxide released by the mouse is absorbed by soda lime, the volume of gas in the chamber decreases. The rate of oxygen use is determined by timing the movement of the soap bubble toward the chamber.

substituted. The apparatus consists of a Plexiglas chamber fitted with a rubber stopper and a 5-ml pipette (0.1-ml graduations). A wire cage for holding the mouse and a thermometer are supported within the chamber. You will be measuring oxygen consumption by measuring the change in gas volume inside the chamber. As you can see in the equation for aerobic cellular respiration, a molecule of CO_2 is released with every molecule of O_2 used. To see a change in the volume of gas in the system, the CO_2 given off must be removed. Absorption of CO_2 is accomplished by enclosing soda lime in the chamber below the wire cage. Now, as the animal uses oxygen, the volume of gas in the system will decrease. Decreases in gas volume are measured by introducing a soap bubble into the open end of the pipette. By timing the movement of the soap bubble, one can calculate the rate of oxygen consumption. The following exercise uses one control in the form of a thermobarometer. The thermobarometer consists of a second respirometer excluding the experimental animal. Changes in atmospheric pressure and environmental temperature can be expected to change the volume of gas in all systems. In analyzing data, adjust changes in the first respirometer by adding or subtracting values obtained from the thermobarometer. After designing your experiment using the information under "Getting Started" on page 47, you will use the following data gathering protocol.

MOUSE (MUS MUSCULUS) DATA GATHERING PROTOCOL

1. Select an experimental mouse that has not already been measured.
2. Determine the mouse number and record the number in worksheet 3.1, table 3.3.
3. Determine the mouse's mass and record it in table 3.3.
4. Record the atmospheric pressure at the time of the experiment in table 3.3. (Use a laboratory barometer or an Internet weather information site. You may need to convert barometric readings from inches of mercury to millimeters of mercury by multiplying by 25.4.)
5. Place the mouse in the respirometer, and snuggly fit the black stopper into the chamber opening. Do not add the soap bubble. Start the stop watch for a 5-minute acclimation period. Do not handle the respirometer during the experiment. Heat from your hands could warm the air in the respirometer and cause gas expansion. Do not disturb the mouse during the experiment. Doing so will alter its rate of respiration.
6. Set up the thermobarometer, adjusting a soap bubble to near the middle of the graduated tube.
7. After 5 minutes of acclimation record the temperature in the experimental apparatus (table 3.3) and add a soap bubble to the end of the graduated tube. Reset the stop watch to 0. When the soap bubble passes the 5-ml mark on the graduated tube, start the stop watch and note and record the position of the soap bubble in the thermobarometer (table 3.4).
8. Record the time required for the mouse to use 3 ml of oxygen (table 3.3). Record the position of the soap bubble in the thermobarometer at the end of the time noted above (table 3.4).
9. Remove the black stopper, shake soap from the graduated tube and repeat steps 5–8 with the same mouse.
10. Repeat steps 1–9 with a control mouse. Record data in tables 3.5 and 3.6.

DATA ANALYSIS

You will analyze data using tables 3.7 and 3.8 in worksheet 3.1. The tables, including the required calculations, can be easily incorporated into a computer spreadsheet if desired. Once you have calculated the rate of oxygen consumption in ml O_2/gram/minute, the data must be corrected to conditions of standard temperature and pressure. This correction makes values comparable to published data. Use the following formula:

$$V_2 = \frac{P_1 (273)}{T_1 (760)} \times V_1$$

P_1 = Atmospheric pressure (mmHg) during the experiment.
T_1 = Temperature (°K) during the experiment (°K = °C + 273).
V_1 = Uncorrected volume of oxygen used per minute.
V_2 = Corrected volume of oxygen used per minute.

PRESENTING YOUR RESULTS

Results of research can be presented in written, poster, and oral formats. Presentation of results is one of the most important parts of science. It communicates information that would otherwise go unnoticed, and it allows others to evaluate the validity of the research and the conclusions drawn. Peer review, the examination of research results by other scientists, is the reason that science works. Properly designed experiments, careful methods, and reasonable conclusions are all confirmed when scientists examine each others' work. Your instructor will decide on the format for the presentation of your results. Regardless of the format, thoroughness, accuracy, conciseness, grammatical correctness, originality, and the absence of plagiarism are all hallmarks of good presentations.

Originality and Plagiarism

Scientific research and writing is more than putting together others' ideas and reporting on them. Original research means that, even though others may have performed similar research and even though the writer has certainly been influenced by previous work, the essence of what is presented is based on research and ideas that are the writer's own.

Format

The format for a scientific report varies slightly depending upon whether it is prepared for a journal article, poster, or oral presentation. A typical format, however, would include the following elements:

- title
- abstract
- introduction
- materials and methods
- results
- discussion
- literature cited

Title A short descriptive title should include three elements: (1) the independent variable being tested (e.g., the effect of altered photoperiod), (2) the process being investigated (i.e., aerobic cellular respiration), and (3) the species name of the experimental animal (e.g., a mouse, *Mus musculus*).

Abstract The abstract is a concise summary of information presented in the paper. It should include a statement of purpose, a brief summary of the results, and a statement of the conclusions drawn. The purpose of an abstract is to familiarize the reader with the content of a paper. The reader can then decide whether to read the entire paper.

Introduction The introduction gives a statement of the hypothesis and information on the questions and observations that led to the formulation of the hypothesis. It is not a complete literature review, but a description of other work that influenced the research being described.

Materials and Methods This section precisely describes all equipment and procedures used in carrying out the research.

The purpose of this section is to present enough detail so that someone else could carry out the same procedures and confirm the results. It allows a critical analysis of the adequacy of the research methodology.

Results The data are presented in the results section. Data should be organized into a clearly understandable form. Data might be presented on a graph or in a table. If so, all information for interpretation should be on the graph or in the table. Each graph or table should be on a separate page, should have a reference in the text (Figure 1 shows . . .), and should appear in the format shown in figures 3.3 and 3.4. Figures and tables should be used when they make results

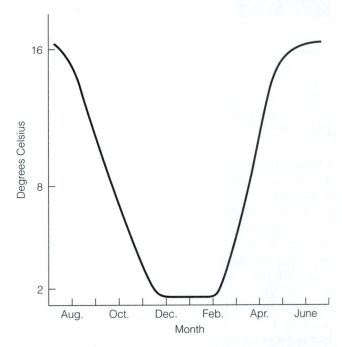

Figure 1 Temperature fluctuations in Parfrey's Glen Creek.

Figure 3.3   Format for a figure.

Table 1.   The monthly weight gain in rats on a diet deficient in iodine.

month	mean weight gain
jan.	4.6 g
feb.	5.2 g
—	—
—	—
—	—
—	—
—	—
—	—

Figure 3.4   Format for a table.

easier to understand. No interpretation of the results is appropriate in the results section.

Discussion In the discussion, results are interpreted and related to other studies. Ideas regarding how procedures might be improved and the need for future research can also be discussed. No new data are presented here.

Citations Citations within the body of the paper are made by using author and date. They may be placed at the end of a sentence or incorporated within the sentence. If there are more than two authors, *et al.* may be used. Following are illustrations of citations.

Pennak (1989) states _____
_____ (statement) _____ (Pennak, 1989).

For more than two authors:
_____ (statement) _____ (Dhar et al., 1997).

Everything cited in the body of the paper must have a reference in the literature cited.

Literature Cited This is an alphabetical listing of all resources mentioned in the paper; nothing should appear in the literature cited that is not mentioned in the paper. Alphabetize by the first author's last name. Do not use *et al.* here. Use the format shown in the following examples. Note spaces to be left between the author's name, the date, and the other items.

Format for citing a journal article:
Dhar, S. R., A. Logan, B. A. McDonald, and J. E. Ward. 1997. Endoscopic investigations of feeding structures and mechanisms in two plectolophous brachiopods. *Invertebrate Biology*, 116 (2): 142–150.

Format for citing a book:
Pennak, R. W. 1989. *Freshwater Invertebrates of the United States.* New York. John Wiley & Sons, Inc. xvi + 628 pp.

Getting Started

▼ The nature of your experiment will be limited by your instructor. You may be asked to work in groups, and your instructor may give you specific suggestions on your project. Read through the sections of your textbook covering aerobic cellular respiration and temperature regulation. As you do, consider the following: What environmental factors might affect the rate of aerobic cellular respiration? What is its role in generating metabolic heat for temperature regulation? Is there any reason that the rate of cellular respiration would vary because of size, age, or sex? Record your observations and questions in worksheet 3.1. Narrow your observations and questions down to a single question and formulate a hypothesis based on that question. Be sure that the hypothesis is in the form of a declarative statement involving a single independent variable. Record your hypothesis in worksheet 3.1. Design a test of your hypothesis

and record the procedure in worksheet 3.1. Be sure to record your independent variable, your dependent variable(s), your controlled variables, and construct a data table for recording results. Consult with your instructor for further instructions and then carry out your experiment. You may be asked to present your results as a formal written report using the guidelines presented on pages 45–47. ▲

3.2 LEARNING OUTCOMES REVIEW—WORKSHEET 3.2

3.3 CELL DIVISION

LEARNING OUTCOMES

1. Describe the significance of mitotic cell division and meiotic cell division in animals.
2. Describe the events of mitotic cell division in animal cells.
3. Recognize stages of mitotic cell division in prepared slides of the whitefish blastula.

Mitosis is the process by which a nucleus divides, and **cytokinesis** is the process by which the cytoplasm divides. Together, they result in the division of the entire cell. From the single-celled zygote arise all of the billions of cells that make up the adult organism. Not only is it a division of one cell to form two, but it is an exact division that results in two cells genetically identical to the original. Each cell of an organism must have a complete set of genetic instructions to function properly.

The **cell cycle** is shown in figure 3.5. Note that the cell spends most of its time in **interphase.** During interphase, the cell is carrying on processes characteristic of that cell. With regard to events related to mitosis, interphase is divided into three stages: **gap 1 (G_1), synthesis (S),** and **gap 2 (G_2).** G_1 stage is a period of active RNA and protein synthesis.

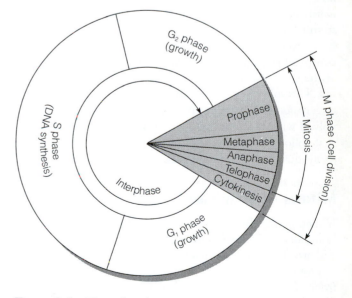

Figure 3.5 The cell cycle.

It immediately follows mitosis. The synthesis phase is a period in which DNA (**chromatin**) is replicated (duplicated) so that each **chromosome** will consist of two chromatids when mitosis begins. The two **chromatids** of a chromosome are joined by a differentiated region of DNA, the **centromere.** The G_2 stage is a period of protein synthesis prior to mitosis. The duration of interphase varies, depending on species and cell type, from hours (embryonic cells) to years (adult bone cells).

Mitosis is divided into four phases that represent artificial points of reference for study of the process: prophase, metaphase, anaphase, and telophase. In reality, mitosis is a smooth, continuous series of changes with no sharp breaks between stages. This can be emphasized by applying the terms "early" or "late" to cells that seem to be between two stages.

MITOTIC CELL DIVISION IN THE WHITEFISH BLASTULA

▼ Obtain a slide of a whitefish blastula. The blastula is an early stage in animal development. It is a hollow ball of rapidly dividing cells. Your slide will show one or more thin sections through that ball of cells. Read through the following descriptions and then examine your slide under low power. Switch to high power, looking for cells representative of each of the stages of mitosis (figs. 3.6 and 3.7). You may need to look at other sections or other slides to see all stages. Remember that mitosis is a continuous process and that you are likely to see intermediate stages. Make drawings of what you see. ▲

Mitosis—Prophase

The beginning of prophase is signaled by the migration of pairs of centrioles to opposite ends (poles) of the cell and the breakdown of the nuclear envelope and nucleolus. Chromatin begins to condense (winding and folding of DNA strands) into chromosomes, and microtubules begin to form from proteins of the cytoplasm. In prophase, chromosomes consist of two chromatids (called sister chromatids) joined at the centromere. Centrioles organize microtubule assembly. The role of centrioles in cell division is still not well understood and is under investigation. Microtubules are involved with the movement of chromosomes in the stages that follow. When viewed under the light microscope, it seems that the microtubules push and pull chromosomes.

Mitosis—Metaphase

In metaphase, chromosomes move to the equator of the cell. Some microtubules (**spindle fibers**) originate at the poles and attach at the centromere of each chromosome at a disk of protein called the kinetochore. Other microtubules run from pole to pole. Microtubules radiating from the poles of the cell form the **aster.** The centrioles and the microtubules

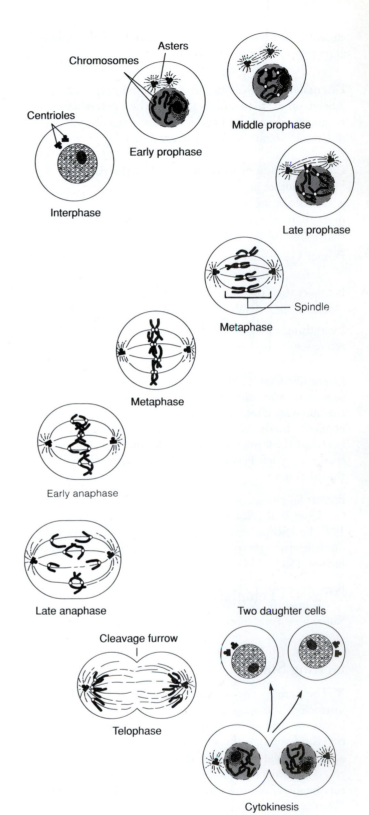

Figure 3.6 Animal mitosis.

radiating between them form the **mitotic spindle.** Late in metaphase, chromosomes are in position along the equator of the cell and the centromeres divide. This latter event frees the sister chromatids from each other (see fig. 3.7a).

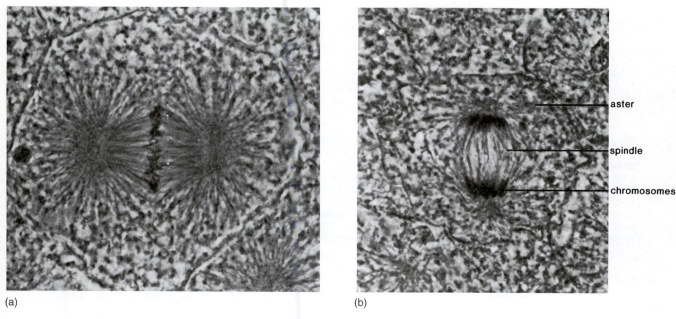

(a)   (b)

Figure 3.7 Mitosis in the whitefish blastula: *(a)* metaphase; *(b)* late anaphase.

Mitosis—Anaphase

Anaphase begins as chromatids of each chromosome separate and move toward opposite poles of the cell. At this point, the sister chromatids are considered full-fledged chromosomes. Depending on the position of the centromere, chromosomes will appear straight, J-shaped, or V-shaped, with the centromere at the apex of each leading the chromosome to the pole. Activities of the pole-to-pole spindle fibers result in an elongation of the entire spindle. Anaphase ends as chromosomes approach the poles of the cell (see fig. 3.7b).

Mitosis—Telophase

Telophase is essentially the reverse of prophase. Chromosomes have reached the poles of the cells, furrowing of the plasma membrane is initiated, spindle fibers disappear, chromosomes "unwind" into chromatin (dispersed DNA and protein strands), and the nuclear envelope and nucleolus reappear. Upon completion of furrowing, cytokinesis is considered to be ended and two new daughter cells result. Each daughter cell enters G1 interphase with chromosomes consisting of one chromatid each and with exactly the same quality of DNA as the parent cell. DNA is duplicated in S phase, so each chromosome will have two chromatids before the next division. A second pair of centrioles is also synthesized during S phase.

CYTOKINESIS IN THE WHITEFISH BLASTULA

Look again at your slide of a whitefish blastula and find a cell in late anaphase of mitosis. As chromosomes approach the poles of a dividing animal cell, cytokinesis begins. Cytokinesis is the division of the cytoplasm by cleavage of the cell at the equator, a process called **furrowing.** Microfilaments are responsible for this gradual constriction of the plasma membrane.

A COMPARISON OF MITOTIC CELL DIVISION AND MEIOTIC CELL DIVISION

Meiosis is a form of cell division that produces the cells involved with sexual reproduction. The products of this form of cell division are usually called eggs and sperm in animals and are referred to as **gametes.** During sexual reproduction, gametes fuse to form a single cell called a **zygote.** The zygote then divides by mitosis to produce the various stages characteristic of an animal's embryonic development.

To maintain a constant number of chromosomes in the next generation, gametes must be produced that have one-half the chromosome number of ordinary body cells. Chromosomes exist in pairs called homologous pairs. Each member of a pair of chromosomes is similar in size and shape and carries genes that code for the same traits of the animal. When a cell contains both members of all homologous pairs, it is diploid (2N). When a cell contains only one member of all pairs of chromosomes, it is haploid (1N). All body cells of most animals, except sperm or eggs, possess a diploid (2N) number of chromosomes. Half of these chromosomes came to the animal through the sperm cell of the father, and half of the chromosomes came in the egg cell of the mother. Meiosis is the form of cell division that produces haploid gametes. Fertilization results in the union of sperm and egg cells and the restoration of the diploid number of chromosomes.

Compare the processes of mitosis and meiosis as shown in figure 3.8. Unlike mitosis, meiosis consists of two nuclear divisions, called meiosis I and meiosis II. During meiosis I,

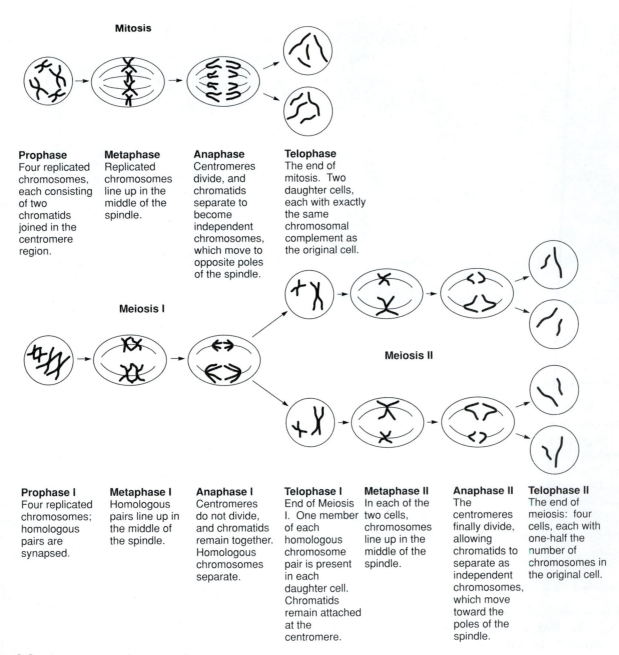

Mitosis

Prophase
Four replicated chromosomes, each consisting of two chromatids joined in the centromere region.

Metaphase
Replicated chromosomes line up in the middle of the spindle.

Anaphase
Centromeres divide, and chromatids separate to become independent chromosomes, which move to opposite poles of the spindle.

Telophase
The end of mitosis. Two daughter cells, each with exactly the same chromosomal complement as the original cell.

Meiosis I

Meiosis II

Prophase I
Four replicated chromosomes; homologous pairs are synapsed.

Metaphase I
Homologous pairs line up in the middle of the spindle.

Anaphase I
Centromeres do not divide, and chromatids remain together. Homologous chromosomes separate.

Telophase I
End of Meiosis I. One member of each homologous chromosome pair is present in each daughter cell. Chromatids remain attached at the centromere.

Metaphase II
In each of the two cells, chromosomes line up in the middle of the spindle.

Anaphase II
The centromeres finally divide, allowing chromatids to separate as independent chromosomes, which move toward the poles of the spindle.

Telophase II
The end of meiosis: four cells, each with one-half the number of chromosomes in the original cell.

Figure 3.8 A comparison of mitosis and meiosis.

homologous pairs of chromosomes pair in a process called synapsis. The pairs then separate and move toward opposite poles of the cell. The two cells produced by this first division contain half of the number of chromosomes present in the parental cell. The chromosomes of these cells consist of two chromatids. Meiosis II is just like mitosis, and the cells produced have chromosomes consisting of one chromatid.

The overall result of mitosis is the production of two cells that are genetically identical to the parent cell. It is the form of cell division that results in the production of all of the cells of an animal's body from a single fertilized egg. In meiosis, chromosome numbers are halved. Members of

a pair of chromosomes sort independently during the first division, and parts of homologous chromosomes can be exchanged in a process called crossing-over. The result is that the cells produced are always genetically different from one another.

Exercise 23 has an activity that helps one understand the processes of spermatogenesis and oogenesis. Your instructor may want you to turn to that exercise at this time.

3.3 LEARNING OUTCOMES REVIEW—WORKSHEET 3.2

WORKSHEET 3.1 Transport Into and Out of Cells

TABLE 3.1	Diffusion of Methylene Blue in Agar
Time (min)	Diameter of stain circle (mm)
0	
15	
30	
45	
60	
75	
90	
105	
120	

1. As regards the diffusion of methylene blue through agarose:
   a. where is the concentration of methylene blue the greatest?

   b. what force caused the movement of methylene blue that you observed?

2. Name two examples of where diffusion of something other than water occurs in an animal. Why is the movement important in each example?

TABLE 3.2	Osmosis Across an Artificial, Selectively Permeable Membrane. After recording your data, use the graph paper provided to plot bag Mass vs. Time. Plot Mass changes for all three bags on the same graph using a unique symbol for the data points associated with each bag		
Time (min)	Mass of bag A (g)	Mass of bag B (g)	Mass of bag C (g)
0			
10			
20			
30			
40			
50			
60			

3. Use one of the following terms to complete each of the following sentences: hyposmotic, hyperosmotic, isosmotic.
   a. The solution in beaker A is _____ to the solution in bag A.
   b. The solution in beaker B is _____ to the solution i n bag B.

4. Write a short description of the events occurring in beaker C. Use the terms osmosis, selectively permeable, net movement, and hypertonic in your answer.

5. If this experiment was allowed to proceed for another two hours, what do you think would happen in each beaker?

Aerobic Cellular Respiration

Observations and Questions from Background Reading

Hypothesis

Materials and Methods
   Identify the variables
      Independent variable

      Dependent variable(s)

      Controlled variables

Describe your experiment

Use an "if/then statement" to predict the results of your experiment.

Results

Use the following tables to record your data or format a spreadsheet for recording data. The values in columns with boldface type will be used in the final calculations of corrected oxygen consumption.

TABLE 3.3   Data from Experimental Animals

Animal number	Mass (g)	T_1 (°K)	P_1 (mmHg)	Trial 1 Time (sec) for 3 ml O_2	Trial 2 Time (sec) for 3 ml O_2	Average time (Trial 1 and Trial 2)

TABLE 3.4   Thermobarometer Readings During Experimental Runs

Animal number	Beginning Reading (ml) Trial 1	Ending Reading (ml) Trial 1	Change 1*	Beginning Reading (ml) Trial 2	Ending Reading (ml) Trial 2	Change 2*	Average Change**

*For thermobarometer readings include a + or − sign. Assign a + if the net movement is toward the respirometer. Assign a − if the net movement is away from the metabolic chamber.
**Add change 1 and change 2 algebraically and divide by 2.

TABLE 3.5   Data from Control Animals

Animal number	Mass (g)	T_1 (°K)	P_1 (mmHg)	Trial 1 Time (sec) for 3 ml O_2	Trial 2 Time (sec) for 3 ml O_2	Average Time (Trial 1 and Trial 2)

TABLE 3.6   Thermobarometer Readings During Control Runs

Animal number	Beginning Reading (ml) Trial 1	Ending Reading (ml) Trial 1	Change 1*	Beginning Reading (ml) Trial 2	Ending Reading (ml) Trial 2	Change 2*	Average Change**

*For thermobarometer readings include a + or − sign. Assign a + if the net movement is toward the respirometer. Assign a − if the net movement is away from the metabolic chamber.
**Add change 1 and change 2 algebraically and divide by 2.

TABLE 3.7   Experimental Run Calculations

Animal Number	Corrected oxygen consumption* (ml O_2)	ml O_2 consumed/ min**	V_1 (ml O_2 consumed/g/min***)	V_2****

*Add or subtract the average thermobarometer change, based on the + or − sign, to 3 ml.
**[Corrected oxygen consumption/ time (seconds)] × 60 seconds
***divide by mass of the animal
****$V_2 = \dfrac{P_1 (273)}{T_1 (760)} \times V_1$

TABLE 3.8 Control Run Calculations

Animal Number	Corrected oxygen consumption* (ml O_2)	ml O_2 consumed/min**	V_1 (ml O_2 consumed/g/ min***)	V_2****

*Add or subtract the average thermobarometer change, based on the + or − sign, to 3 ml.

**[Corrected oxygen consumption/ time (seconds)] × 60 seconds

***divide by mass of the animal

****$V_2 = \dfrac{P_1\,(273)}{T_1\,(760)} \times V_1$

WORKSHEET 3.2   Aspects of Cell Function

3.1   Transport Into and Out of Cells—Learning Outcomes Review

1. A red blood cell has a salt concentration of 0.9%. A salt solution of 1.5% would be
   a. isosmotic to the cell.
   b. hyposmotic to the cell.
   c. hyperosmotic to the cell.

2. The cell in question 1 would
   a. shrivel.
   b. swell.
   c. remain the same.

3. A difference in concentration of a substance between two points of reference
   a. may result in diffusion of the substance toward the point of lower concentration.
   b. is called a concentration gradient.
   c. will result in osmosis if the substance is water and a selectively permeable membrane separates the points of reference.
   d. All of the above are correct.

4. Cells may take up water in which of the following ways?
   a. exocytosis
   b. osmosis
   c. pinocytosis
   d. active transport
   e. Both b and c are correct.

5. A primary difference between active and passive transport is that
   a. active transport moves materials across cell membranes, passive transport does not.
   b. active transport moves materials along concentration gradients, passive transport moves materials against concentration gradients.
   c. passive transport requires carrier molecules, active transport does not.
   d. active transport requires energy expenditure by a cell, passive transport does not.

3.2   Aerobic Cellular Respiration—Learning Outcomes Review

6. The purpose of soda lime in the respirometer used in this exercise is to
   a. react with O_2 in the chamber.
   b. remove CO_2 from the chamber.
   c. stabilize the temperature in the chamber.
   d. provide a source of energy for the experimental animal.

7. One goal of this exercise was to standardize the results of your exercise in order to make your results comparable to the results of others. Select the item in the following list that does not accomplish that goal.
   a. weighing your experimental and control animals
   b. recording temperature within the chamber during the experiment
   c. recording atmospheric pressure during the experiment
   d. All of the above procedures accomplish the goal of making my results comparable to the results of others.

8. The effects of changes in room temperature and atmospheric pressure during the experiment are monitored by
   a. checking atmospheric pressure and room temperature every 10 minutes during the experiment.
   b. ignoring temperature and pressure changes. They cannot influence the results of the experiment.
   c. mathematically correcting for conditions of standard temperature and pressure at the conclusion of the experiment.
   d. recording changes in a control apparatus that does not contain the experimental animal, the thermobarometer.

9. Cellular respiration is monitored by observing the movement of a soap bubble in a pipette attached to the respirometer. During the experiment, the soap bubble
   a. moved toward the respirometer as O_2 was used by the animal.
   b. moved away from the respirometer as CO_2 was given off by the animal.
   c. moved toward the respirometer as CO_2 was used by the animal.
   d. moved away from the respirometer as O_2 was given off by the animal.

10. The reactions of aerobic cellular respiration occur in
    a. ribosomes of the cell.
    b. the Golgi apparatus of the cell.
    c. the cytoplasm and mitochondria of the cell.
    d. the endoplasmic reticulum.

11. The purpose of the reactions of aerobic cellular respiration is to
    a. convert energy into a form useable by the animal, namely ATP.
    b. produce energy.
    c. deliver O_2 to the tissues of the animal where O_2 functions as a cofactor for enzymes.
    d. All of the above are functions of aerobic cellular respiration.

3.3 Cell Division—Learning Outcomes Review

12. Chromatin condenses into chromosomes, the nuclear envelope breaks down, and centrioles begin to organize microtubules of the spindle during
    a. prophase of mitosis.
    b. metaphase of mitosis.
    c. anaphase of mitosis.
    d. telophase of mitosis.
    e. cytokinesis.
    f. S-phase of interphase.

13. Chromatids of a chromosome separate and begin moving to opposite poles of a cell during
    a. prophase of mitosis.
    b. metaphase of mitosis.
    c. anaphase of mitosis.
    d. telophase of mitosis.
    e. cytokinesis.
    f. S-phase of interphase.

14. Chromosomes before _____ consist of one highly dispersed chromatid. After _____ chromosomes consist of two highly dispersed chromatids that are joined at the centromere.
    a. prophase of mitosis.
    b. metaphase of mitosis.
    c. anaphase of mitosis.
    d. telophase of mitosis.
    e. cytokinesis.
    f. S-phase of interphase.

    Do the following statements describe mitosis, meiosis, or both mitosis and meiosis?

15. Daughter cells are haploid (1N).
    a. mitotic cell division
    b. meiotic cell division
    c. both

16. Daughter cells are identical to each other and the parent cell.
    a. mitotic cell division
    b. meiotic cell division
    c. both

17. Chromatids separate at the centromere.
    a. mitotic cell division
    b. meiotic cell division
    c. both

18. Independent assortment of homologous chromosomes and crossing-over occurs.
    a. mitotic cell division
    b. meiotic cell division
    c. both

3.1 Transport Into and Out of Cells—Analytical Thinking

1. Chinook salmon, *Oncorhynchus tshawytscha*, are anadromous. They hatch from eggs in rivers of the Pacific Northwest. Young salmon remain in the rivers for about a year. They then migrate to estuaries and then to the ocean where they spend up to 8 years. Adults then return to their home river for breeding. After breeding the fish die. Describe the osmotic changes encountered by this fish during its lifecycle. Use the terms hyposmotic, hyperosmotic, and isosmotic in your answer. Compare conditions experienced by these fish to the experimental set-up used in the lab to demonstrate osmosis.

2. Fish must be able to maintain relatively constant ion concentrations in their body fluid. Despite cellular membranes that are relatively impermeable to some ions, ion losses in freshwater can occur through gill surfaces, excretion, and defecation. In the sea, salt may accumulate during feeding and by diffusion across soft tissues. To make up for these exchanges, ions are actively transported across gill, and other, surfaces. Describe the direction of active ion transport when the Chinook salmon is in the river vs. the ocean.

3.2 Aerobic Cellular Respiration—Analytical Thinking

3. The rate of aerobic cellular respiration varies in animals based on a variety of environmental factors. In addition to providing ATP to maintain basic body functions, it provides energy for growth, producing young, caring for young and may other productive functions. Much of the energy processed in aerobic cellular respiration is eventually lost to the environment as heat. Evaluate the energy needs and metabolic rates of the following pairs of animals.
    a. arctic hare (*Lepus arcticus*) in winter vs. arctic hare in midsummer
    b. bull elk (*Cervus canadensis*) during mating season in autumn vs. bull elk in late spring
    c. ruby throated hummingbird (*Archilochus colubris*) vs. barnyard chicken (*Gallus gallus domesticus*)

3.3 Cell Division—Analytical Thinking

4. Draw an animal cell with 6 chromosomes in anaphase of mitosis. Draw that same cell in anaphase I of meiosis. Draw the chromosomes differently so that you can distinguish three homologous pairs of chromosomes. Refer to the diagrams as you explain the role of mitotic cell division and meiotic cell division in animals.

Exercise 4

Genetics

Prelaboratory Quiz

Study this week's laboratory exercise and then complete the following quiz to assess your preparation for the laboratory.

1. Two chromosomes containing genes coding for the same group of traits are called
   a. allelic chromosomes.
   b. homologous chromosomes.
   c. homozygous chromosomes.
   d. heterozygous chromosomes.

2. A gene that, when present, is always expressed is said to be
   a. dominant.
   b. recessive.
   c. homozygous.
   d. heterozygous.

3. The genes that code for alternate expressions of a trait are called
   a. alleles.
   b. homologous genes.
   c. homozygous genes.
   d. recessive genes.

4. Use the word "alleles" or "genes" to fill in the following blanks.
   a. The two _____ for body color studied in this week's laboratory are ebony and wild.
   b. Diploid organisms possess two sets of chromosomes containing _____ that code for the same traits, but not necessarily the same expression of those traits.
   c. When a fly has the recessive ebony phenotype, both _____ coding for the body color trait must be the same.

5. True/False   A fruit fly that has sex combs on its prothoracic legs; a genital plate; and a dark, wide band at the tip of the abdomen is a female.

6. True/False   Fruit flies have four larval instar stages and a pupal stage.

7. True/False   In fruit flies, the genetic symbol used to represent a trait is usually taken from the mutant expression of the trait.

8. True/False   The expression "e^+" is indicating a recessive allele derived by mutation from the wild allele, e.

9. True/False The expression "B" is indicating a dominant mutant derived from a recessive wild allele indicated by B^+.

10. True/False   The expression "dp^+" is indicating the dominant wild allele of dp.

11. True/False   The chi-square test is used in this laboratory exercise to determine whether or not the results of your crosses approximate the results expected based on Mendelian inheritance patterns.

12. True/False   The time from egg-laying to emergence of fruit fly adults spans approximately two weeks.

13. True/False   The sex-linked trait used in this week's laboratory exercise is a wing-shape trait.

14. True/False   The second fruit fly cross in this week's laboratory exercise involves a demonstration of Mendel's principle of independent assortment and involves inheritance of traits carried by genes located on a single pair of homologous chromosomes.

Genetics is the study of the mechanisms of transmission of genes from parents to offspring. These mechanisms are dependent on the behavior of chromosomes during the formation of gametes and the way in which gametes are brought together in fertilization. The processes occurring at the cellular level are those that take place during meiotic cell division. In this exercise, it is assumed that the student is thoroughly familiar with meiotic cell division.

Diploid organisms possess two sets of chromosomes containing genes that code for the same traits. Chromosomes containing genes that code for the same traits are referred to as **homologous chromosomes,** and the genes that code for alternate expressions of a trait are referred to as **alleles.** Genes are often either dominant or recessive. A **dominant** gene is one that, when it is present, is always expressed. A **recessive** gene is one that is expressed only when a similar recessive gene is on the homologous chromosome. When the two genes that determine a trait are alike, the organism is **homozygous** for that trait. When the two genes are different, the organism is **heterozygous** for that trait. Thus, a dominant trait is expressed in either the homozygous or heterozygous state; a recessive trait is expressed only in the homozygous state.

4.1 INHERITANCE IN DROSOPHILA—MATERIALS AND METHODS

LEARNING OUTCOMES

1. Demonstrate the procedure for anesthetizing fruit flies.
2. Determine the sex of a fruit fly.
3. Distinguish phenotypes of a fruit fly and write fruit fly genotypes using proper genetic symbols.

In this exercise you will be making crosses involving the fruit fly, *Drosophila melanogaster*. The development from egg to adult spans approximately two weeks, as shown in figure 4.1. Because of this, the following exercise will span the next few

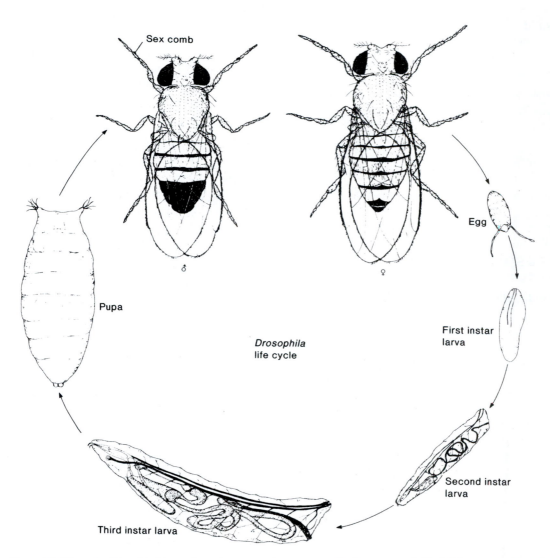

Figure 4.1 *Drosophila* life cycle. The third instar larva is present about one week after eggs are laid. Adults emerge from the pupal case two weeks after eggs are laid.

laboratory periods. You should be prepared to come to the laboratory outside of the normal class period to analyze your results.

Anesthetizing Fruit Flies

▼ Obtain a vial of mixed fruit flies. Charge an anesthetizer or an anesthetizing wand with anesthetic. Your instructor will describe this process, as it differs depending on the anesthetizing method used. The following procedure assumes you are using a plastic anesthetizer. Gently tap the flies to the bottom of the vial. Remove the plug from the vial and tap the flies into the anesthetizer. Allow the anesthetizer to sit until the activity has ceased. ▲

Determination of Sex

▼ When flies are anesthetized, empty them onto a white 3 × 5 card. Examine them under a dissecting microscope. Table 4.1 and figure 4.2 show five characteristics that can

TABLE 4.1	Distinguishing Sexes in *Drosophila Melanogaster*	
	Male	**Female**
Overall size	Smaller	Larger
Sex combs on prothoracic legs	Present	Absent
Abdomen shape	Rounded	Tapering
Abdomen color pattern	Large terminal dark band	More, narrower bands
Genitalia	Sclerotized	Nonsclerotized

be used in distinguishing sexes. All characteristics are easily visible under the dissecting microscope. To see clearly the sex combs on the prothoracic legs (front legs), use the high power of the dissecting microscope. In manipulating flies, use a fine-bristled brush. Separate flies into males and females. You will find the banding pattern on the abdomen and the appearance of the genitalia easiest and quickest to use. Have the instructor confirm your decisions. ▲

Distinguishing Phenotypes

▼ Notice that the flies are of a number of different body forms. Wild-type flies have long wings, gray-brown bodies, and red eyes. A mutation called "ebony" is expressed as a black-bodied fly. A mutation called "dumpy" is expressed in wings that just reach the tip of the abdomen or have a notch along the inside terminal-margin of the wing. Other mutations may have been selected by your instructor. If so, they will be described. Separate all flies by phenotype and then have your instructor confirm your decisions. ▲

GENETIC SYMBOLS

The conventional method of representing alleles for a particular locus is based on the name for the dominant expression of the trait. The dominant allele is represented by an uppercase letter from the name of the dominant expression. The recessive allele is represented by the lowercase of the same letter. In the fruit fly, all traits are derived from mutations of wild-type genes. The wild-type expressions of traits are usually dominant. Using an uppercase W to represent many different wild-type traits would be impossible. The alternative convention with the fruit fly is to use a symbol derived from the mutant trait to represent the mutant allele. If the mutant allele is recessive, it is represented by a

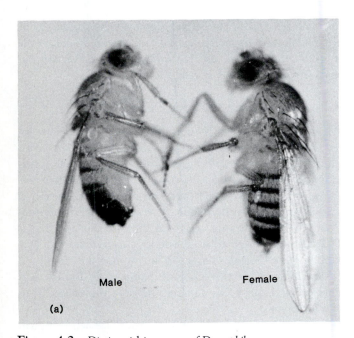

Male   Female

(a)

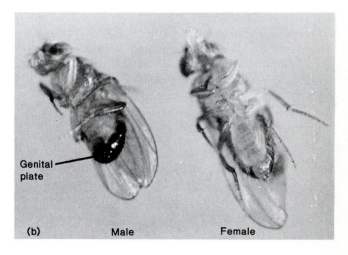

Genital plate

(b)   Male   Female

Figure 4.2   Distinguishing sexes of *Drosophila*.

lowercase letter; if it is dominant, the mutant allele is represented by an uppercase letter. The wild-type allele is represented by the same upper- or lowercase letter followed by a superscript [+].

4.1 LEARNING OUTCOMES REVIEW—WORKSHEET 4.2

4.2 EXPERIMENTAL PROCEDURE

LEARNING OUTCOMES

1. Predict the results of crosses involving one or two autosomally linked traits.
2. Predict the results of crosses involving single X-linked traits.
3. Perform crosses using the classical genetic organism *Drosophila melanogaster*.
4. Analyze data using a simple statistical test.
5. Describe the principles of segregation and independent assortment and use these principles in solving problems involving monohybrid, dihybrid, and X-linked crosses.

Depending on the amount of time available, you will be provided either with parental flies or progeny of the parental flies (first filial or F_1 generation). If the former, your crosses will span four or five weeks; if the latter, two or three weeks. The timing of the manipulations involved is indicated in table 4.2 in worksheet 4.1. Fill in the dates for later reference.

Cross Involving One Pair of Characters— The Principle of Segregation

Mendel's principle of segregation states that pairs of genes are distributed between gametes during gamete formation. Segregation occurs during meiotic cell division when genes carried on one member of a pair of homologous chromosomes end up in one gamete, and genes carried on the other member are segregated into a different gamete.

The following cross involves one pair of fruit fly characteristics and illustrates the principle of segregation. Cross I is as follows:

Parental Generation (P)
Homozygous Wild × Homozygous Ebony
↓
First Filial Generation (F_1)
F_1 × F_1 (Brother–Sister mating)
↓
Second Filial Generation (F_2)

Determine the genotypes of all flies involved in this cross, and complete the Punnett squares for Cross I in worksheet 4.1.

▼ Place two anesthetized males and two anesthetized virgin females of appropriate genotypes in each of two vials and then plug the vials. Keep the vials on their sides until the flies become active again. This prevents flies from sticking in the media. Label the vials with the cross (Cross I), the generation (P or F_1), the date, and your initials. You should score at least 50 flies in the F1 generation and 100 flies in the F_2 generation. Record your data in table 4.3 in worksheet 4.1. ▲

Cross Involving Two Pairs of Characters— The Principle of Independent Assortment

Mendel's principle of independent assortment states that during gamete formation, pairs of genes carried on different pairs of homologous chromosomes segregate independently of one another. Independent assortment means that during meiosis, when homologous pairs of chromosomes line up at metaphase I and then segregate, the behavior of one pair of chromosomes does not influence the behavior of any other pair. Offspring of a cross between individuals carrying genes for two traits should show random combinations of the expressions of the two traits.

The following cross involves two pairs of fruit fly characteristics and illustrates the principle of independent assortment. Cross II is as follows:

Parental Generation (P)
Homozygous Wild × Homozygous Dumpy Ebony
↓
F_1 Generation
F_1 × F_1 (Brother–Sister mating)
↓
F_2 Generation

Determine the genotypes of all flies involved in this cross, and complete the Punnett squares for Cross II in worksheet 4.1.

▼ Make your crosses as before using two males and two virgin females in each of two vials. Score at least 50 flies in the F_1 generation and 200 flies in the F_2 generation. Record your data in table 4.4 in worksheet 4.1. ▲

Cross Involving X (Sex) Linkage

Genes carried by sex (X) chromosomes show different inheritance patterns depending on whether the male or female introduces a particular expression into a cross. When a trait is determined by genes carried by the X chromosome, it is said to be X-linked, and the inheritance pattern is called sex-linked inheritance. The symbols used to represent X-linked traits are an X—to indicate that the trait is X-linked—followed by a superscript—to indicate the particular allele. Females have two X chromosomes, and both Xs are indicated in the genotype. In the male, a single X is present, and the Y chromosome replaces the second X in the genotype.

You should work in teams in Cross III. Each team will receive flies for the following crosses.

Cross A		Cross B
White-Eyed × Wild Male	(P)	Wild × White-Eyed
Female		Female Male

$$\downarrow \qquad\qquad\qquad\qquad \downarrow$$

F₁ Generation   Brother–   F₁ Generation
                    Sister
$F_1 \times F_1$ matings $F_1 \times F_1$
$\downarrow$ $\downarrow$
F_2 F_2

Determine the genotypes of all flies involved in these crosses and complete the Punnett squares for Cross III in worksheet 4.1.

▼ Each half of a team should be responsible for the care and scoring of cultures for either Cross IIIA or IIIB. At the end of the exercise, share results. Members of each team should make crosses as before using two males and two females in each of two vials. Score 200 flies in both the F_1 and F_2 generations. Be sure to record both sex and eye color (tables 4.5 and 4.6 in worksheet 4.1). ▲

Written Report

▼ An informal written report is due one week after the F_2 generations are scored. Summarize the results of your crosses by answering the following for each cross.

1. Which character is dominant and which is recessive?
2. How do you know?
3. What is the ratio of dominant to recessive in the F_2 generation?
4. What ratio would you expect (see worksheet 4.1)? How many flies of each phenotype would you expect assuming a total equal to the total number of flies you counted?
5. To test whether your results conform to what you expected (from question 4), perform chi-square tests. The chi-square test is a statistical tool that can be used to determine the probability of obtaining a set of results, assuming some theoretical expectation. You will be testing your results based on the predictions of Mendelian genetics. If the probability of obtaining a set

of results (given the theoretical expectation) is less than 0.05 (5%), the results are said to be significantly different than expected. If the probability of obtaining a set of results is greater than 0.05, the results are said to conform to the theoretical expectation. Rather than calculating a probability, you will calculate a unitless number, the chi-square value. The chi-square formula is as follows:

$$X^2 = \Sigma \frac{(O - E)^2}{E}$$

X^2 = Chi-square value

O = Observed number of offspring in one phenotypic class

E = Expected number of offspring in the same phenotypic class

Σ = The summation of $\frac{(O - E)^2}{E}$ for all phenotypic classes

The chi-square value becomes greater as the difference between observed and expected increases. If this chi-square value is greater than the chi-square value associated with a 0.05 probability, the results are interpreted as being different than expected. If the calculated chi-square value is less than that associated with a 0.05 probability, the results conform to the expectation. Normally, the chi-square value associated with 0.05 probability is taken from a table of chi-square values. Tables 4.7–4.9 in worksheet 4.1 give instructions for calculating the chi-square value and give the chi-square value associated with the 0.05 probability for each of your crosses.

6. Explain any results that deviate from what you expected.
7. In Cross III, explain the differences between crosses A and B. ▲

4.2 LEARNING OUTCOMES REVIEW—WORKSHEET 4.2

WORKSHEET 4.1   Genetics

TABLE 4.2   Dates of Fruit Fly Manipulations

	Approximate Elapsed Time	Cross I	Cross II	Cross III
Parental cross	(Day 0)	_____	_____	_____
Parents removed	(Day 7)	_____	_____	_____
F_1 Scored	(Day 14)	_____	_____	_____
F_1 Cross performed	(Day 14)	_____	_____	_____
F_1 Parents removed	(Day 21)	_____	_____	_____
F_2 Scored	(Day 28)	_____	_____	_____

1. Expected Results from Cross I
   Symbols used to represent the alleles involved in this cross:

   ebony—e
   wild—e^+

   Genotypes and phenotypes of individuals and gametes:
   a. Parental generation

   homozygous ebony   homozygous wild

   b. Parental Gametes

   c. F_1 generation genotype   d. F_1 generation phenotype

   e. F_1 gametes   f. Punnett square for F_2 progeny.

   g. Expected phenotypic ratio for the F_2 generation

2. Expected Results from Cross II
   Symbols used to represent the alleles involved in this cross:

   dumpy—*dp*   ebony—*e*
   wild—*dp*$^+$ wild—*e*$^+$

   Genotypes and phenotypes of individuals and gametes:
   a. Parental generation

   homozygous dumpy ebony   homozygous wild

   b. Parental gametes

   c. F$_1$ generation genotype   d. F$_1$ generation phenotype

   e. F$_1$ gametes

   f. Punnett square for F$_2$ progeny

   g. Expected phenotypic ratio for the F$_2$ generation

3. Expected Results from Cross III
   Symbols used to represent the alleles involved in this cross:

   wild eye—X^{w^+}
   white eye—X^w

   Genotypes and phenotypes of individuals and gametes:
   a. Parental generation
      Cross A

wild male	white-eyed female

      Cross B

white-eyed male	wild female

   b. Parental gametes
      Cross A

male	female

      Cross B

male	female

   c. F_1 generation genotype
      Cross A

male	female

      Cross B

male	female

   d. F_1 generation phenotype
      Cross A

male	female

      Cross B

male	female

   e. F_1 gametes
      Cross A

male	female

      Cross B

male	female

f. Punnett square for cross A F_2 progeny

_____ _____

_____		
_____ | | |

g. Phenotypic ratio for cross A F_2 progeny

h. Punnett square for cross B F_2 progeny

_____ _____

_____		
_____ | | |

i. Phenotypic ratio for cross B F_2 progeny

TABLE 4.3   Data from Cross I

	Wild Flies	Ebony Flies
F_1		
F_2		

TABLE 4.4   Data from Cross II

	Wild Wild	Wild Ebony	Dumpy Wild	Dumpy Ebony
F_1				
F_2				

TABLE 4.5   Data from Cross IIIA

Cross A	White-Eyed Female	×	Wild Male	
	White-Eyed		Wild	
	Male	Female	Male	Female
F_1				
F_2				

TABLE 4.6   Data from Cross IIIB

Cross B	Wild Female	×	White-Eyed Male	
	White-Eyed		Wild	
	Male	Female	Male	Female
F_1				
F_2				

TABLE 4.7 Chi-square Calculations for Cross I, F$_2$ Generation

	Ebony Flies	Wild Flies	Total
Observed number (O)			
Expected number (E)			
$(O - E)^2$			*
$\dfrac{(O - E)^2}{E}$			

$$X^2 = \Sigma \frac{(O - E)^2}{E}$$

$$X^2 = \text{_____}$$

X^2 at 0.05 probability = 3.84
*Total is not needed here.

TABLE 4.8 Chi-square Calculations for Cross II, F$_2$ Generation

	Wild Wild	Wild Ebony	Dumpy Wild	Dumpy Ebony	Total
Observed number (O)					
Expected number (E)					
$(O - E)^2$					*
$\dfrac{(O - E)^2}{E}$					

$$X^2 = \Sigma \frac{(O - E)^2}{E}$$

$$X^2 = \text{_____}$$

X^2 at 0.05 probability = 7.81
*Total is not needed here.

TABLE 4.9   Chi-square Calculations for Cross III, F_2 Generations (Test the cross you scored.)

	White-Eyed Males	Wild Males	White-Eyed Females	Wild Females	Total
Observed number (O)					
Expected number (E)					
$(O - E)^2$					*
$\dfrac{(O - E)^2}{E}$					

$$X^2 = \Sigma \frac{(O - E)^2}{E}$$

$$X^2 = \underline{\hspace{3cm}}$$

X^2 at 0.05 probability = 7.81
*Total is not needed here.

WORKSHEET 4.2

4.1 Inheritance in *Drosophila*—Learning Outcomes Review

1. Chromosomes that have genes coding for the same trait, but not necessarily for the same expression of the trait are called
   a. recessive chromosomes.
   b. daughter chromosomes.
   c. homologous chromosomes.
   d. dominant chromosomes.

2. A gene that is expressed only when a similar gene is present on the homologous chromosome is a
   a. recessive gene.
   b. dominant gene.
   c. codominant gene.
   d. sex-linked gene.

3. When an organism possesses two genes for a particular trait, and the genes are coding for different expressions of the trait, the organism is
   a. homozygous for the trait.
   b. heterozygous for the trait.
   c. carrying two alleles for the trait.
   d. Both b and c are correct.

4. All of the following are characteristics of male fruit flies (*Drosophila melanogaster*) except one. Select the exception.
   a. sex combs
   b. larger size
   c. large terminal dark band
   d. sclerotized genitalia

The following true/false questions pertain to a fruit fly (*Drosophila melanogaster*) that has the following genotype: $B B^+$; *se se*. *B* codes for a bar-eye mutant and *se* codes for a brownish-eye-color mutant called sepia.

5. The B^+ allele codes for the wild eye.
   a. true
   b. false

6. Because these traits are both related to eye characteristics, B, B^+, *se*, and se^+ must all be alleles of each other.
   a. true
   b. false

7. The bar mutant is dominant and the sepia mutant is recessive.
   a. true
   b. false

4.2 Experimental Procedure—Learning Outcomes Review

8. The separation of homologous chromosomes during meiotic cell division results in pairs of genes coding for the same trait being distributed into different gametes. This process is known as
   a. the principle of segregation.
   b. the principle of independent assortment.
   c. the principle of homologous chromosomes.
   d. crossing over.

9. The offspring of two parents that are each homozygous for a different allele of single trait are
   a. homozygous for the trait.
   b. heterozygous for the trait.
   c. members of the F_2 generation.
   d. phenotypically recessive.

10. The principle of independent assortment applies when one is considering
    a. a single trait carried by one pair of homologous chromosomes.
    b. a single trait carried by two pairs of homologous chromosomes.
    c. two traits carried by different pairs of homologous chromosomes.
    d. two traits carried by one pair of homologous chromosomes.

11. The principle of independent assortment reflects events of
    a. mitotic cell division.
    b. meiosis I of meiotic cell division.
    c. meiosis II of meiotic cell division.
    d. fertilization.

12. In X linkage,
    a. males have two copies of a gene for a particular trait and females have one copy.
    b. male offspring usually resemble their female parent, regardless of whether the trait is the dominant or recessive expression.
    c. male offspring resemble their female parent only if the female shows the dominant phenotype.
    d. both male and female parents have only one copy of a gene for a particular trait.

13. In a cross between a white-eyed female fruit fly and a red-eyed male fruit fly,
    a. all male offspring would be red-eyed and all female offspring would be white-eyed.
    b. all male offspring would be white-eyed and all female offspring would be red-eyed.
    c. one-half of male and female offspring would be red-eyed and one-half would be white-eyed.
    d. all offspring, male and female, would be red-eyed.

4.1 Inheritance in *Drosophila*—Analytical Thinking

1. A geneticist crossed a homozygous ebony-bodied fruit fly with one that is heterozygous for the ebony body color. What are the genotypes of both flies and phenotypic and genotypic ratios of flies in the next generation?

2. A gene that is expressed only when a similar gene is present on the homologous chromosome is a

	Parents		Progeny	
a.	_____ ebony ×	_____ wild	82 ebony	78 wild
b.	_____ ebony ×	_____ wild	0 ebony	74 wild
a.	_____ wild ×	_____ wild	39 ebony	118 wild

3. A geneticist crossed a wild-bodied, ebony fruit fly with one that is dumpy-bodied and wild in color. Assuming that both flies are heterozygous for the wild traits, what are the phenotypes and ratios expected for the next generation of fruit flies?

4. A geneticist crossed a dumpy ebony fruit fly to a wild fly that is heterozygous for both traits. What phenotypic ratio of offspring would you expect to see in the next generation?

4. A geneticist crossed a dumpy ebony fruit fly to a wild fly that is heterozygous for both traits. What phenotypic ratio of offspring would you expect to see in the next generation?

5. Supply the genotypes of the fruit flies described below.

   a. A wild fly that had one dumpy ebony parent:
      Dumpy ebony parent   Wild fly

   b. A dumpy-bodied, wild-colored fruit fly that had a wild-bodied, ebony offspring:
      Dumpy wild fly   Wild ebony offspring

   c. A fly that produced all of the following kinds of gametes: dpe^+, dp^+e^+, dpe, dp^+e

6. Describe the results of a cross between a white-eyed female fruit fly and a wild-eyed male fruit fly.

7. Describe the results of a cross between a female that is heterozygous for the white-eye trait and a white-eyed male.

8. Explain why all of the male offspring of a white-eyed female will have white eyes, while none of the male offspring of a white-eyed male will have white eyes unless the mother also carries a white-eye allele.

Adaptations of Stream Invertebrates— A Scavenger Hunt

Prelaboratory Quiz

Study this week's laboratory exercise and then complete the following quiz to assess your preparation for the laboratory.

1. The term that refers to all organisms living in a certain area and the organisms' physical environment is
   a. population.
   b. ecosystem.
   c. community.
   d. niche.

2. Which of the following is an accurate definition of "evolutionary adaptation"?
   a. any genetic change than improves the organism's ability to survive
   b. any genetic change that improves the organism's ability to exploit food resources
   c. any genetic change that improves the organism's reproductive success in a specified environment
   d. any genetic change that improves the organism's reproductive success in any environment

3. All of the following statements are teleological except one. Select the exception.
   a. Genetic change and evolution in a species occur because the species needs to change in order to survive.
   b. Evolution occurs as a result of random genetic changes in a species that may or may not be adaptive.
   c. Evolution directs change in a species.
   d. Evolution always results in progress. Species improve through evolutionary change.

4. A thin, 2- to 3-mm layer of nonmoving water at the surface of objects in a stream is called the
   a. neuston.
   b. benthos.
   c. boundary layer.
   d. hyporheos.

5. Current velocity in a stream is usually greatest
   a. near the bottom substrate.
   b. near the shore.
   c. throughout the stream cross section.
   d. near the center of the stream.

6. When a chemical parameter of a stream falls outside of the range that supports an animal and the animal dies, this parameter is said to be outside of the animal's
   a. range of the optimum.
   b. tolerance limit.
   c. trophic level.

7. The source of most primary production for streams is
   a. vegetation growing along the stream bank— especially autumn leaf fall.
   b. algae growing in the stream.
   c. attached aquatic vegetation.
   d. All of the above contribute equally to primary production for stream food webs.

8. Herbivory in streams begins with animals classified as
   a. shredders.
   b. filterers.
   c. gatherers.
   d. predators.

9. Animals, like the larvae of the water penny beetle, that feed on fine organic material attached to or settled on substrate materials are classified as
   a. shredders.
   b. filterers.
   c. scrapers and gatherers.
   d. predators.

10. Animals living in or on the streambed are living in the
    a. neuston.
    b. boundary layer.
    c. neckton.
    d. benthos.

11. True/False   A pH of 5.0 is harmful to most aquatic animals.

12. True/False   Very soft water is required for the formation of exoskeletons of arthropods and the shells of molluscs.

13. True/False   Dissolved oxygen levels in warm streams tend to be higher than those in cold streams.

14. True/False   Flattening of some stream insects allows them to live in the neuston of a stream.

15. True/False   Some stream insects use silk to attach themselves to the substrate of a stream.

5.1   EVOLUTIONARY ADAPTATION

LEARNING OUTCOMES

1. Describe the concept of evolutionary adaptation.
2. Explain why teleological explanations of evolutionary occurrences are common, but wrong. Give examples of teleological thinking.

Evolution has resulted in the great variety of living organisms that we see today. A goal of the next unit of this laboratory manual is to study that diversity in a survey of the animal phyla. The survey approach helps one appreciate common themes of animal organization that result from common evolutionary ancestry. Another approach to studying animal diversity is to study particular ecosystems. An **ecosystem** consists of all of the organisms living in a certain area and the organisms' physical environment. The ecosystem approach helps one appreciate the interactions that occur between animals and their environment. The goal of this exercise is to study a freshwater stream in order to appreciate the evolutionary adaptations of invertebrates to the stream ecosystem.

An **evolutionary adaptation** is any change in an animal's genetic makeup that increases the animal's chance of successful reproduction in a specified environment. Changes that could be considered adaptations can directly affect an animal's ability to produce large numbers of eggs, care for large numbers of offspring, or reproduce more often. Other adaptations can indirectly increase an animal's reproductive potential. These changes can increase an animal's ability to feed more efficiently and store more energy for egg production. Other changes allow animals to live in places where other animals are unable to live, thus reducing competition for resources and making it more likely that young will survive. Still other changes allow an animal to withstand harsh climatic conditions or potentially harmful interactions with other individuals. These changes allow an animal to survive and reproduce in the future.

Notice that the definition of adaptation specifies a particular environment. This qualification is very important. A trait that is adaptive in one environment may result in decreased reproductive success in another environment. If an animal migrates to another location or if the environment changes, what were formerly adaptive traits could result in extinction.

When studying animal adaptations, it is tempting to think of evolution as a directed process that produces change in response to the needs of an animal. This kind of goal-oriented thinking is said to be **teleological.** Evolution is simply a result of random genetic changes (mutations) that result in some individuals being more effective at reproducing than others in the population. There is no force that directs when a mutation will occur or what the nature of the change will be.

A second pitfall in thinking is to attribute adaptive significance to every process or structure that an animal possesses. Mutations are random occurrences. They are usually neutral rather than harmful or beneficial for an animal. This means that evolution may result in structures that are useless for an animal. These structures may be retained because they are coded by genes that may be linked to chromosomes containing genes that code for adaptive traits.

5.1 LEARNING OUTCOMES REVIEW—WORKSHEET 5.3

5.2   STREAM ECOLOGY

LEARNING OUTCOMES

1. Give examples of physical characteristics of stream ecosystems, and describe how these characteristics influence animals that live in streams.
2. Give examples of chemical characteristics of stream ecosystems, and describe how these characteristics influence animals that live in streams.
3. Describe the major source of energy for animals living in streams and explain how that energy is processed as it moves through a stream food web.

Freshwater stream ecosystems are a part of nearly all landscapes. Their physical and chemical characteristics are relatively restrictive, and animals that live in streams often show body structures that are adaptations to stream habitats. Many of these structural (morphological) adaptations are easily recognized by careful observation. Before one can understand the adaptive significance of the morphological traits, one needs to know something about the stream ecosystem. While streams can be very different from one another in terms of depth, type of substrate, current velocity, and so forth, there are certain characteristics that streams share. Some of these are discussed next.

PHYSICAL CHARACTERISTICS

The most obvious characteristic of streams is the current. Animals that live in streams must be able to maintain themselves in, or seek shelter from, moving water. Some members of a species must withstand periods of flooding, in which current velocity increases to the point that it threatens to scour animals from the streambed. On the other hand, many stream animals have an inherent requirement for moving water. Moving water delivers nutrients and dissolved oxygen to stream animals and aids in animal dispersal. The current that we experience when we stand in a stream, however, is not necessarily what is experienced by an insect in the stream (fig 5.1). As one moves away from the center of a stream, current velocity decreases because of friction between the water and the stream bank or streambed. Within 2–3 mm of the surface of a rock, or other object on the substrate, the current disappears and water is stationary. This layer of standing water is called the **boundary layer.**

Many invertebrates that live on the streambed **(benthos)** are adapted to living in this layer of stationary water.

Other physical characteristics that influence animal life in streams include water temperature, substrate type, and turbidity. Temperature influences where an animal lives, the stage of its life cycle, and the availability of oxygen for the animal. Substrate type also influences stream animals. The particle size of substrate materials may vary from meters in diameter (boulders) to fractions of a millimeter (silt). The color of substrates varies from very light to very dark. Although the reasons are not always entirely clear, a certain species may be consistently found associated with a particular substrate. Streams are often fairly turbid, especially larger rivers with silty substrates. Fine particles are often suspended in moving water. Animals adapted to larger rivers are usually not affected by turbid conditions. On the other hand, the runoff of topsoil from construction sites and farming operations can result in sedimentation that may foul gills, cover habitats in the spaces between rocks and gravel, or foul food supplies.

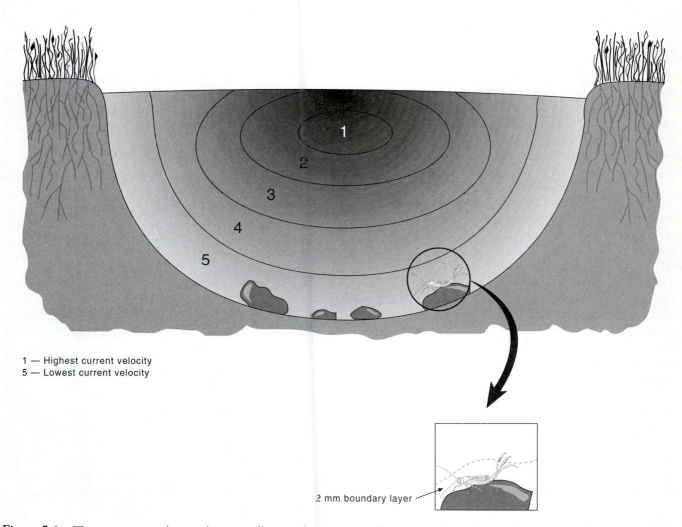

1 — Highest current velocity
5 — Lowest current velocity

2 mm boundary layer

Figure 5.1   This cross-sectional view of a stream illustrates how current velocity varies depending on where the measurements are taken. As one moves from the center of the stream toward the banks or bottom substrate, current velocity is reduced because of friction between the water and the streambed. The inset shows the boundary layer, which is a small layer of still water on the surface of objects in the streambed. Many benthic stream invertebrates are adapted for life in the boundary layer.

CHEMICAL CHARACTERISTICS

Streams also have chemical characteristics that influence animal life. Most important among these are the amount of dissolved oxygen, pH, and hardness. Animals are most successful when these chemical characteristics are within certain ranges of values. These ranges are species-specific and are called **ranges of the optimum.** Animals can survive when a chemical parameter falls out of the range of the optimum. Extreme variations in the parameter, however, make conditions unsuitable for life, and the animal dies. This is referred to as the **tolerance limit** of the animal. Different species of animals have different ranges of optimum for various chemical characteristics of streams.

Oxygen levels of streams vary from low values of two or three parts per million (p.p.m.) to saturation levels of 14 p.p.m. The amount of oxygen present in a stream depends partly on water temperature. The solubility of oxygen in water is greater at lower temperatures, so cold water usually has more dissolved oxygen than warm water. Oxygen is rarely a limiting factor in clean streams. Streams have a larger surface area relative to their volume than lakes. This large surface area provides for adequate exchange of gases between the water and the air. Gas exchange is enhanced by the fact that water is moving in streams. Water rushing over boulders or tree limbs causes turbulence that mixes air and water, providing faster rates of gas exchange between air and water. Streams can be naturally low in oxygen for two reasons: (1) heavy leaf fall in autumn can enhance rates of decomposition, and oxygen in the stream can become depleted and (2) continuous ice cover can limit the exchange of oxygen between the air and the water and can result in low oxygen levels. Pollution, on the other hand, can deplete streams of oxygen. Heated discharge from factories can raise water temperature and reduce the solubility of oxygen in water, and organic pollutants enhance rates of decomposition and deplete the oxygen supply.

The pH (acidity) also influences animals of a stream. The pH of a stream is determined by characteristics of the surrounding landscape. Streams that run through regions where the bedrock is limestone (calcium carbonate) will normally have a pH close to neutral. In other regions where there is little calcium in the surrounding soils, the pH can be naturally lower. Acid deposition and nitrogen loading from organic pollutants have lowered the pH of many streams. Although there is much variation in the tolerance levels of stream animals, most evidence suggests that pH levels below 5.5 are harmful to many common stream animals.

The hardness of a stream is a measure of the dissolved minerals in the water. Dissolved calcium, magnesium, chloride, sulfate, and sodium ions are important for at least three reasons. First, these ions are used by an animal for various physiological processes. Second, the presence of these ions in the water reduces the osmotic stress on a freshwater animal. This means that the animal will take on less water by osmosis and expend less energy expelling excess water. Lastly, in molluscs and many arthropods, calcium is used to harden the shell (or exoskeleton in arthropods). Reduced availability of these ions (soft water) may cause problems for an animal in any of these three areas. Hardness often varies between 50 and 300 p.p.m.

STREAM FOOD WEBS

The sequence of organisms through which energy moves in an ecosystem is a **food chain.** In most ecosystems, it is more accurate to envision complexly interconnected food chains called **food webs.** There are relatively few attached aquatic plants in most streams and very little floating algae (phytoplankton). The source of most plant material (primary production) that provides energy for stream animals comes from outside the stream. Plants on the stream bank supply primary production in the form of leaves that fall into the stream, especially in autumn (fig. 5.2).

The feeding (trophic) levels of a stream can be classified as follows. Herbivory begins with animals that serve as **shredders.** Many mayfly nymphs, some caddisfly nymphs, and crustaceans (amphipods), for example, feed on entire leaves. As they feed, smaller pieces of leaves are freed and carried downstream by the current. This shredding continues until the particle size becomes very small and the food is prepared for the next set of users. **Filterers** are animals with specializations for trapping suspended food particles. Some mayfly nymphs have fine hairlike setae on their forelegs that are used as filters. Some caddisfly nymphs spin silk nets that are used to trap food particles. Black fly larvae possess fanlike appendages that trap food. Bivalves filter food using gill filaments. Some filtering animals are specialized for filtering coarse food particles from the water. The smallest particles that they miss are filtered by animals with finer filtering surfaces. Other animals, called **scrapers and gatherers,** scrape fine organic material that has settled on, or is attached to, substrate materials. Scrapers and gatherers include snails, some mayflies, and beetle larvae such as water penny. There are also many predators in freshwater streams. Some are

Figure 5.2 Primary production for most streams comes from vegetation surrounding the stream. Leaf fall in autumn may supply enough nutrients for stream food webs for the entire year.

invertebrates that prey on other invertebrates. Stone fly larvae and helgramites are invertebrate predators. Fishes and salamanders are examples of vertebrate predators. Animals such as crayfish are **detritivores** and feed on a variety of dead and decaying animal and plant matter.

5.2 LEARNING OUTCOMES REVIEW—WORKSHEET 5.3

5.3 A STUDY OF A STREAM ECOSYSTEM

LEARNING OUTCOMES

1. Evaluate the physical and chemical parameters of a stream.
2. Make a collection of stream invertebrates, examining them for adaptations they may possess for living in a stream.
3. Assign common stream animals to their position in stream food webs.

The purpose of the following activities is to study the ecology of a freshwater stream. You will first study the physical and chemical characteristics of the stream you are visiting. Following that exercise, you will be using collecting equipment to study adaptations of animals living in the stream. Before going into the stream, a few safety precautions should be followed.

1. Use a buddy system. Work in groups of two or more and stay with your group. If one member of the group encounters trouble, the other members will be there to summon help.
2. Never get out of visual and calling range of other groups.
3. Do not exceed knee depth when wading.
4. Watch for underwater obstacles. Use the handle of your net to steady yourself and explore the substrate.
5. Rocky substrates are often very slippery. Be very careful when walking across them.
6. Avoid extremely fast currents.

PHYSICAL AND CHEMICAL CHARACTERISTICS

▼ When you arrive at the stream you will be studying, take some time to look around. How wide is the stream? How deep is the stream in your location? Look up and down the length of the stream. What do you think is the primary source of energy for animals living in the stream? What makes up the substrate of this stream? What other factors are there that may influence the lives of animals in the stream (e.g., farm pastures, industrial establishments, highways, subdivisions)? Record the date and location of your field trip and your observations on worksheet 5.1.

Determine the current velocity in the stream. Your instructor will give you directions for using a current meter if one is available. Alternatively, measure a convenient distance (perhaps 15 m) and time the movement of a float across that distance. Repeat your measurement three times and calculate an average. Express your results in meters per second and record the results on worksheet 5.1. ▲

Various chemical tests that can be performed with stream water. The three parameters described in this exercise—dissolved oxygen, pH, and hardness—are important for the reasons described. Your instructor will provide you with reagents from a water analysis kit and give you specific instructions on their use. Perform the chemical tests and record your results in worksheet 5.1. If additional reagents are available, your instructor may request that more tests be performed. Record your conclusions regarding the chemical characteristics of your stream on worksheet 5.1.

STREAM INVERTEBRATES— A SCAVENGER HUNT

Invertebrates are used in this exercise because they are easily collected, usually quite numerous, and show remarkable diversity. They are not often observed in casual visits to streams, however, because they are quite small. Stream invertebrates show a number of morphological adaptations to the physical characteristics of the stream ecosystem. Table 5.1 lists adaptations that are easily observed in stream invertebrates. Many of these are adaptations to living in moving water. Others are behavioral traits that allow invertebrates to avoid the current (figs. 5.3–5.10).

▼ Each group of students will be given a D-frame net, a collecting pan, vials, forceps, a squeeze bottle of 70% ethanol, a pencil, and vial labels. Your goal is to collect as many different kinds of invertebrates as you can find in the time allotted. Approach your collecting site from downstream so you do not disturb the substrate where you will be working. Collecting simply involves sweeping your net through vegetation along the stream bank or through debris on the bottom of the stream. After collecting in sandy or muddy substrates, wash the contents of the net by gently dipping the net into the water. In swift water with rocky substrate, place the flat edge of the net on the stream bottom so that the net opens upstream. Disturb the substrate in front of the net with your foot and allow the current to carry debris into the net. Fill your collecting pan about one-half full of water and empty the contents of the net into the pan. Examine the net and the pan for invertebrates. Preserve one or two specimens of each kind of animal in a separate vial of ethanol. Use a pencil and a vial label to number the animals in your collection. Place the label inside the vial with the specimen. Determine whether or not any of the adaptations listed in table 5.1 apply to the animal. A specimen may have more than one adaptation. If so, indicate all that apply. Other specimens may have no adaptations visible. Indicate that as well. Use

TABLE 5.1 Adaptations of Stream Insects

Adaptation	Adaptive Significance
Morphological adaptations	
Flattening	Flattening of the body allows an animal to utilize the boundary layer, thus avoiding current. It also allows animals to crawl under objects and within the spaces between substrate particles (figs. 5.3 and 5.7).
Streamlining	Streamlining lessens resistance to fluids moving around the animal. The animal tends to orient upstream (figs. 5.3 and 5.8).
Suckers	Suckers help an animal adhere to the substrate. They are not common in stream animals except for leeches and a few fly larvae.
Friction pads and marginal contact	A flattened ventral surface and domed upper surface increase frictional resistance with the substrate and force the margin of the animal against the substrate. This seals the ventral side and prevents moving water from getting under the animal (fig. 5.7).
Hooks and grapples	Claws on the legs or other appendages of animals are used to hold onto the roughness of stone surfaces (figs. 5.5 and 5.6).
Small size	Small animals can utilize the boundary layer and spaces between substrate particles.
Silk and sticky secretions	Silk is used to attach animals to their substrate (figs. 5.6 and 5.9).
Ballast	Ballast may be in the form of stone cases constructed with the help of silk (fig. 5.4).
Neuston	Some invertebrates live on top of the water, supported by water's surface tension. Hairlike setae that repel water may be present on legs or other body parts and help prevent the animal from breaking through the water surface (fig. 5.10).
Behavioral adaptations	
Avoidance	Many animals seek refuge from the current by living under stones or under the banks of streams.
Living in the hyporheos	Some stream animals burrow below the surface of the substrate. Some spend their entire life histories in this way—completely hidden from the current.

worksheet 5.2 to record the specimen by number and location. The location record should include information on substrate type, water depth, current velocity, and location within the substrate. Your instructor may have set up dissecting microscopes so that you can view specimens more closely. It is not necessary to know the names of the animals you are collecting to complete this exercise. However, your instructor may want to take time to help you learn to recognize a few of your specimens by name. Record the name in worksheet 5.2 if the specimen has been identified.

At the end of today's laboratory period, you will hand in your collection and worksheets 5.1 and 5.2. Your instructor will grade your worksheets and assign points based on the number of different adaptations that you found. Based on your observations in the stream and the information in this manual, fill in the food web in figure 5.11 in worksheet 5.3. ▲

5.3 LEARNING OUTCOMES REVIEW—WORKSHEET 5.3

(a)

(b)

Figure 5.3 (*a*) Mayfly nymphs (order Ephemeroptera) are recognized by the presence of two or three tail filaments and dorsal or lateral abdominal gills. Mayflies usually serve as shredders or filterers in stream ecosystems and are often streamlined or flattened. A member of the family Heptageniidae is shown here. (*b*) An adult mayfly, (*Hexagenia*). The adult mayfly's lifespan ranges from only a few hours to one to two days. They do not feed. They mate, the females lay eggs on the water's surface, and they die.

Figure 5.4 Caddisfly larvae (order Trichoptera) often build cases of stone or twigs. These materials are cemented together with silk secretions. *Neotrichia* is shown here.

Figure 5.5 Helgramites or dobson fly larvae (*Corydalus cornuta*) are predators that feed on other invertebrates. They are often found under, or clinging to, rocks in swift currents. Hooks on prolegs at the posterior tip of the abdomen help the animal to maintain its position in the current.

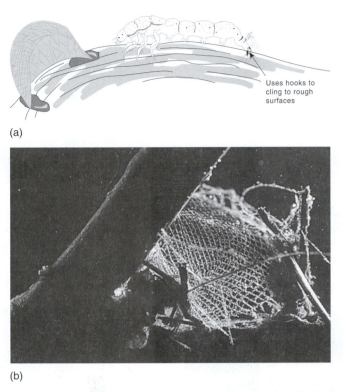

(a)

(b)

Figure 5.6 Some caddisfly larvae (order Trichoptera) do not build cases but may construct silken retreats on the undersurface of rocks or nets used to capture debris drifting downstream. *(a) Hydropsychae* is shown here. *(b)* Food-filtering net of *Hydropsychae* (×8).

Figure 5.7 This flattened, scale-like beetle larva (order Coleoptera) is a water penny (*Psepheneus lecontel*). Flattening allows it to live within the boundary layer. The doming of its dorsal surface and the tight marginal contact of its body with rocks that it crawls over prevent it from being swept away. Water pennies are scrapers and gatherers (×10).

Figure 5.8 This insect is a stone fly nymph (order Plecoptera). Stone flies are recognized by two tail filaments and gills that attach at the ventral surface of the thorax. They are often streamlined insects and prey on other stream invertebrates

Figure 5.9 These black fly larvae (*Simulium*) secrete silk from their salivary glands. The silk is attached to rocks or other firm substrates. Hooks at the tip of the abdomen allow the black fly larvae to cling to the silk and the substrate in spite of very fast currents. The mouthparts of black fly larvae are fan shaped and filter their food—fragments of plant material—from the water. Adult black flies are annoying biters.

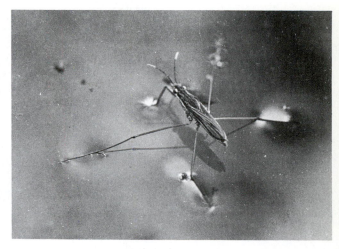

Figure 5.10 This water strider (*Gerris*) is a member of the insect order Hemiptera. The water-repelling setae on its legs and the surface tension of the water allow water striders to live on top of the water (neuston). Note the "dents" in the water's surface where the insect's legs contact the water. Water striders are usually found in quiet water in pool areas or along the bank. Water striders prey on invertebrates swimming just below the water surface.

WORKSHEET 5.1   Physical and Chemical Characteristics

Name and location of the stream _____

Date of the scavenger hunt _____

1. Summary of the physical characteristics of the stream
    a. The approximate width of this stream is

    b. The average depth of this stream is about

    c. The current velocity is

    d. The major sources of energy for animals living in this stream are

    e. The substrate of this stream consists of

    f. Other factors that may influence animal life in this stream (list)

2. Summarize the results of your physical and chemical measurements. Are there any parameters that you measured that seem to be out of the ordinary?

WORKSHEET 5.2   Adaptations of Stream Animals

1. Fill in the following table by placing the number of the specimen beside the adaptation. Indicate where each specimen was collected. Specific information on location requires careful observation during the collecting process. Make notes on current, location in the stream (e.g., midstream vs. under bank), substrate surface (e.g., on top of, or under, a rock), and water depth.

Adaptation	Specimen Number	Location	Specimen Name (optional)
Flattening			
Streamlining			
Suckers			
Friction pads and marginal contact			
Hooks and grapples			
Small size			
Silk and sticky secretions			
Ballast			
Neuston			
Avoidance			
Hyporheos			
No apparent adaptations present			

WORKSHEET 5.3   Adaptations of Stream Invertebrates—A Scavenger Hunt

5.1   Evolutionary Adaptation—Learning Outcomes Review

1. An evolutionary adaptation will always
   a. promote survival of an animal in all environments.
   b. promote reproduction of an animal in all environments.
   c. promote reproduction of an animal in a specific environment.
   d. reduce competition for resources within a particular population.

2. Which of the following statements is teleological?
   a. Mutations are random events and usually do not produce adaptive changes.
   b. Giraffes acquired their long necks because of a need to reach higher and higher into trees to feed.
   c. Streamlining of an insect is an adaptation in some, but not all, environments.
   d. Evolution does not always lead to adaptation. It can result in extinction.

5.2   Stream Ecology—Learning Outcomes Review

3. All of the following are physical characteristics of streams except one. Select the exception.
   a. current velocity
   b. substrate type
   c. dissolved oxygen
   d. water temperature

4. Hardness is an important chemical characteristic of a stream. When hardness is too low
   a. animals will be unable to feed.
   b. many arthropods and molluscs will have a difficult time forming exoskeletons and shells.
   c. enzymatic reactions that support life will not occur.
   d. gas exchange will be difficult.

5. Dissolved oxygen may be reduced in a stream because of
   a. continuous ice cover in winter.
   b. decomposition of organic matter after a heavy leaf fall in autumn.
   c. pollution by organic runoff.
   d. All of the above may reduce dissolved oxygen in a stream.

6. Animals with specializations for trapping suspended food particles from moving water are called
   a. shredders.
   b. filterers.
   c. scrapers.
   d. gatherers.

7. An example of an animal that traps suspended food particles from moving water is
   a. a water penny larva.
   b. a water strider.
   c. some caddisfly larvae.
   d. a helgramite.

8. Animals like crayfish feed on a variety of dead and decaying plant and animal matter. These animals are called
   a. detritivores.
   b. gatherers.
   c. scrapers.
   d. predators.

5.3 A Study of a Stream Ecosystem—Learning Outcomes

9. Match the description to the adaptation.

_____ Neuston	A. These are small invertebrates that do not experience current. They spend most of their lives below the surface of the substrate. These invertebrates include some mayfly nymphs and fly larvae.
_____ Friction pads and marginal contact	B. Stone cases permit many caddisfly larvae to withstand current by weighting the larvae down and preventing them from being washed downstream.
_____ Ballast	C. Water striders avoid current by living on top of the water, usually near the shoreline of the stream.
_____ Living in the hyporheos	D. Some mayfly nymphs can utilize the boundary layer because of this morphological adaptation.
_____ Flattening	E. Mayfly nymphs that show this adaptation may be minnow-like and are probably collected swimming in the water and are very active swimmers in the collecting pan.
_____ Streamlining	F. Water penny larvae are dome-shaped. Tight contact with the substrate around the edge of this insect prevents it from being washed downstream.

5.1 Evolutionary Adaptation—Analytical Thinking

1. Many mayfly nymphs display flattening as an adaptation to living in a stream. Write two statements that could explain the origin and significance of this adaptation. One statement should be teleological and the other statement should reflect a proper understanding of how adaptations arise. Explain why the teleological statement misrepresents evolutionary mechanisms.

5.2 Stream Ecology—Analytical Thinking

2. The following table summarizes measurements of physical and chemical parameters of a stream. Would you expect organisms living in this stream to experience difficult living conditions? Explain.

Current velocity	0.25 M/sec.
Average depth	0.5 M
Substrate type	Sand and small gravel
Temperature	20°C
Dissolved oxygen	4 mg/L
Hardness	80 mg/L
pH	5
Other observations	Students noticed that there was a considerable algal growth on rocks and logs in slower current areas. There was a strong "fishy" odor in their collecting site. On the drive to the collecting site students noticed cattle grazing in fields upstream from the collecting site. There was very little shoreline vegetation and cattle had apparently trampled the shoreline when coming to the stream for water. Collections revealed a preponderance of fly larvae and annelid worms. There were few mayfly and stonefly nymphs, or other stream invertebrates that students hoped to find.

5.3 A Study of a Stream Ecosystem—Analytical Thinking

3. Examine your collection records in worksheet 6.2. Pay particular attention to the adaptations and the location descriptions that you provided. Realizing that habitats for stream animals overlap, do you see any correlation between an adaptation and the location in the stream where the adaption was observed? Explain any correlations suggested by your results.

4. A stream food web. Fill in the following food web by matching the organism with its location in the food web. Did you collect any organisms not listed here? Consult with your laboratory instructor and additional references to determine where they fit in this food web. How many trophic levels are present in this food web?

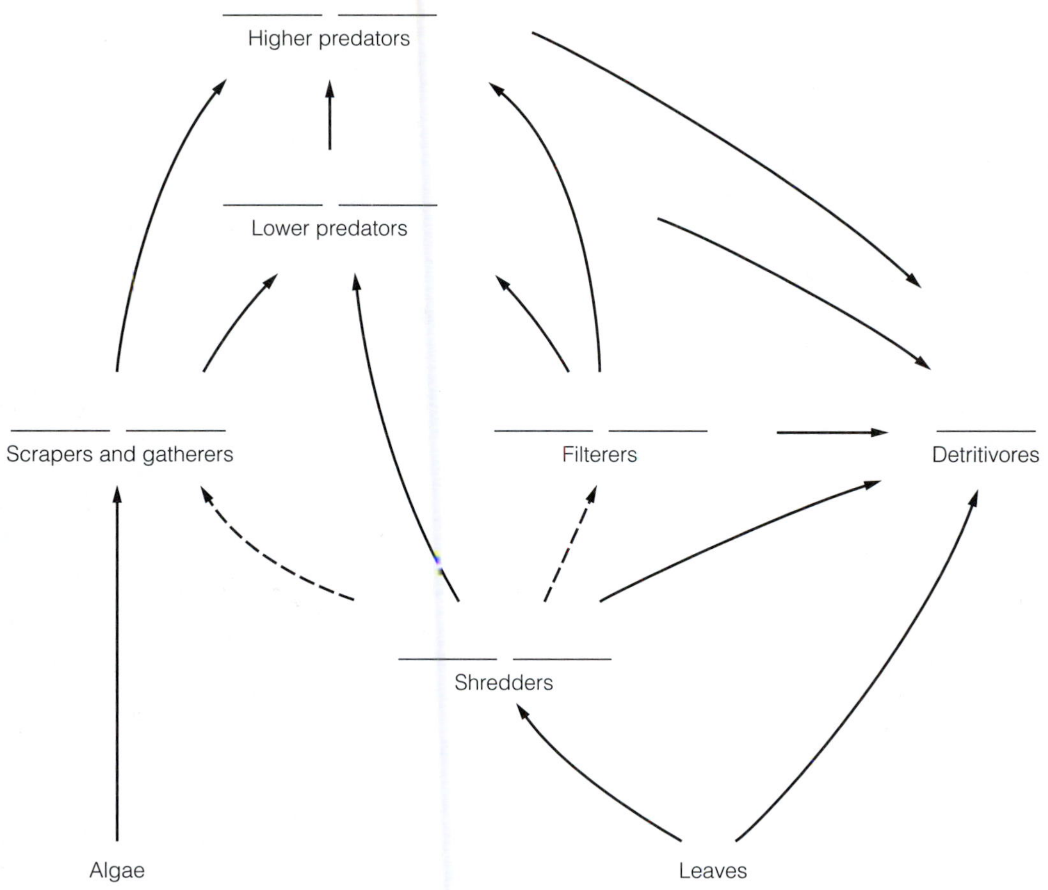
Higher predators
Lower predators
Scrapers and gatherers
Filterers
Detritivores
Shredders
Algae
Leaves

A. Water penny larva
B. Mayfly
C. Caddis fly
D. Helgramite
E. Snail
F. Crayfish
G. Amphipod
H. Stone fly
I. Trout
J. Salamander
K. Black fly larva

Figure 5.11 A simplified food web of a hypothetical stream. Dashed lines going through the shredders to scrapers, gatherers, and filterers indicate that the shredders tear up plant material and the smaller pieces that are left are used by other classes of herbivores. Solid lines indicate energy flow between different trophic levels.

UNIT II

An Evolutionary Approach to the Animal Phyla

Unit II is similar to the traditional survey approach of generations of zoology courses. The goal of this unit is to give you a broad exposure to the protozoans, the animal phyla, and the evolutionary relationships that unite the diversity found in the animal kingdom. Unit II begins with exercise 6—an introduction to animal classification and cladistics. The "Evolutionary Perspective" at the beginning of exercises 7 through 16 is meant to help you relate the phylum being studied to those already discussed and those that will be studied in subsequent laboratory exercises. The introduction in exercises 7 through 16 is designed to introduce you to the unique characteristics of the phylum (phyla) studied in each exercise. That is followed by studies of class representatives and processes unique to particular animal groups. Most exercises in unit II end with an exercise on class-level taxonomy of the phylum studied.

Exercise 6

The Classification of Organisms

Prelaboratory Quiz

Study this week's laboratory exercise and then complete the following quiz to assess your preparation for the laboratory.

1. An inherited trait used in determining taxonomic relationships among a group of animals is a(n)
   a. characteristic.
   b. character.
   c. anamorphy.
   d. symplesiomorphy.

2. A character that has been derived since a group diverged from a common ancestor is called a(n)
   a. symplesiomorphy.
   b. anamorphy.
   c. synapomorphy.
   d. anapomorphy.

3. The next level of taxonomy more inclusive than order is
   a. class.
   b. family.
   c. genus.
   d. phylum.

4. A subset of taxa that shares a certain derived character is called a
   a. polyphyletic group.
   b. paraphyletic group.
   c. species group.
   d. clade.

5. Which of the following is the correct way of designating a species name?
   a. Musca domestica
   b. *Musca domestica*
   c. *domestica*
   d. *Musca Domestica*

6. True/False   If all information about the evolution of a group is known, the group will always be represented as a monophyletic group.

7. True/False   A paraphyletic group can be traced to more than one ancestral species.

8. True/False   The simplest, most economical explanation for an evolutionary lineage is a parsimonious explanation.

9. True/False   A group of organisms that is used to help decide what character is ancestral for a group being studied is called an ingroup.

10. True/False   The goal of animal systematics is to study the diversity of animals and organize them into groupings that reflect evolutionary relationships among the groups.

Biologists estimate that there are in excess of 30 million different kinds of organisms on the earth. Most of these different organisms are undescribed and reside in tropical rain forests. As scientists work to find these undescribed organisms, the organisms must be named. Our need to name organisms stems from the fact that we have a language, and that language allows us to encode and classify concepts. This need to encode and classify is a natural consequence of being human. For example, the first question that any young child asks when seeing an unfamiliar animal in a zoo is, "What is that?"

Scientists go about naming and classifying organisms in ways that have helped to make sense out of the diversity in nature. A potpourri of millions of names is of little use to anyone. To be useful, a naming system must reflect the order that is present in nature. This means that before names are assigned, a scientist must study the organism to determine the relationships of the organism to other previously described organisms. These interrelationships are reflections of the evolutionary pathways that lead to modern organisms. The study of the kinds and diversity of organisms and the evolutionary relationships among them is called systematics, or **taxonomy.** (Some biologists distinguish between taxonomy and systematics. Taxonomy is often thought of as the work involved with the original description of species, and systematics as the assignment of species into evolutionary groups.) The assignment of a distinctive name to each species is called **nomenclature.**

6.1 A TAXONOMIC HIERARCHY

LEARNING OUTCOMES

1. Describe the binomial system of classification.
2. Explain how the binomial system of classification reflects the order present in nature.
3. Discuss the species concept and problems associated with it.
4. Describe the taxonomic hierarchy.

Our modern classification system is founded on the work of Karl von Linne (Carolus Linnaeus, 1707–1778). In this classification system, each different kind of organism is called a **species.** The species concept is actually quite difficult to define precisely. In fact, there is no single definition that can be used in all cases. It will be sufficient for our purposes to think of a species as a group of organisms that produces more of its own kind through sexual reproduction. Realize, however, that there are many instances in which this definition does not work well.

Karl von Linne used a two-part name for each species; therefore, our naming system is referred to as a **binomial system.** The first part of a species name is the **genus** name. It always begins with a capital letter. The second part of a species name is the **species epithet.** It always begins with a lowercase letter and is never used without the genus name. The entire name is written in italics, or underlined, because it is Latinized. For example, the species name (often called the scientific name) of humans is *Homo sapiens,* and the species name of the housefly is *Musca domestica.* When the genus name is understood, it can be abbreviated (*H. sapiens*). An abbreviation is used only after the entire name has been written earlier in a paper.

Taxonomic levels above the species are used to describe evolutionary relationships. These taxonomic levels, or taxa (sing. taxon), are arranged hierarchically (from more specific to broader): **species, genus, family, order, class, phylum, kingdom,** and **domain** (table 6.1). These groupings are based upon increasingly divergent characteristics and, ideally, reflect evolutionary relationships. The housefly (*Musca domestica*), for example, has its own set of unique characteristics that makes it a distinctive species. It also shares certain characteristics with other true flies—the horsefly, for example (fig. 6.1). The most important of these shared characteristics is a single pair of wings. This shared characteristic is a reflection of the relatively close

TABLE 6.1   Taxonomic Categories of a Human and a Housefly

Taxon	Human	Housefly
Domain	Eukarya	Eukarya
Kingdom	Animalia	Animalia
Phylum	Chordata	Arthropoda
Class	Mammalia	Insecta
Order	Primates	Diptera
Family	Hominidae	Muscidae
Genus	*Homo*	*Musca*
Species	*Homo sapiens*	*Musca domestica*

evolutionary ties between these species. They are both members of the order Diptera. These flies also share characteristics with a broader group of animals—butterflies, beetles, and wasps. These animals all have three pairs of legs and are called insects (class Insecta). Insects are all related, but not all insects are as closely related as houseflies are to horseflies. Insects, in turn, share important characteristics with other animals—ticks, spiders, crayfish, and crabs. They all possess a jointed exoskeleton. Thus, the housefly and all of its relatives belong to the phylum Arthropoda.

Above the species level, there are no precise definitions of what makes up a particular taxon. Disagreements among scientists as to whether two species should be included in the same or different taxonomic groups are common. Disagreements like these are the fuel that keeps taxonomy a vital, lively scientific field of study.

Students often ask the question, "Why are scientific names more useful than common names?" The answer to this question is easily understood. As pointed out previously, scientific names reflect evolutionary relationships, common names do not. Just as importantly, scientific names make sense out of a chaotic world of common names. Many species have common names that vary from region to region and country to country. For example, a common sparrow is called an English sparrow, barn sparrow, and house sparrow in different regions of the country. It has a single scientific name that is recognized by scientists throughout the world—*Passer domesticus.* In addition, common names often refer to more than one species. It is impossible to distinguish between species of mosquitoes or crayfish with superficial examination. Common names, therefore, often do not designate a particular species. On the other hand, species names refer to only one kind of organism. No two species have the same scientific name.

6.1 LEARNING OUTCOMES REVIEW—WORKSHEET 6.2

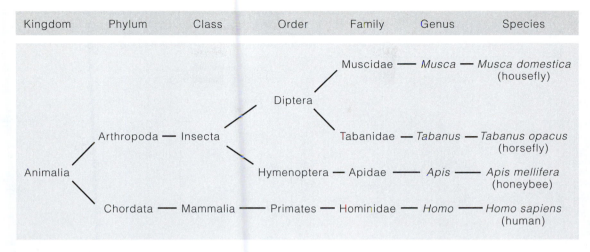

Kingdom	Phylum	Class	Order	Family	Genus	Species

Figure 6.1 A taxonomic hierarchy. The classification of a housefly, horsefly, honeybee, and human.

6.2 EVOLUTIONARY TREE DIAGRAMS

LEARNING OUTCOMES

1. Describe the goal of animal systematics.
2. Interpret the information presented in a cladogram.

The goal of systematic studies is to arrange animals into **monophyletic groups,** which include a single ancestor and all of its descendants. Insufficient knowledge of a particular group of animals sometimes results in lineages that appear to have more than one ancestor. This kind of lineage is called a **polyphyletic group.** Polyphyletic interpretation is always considered erroneous, and a search for additional data usually ensues. Sometimes a lineage includes some, but not all members of the group. This kind of lineage is called a **paraphyletic group** and also reflects insufficient knowledge of the group.

Evolutionary tree diagrams are used to depict ideas regarding family relationships. One of the most commonly used tree diagrams is called a **cladogram.** Cladograms are a result of one approach to systematics called **phylogenetic systematics (cladistics).** Cladograms depict a sequence in the origin of taxonomic characteristics. Figure 6.2 is a cladogram that shows five hypothetical taxa (1–5) and the characteristics (A–H) used in deriving the taxonomic relationships. Characteristics that have a genetic basis, that can be measured, and that indicate relatedness are called **characters.** To decide what character is ancestral for a group of organisms, cladists look for a related group of organisms that is not included in the study group. This related group is called an **outgroup.** Notice that taxon 5 is the outgroup for the other four taxa. Character A is common to all members of the lineage. This character is called a **symplesiomorphy.** It indicates that all members of the study group are related because it is present in ancestors of all members of the study

group. It cannot be used to distinguish between members of the study group.

A character that has arisen since common ancestry with the outgroup is called a **derived character,** or **synapomorphy.** Taxa 1–4 in figure 6.2 share character B. Character B separates taxa 1–4 from the outgroup and is a synapomorphy. Similarly, characters C and D arose more recently than B. Character C is shared by closely related taxa 1 and 2. Character D is shared by closely related taxa 3 and 4. Taxa that share a certain synapomorphy form a subset called a **clade.** Taxa 1 and 2 form a clade characterized by C. The entire cladogram can be considered a hypothesis regarding a monophyletic lineage. New data (newly investigated

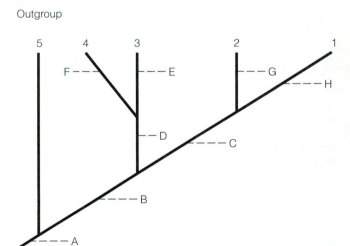

Figure 6.2 Interpreting cladograms. This hypothetical cladogram shows five taxa (1–5) and the characters (A–H) used in deriving the taxonomic relationships. Character A is symplesiomorphic for the entire group. Taxon 5 is the outgroup because it shares only that ancestral character with taxa 1–4. All other characters are more recently derived.

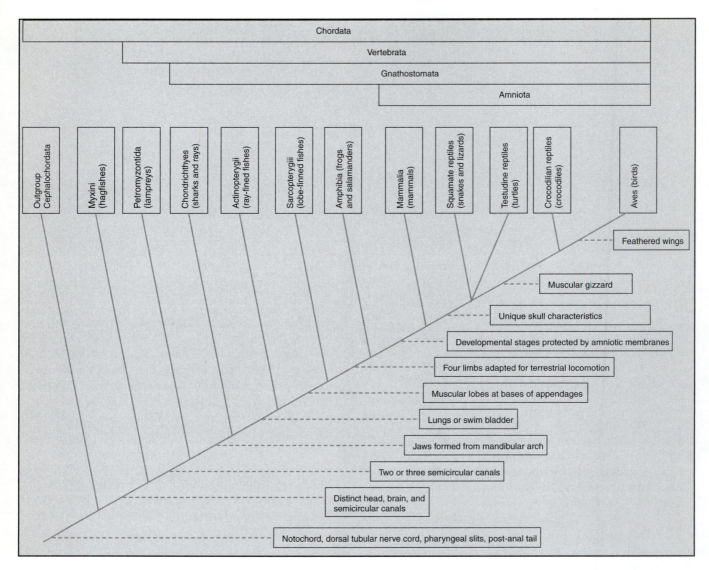

Figure 6.3 A cladogram showing vertebrate phylogeny. Numerous synapomorphies that distinguish vertebrate groups have been omitted to simplify the interpretation of this figure.

characters) are used to test the hypothesis. Figure 6.3 is an actual cladogram showing the taxonomic relationships of the vertebrates (fishes, amphibians, reptiles, birds, and mammals).

6.2 LEARNING OUTCOMES REVIEW—WORKSHEET 6.2

6.3 CONSTRUCTING A CLADOGRAM

LEARNING OUTCOMES

1. Construct a cladogram using simple block animals.
2. Assign names to simple block animals using a few basic rules of zoological nomenclature.

This exercise will help you understand the hierarchical nature of the classification system and the cladograms that you will be working with in unit II. In the first part of this exercise, you will use a set of wooden block animals to construct a hypothetical monophyletic lineage. You will use the wooden body as a sympleisomorphic character that unites all members of this lineage. In practice, the character that unites the outgroup and the study group must be ancestral for both groups. The choice of this ancestral character is based on evidence from other studies. It is not an arbitrary character selection, as the wooden-body choice seems to be in this exercise. Unlike the characteristics of wooden animals constructed for this exercise, the characters used in cladistics are phylogenetically significant.

▼ Examine the block animals on display (fig. 6.4). Complete the following observations:

1. One of the animals has a character that clearly sets it apart from all other animals present. This animal is a

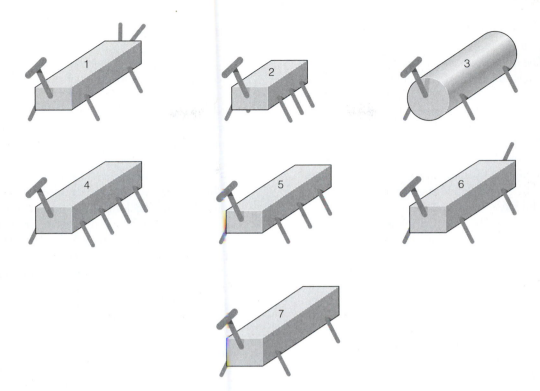

Figure 6.4 Block animals.

member of the outgroup that we will use to establish the ancestral character for the study group. The simplesiomorphic character that unites the outgroup with all other animals is "wooden body." Record "wooden body" in blank A of the cladogram (fig. 6.5) in worksheet 6.1.

2. Group animals other than the outgroup together on the table. What character distinguishes these taxa from the outgroup? Record that character in blank B1 and the contrasting outgroup character in blank B2 in figure 6.5. Write the number 3 in the outgroup blank at the tip of the outgroup lineage of the cladogram.

3. Group animals 1, 6, and 7 together on the table in front of you. They constitute a clade. What character distinguishes this clade from animals 2, 4, and 5? Record this character in blank C1, and the contrasting character in blank C2, of figure 6.5.

4. Animal 7 can be easily distinguished from animals 1 and 6. Record the pair of contrasting characters in blanks D1 and D2 of figure 6.5.

5. In a similar way, distinguish between animals 1 and 6 and record the contrasting characters. Write the number of these animals in their positions at the tips of the branches of this clade.

6. Now return to animals 2, 4, and 5. What are the characters that can be used to group these animals? Record the characters and the numbers designating animals 2, 4, and 5 at the ends of their branches. ▲

You have now constructed one hypothesis that depicts a monophyletic grouping. In completing the following exercise, you will assign names to the six animals in the study group. There are rules of zoological nomenclature that are used in assigning scientific names. The rules are very lengthy, and we will mention only a few.

1. Species names are Latinized by adding Latin endings. A few of these endings are shown in table 6.2. Notice that some of the endings are feminine and some are masculine. In naming a species, both the genus name and the species epithet should reflect the same gender.

2. The name of a genus is one word, which is a singular Latinized noun. It begins with a capital letter, and the word is italicized. The species epithet is also a singular Latinized noun. It begins with a lowercase letter and is also italicized.

3. Above the genus level, names are Latinized, capitalized, plural nouns. They are not italicized.

TABLE 6.2   Case Endings of a Few Latin Nouns

	Feminine	Masculine
Singular	-a	-us
	-am	-er
	-ae (possessive)	-um
Plural	-ae	-i
	-as	-a
		-os
	-arum (possessive)	-orum (possessive)

4. There are conventional endings for superfamily (-oidea), family (-idae), and subfamily (-inae) taxa. You will be using the family-level taxonomic category in the exercise that follows. Higher taxonomic categories do not officially have standard endings for their name.

5. Scientific names are often descriptive of a character used in classifying the animal, although this is not required. It is impossible for a single class or order name to be descriptive of all animals in the class or order. In the past, species names have been given in honor of the person describing the species or another person who has contributed to knowledge of the animal or group. This practice is less common today. The purpose of a name is not to describe a taxon or to honor a person. It is only a label.

6. A taxon can bear only one correct name.

7. The correct name for an animal is the first correctly published name.

▼ Assume that all animals, 1–7, are separate species and members of a single class united by the common ancestral character—a wooden body. We will assign the class name "Wooda" to the class. We will also assume that animals other than the outgroup do not constitute a paraphyletic grouping—there are no species missing from the group. Complete the following observations:

1. Species that are most closely related are more recently diverged from a common ancestor, and their branch points will be nearer the top of the cladogram. Consider animals 1, 6, and 7. There are two animals that are more closely related to each other than to the third. What are the numbers designating these two closely related animals?_____ This pair of animals should be named as two species within a common genus. The other member of this clade should be given a separate genus name and species epithet. Notice, however, that there is no other species in the genus

containing animal 7. In this case, any species epithet can be used. Using the naming rules described previously, assign species names to animals 1, 6, and 7. Record these names in the taxonomic hierarchy in worksheet 6.1 (fig. 6.6).

2. All three species of the 1, 6, 7 clade are united by a single character. Assign a family name to the clade and record it in figure 6.6. Remember that family names use the "-idae" suffix.

3. Repeat the naming procedure for animals 2, 4, and 5. In this case, you need to decide whether or not animals 2 and 5 belong to the same family as animal 4. Unfortunately, above the species level, assignment of taxonomic names can be somewhat subjective. Taxonomists often struggle with these kinds of decisions. Once you have made your decision, write your reasoning for it at the bottom of figure 6.6. Add to the hierarchy in figure 6.6 by drawing lines connecting the genus level with the order level, including either one or two additional family designations. Write in the family name(s).

4. Finally, complete the hierarchy by adding an order name for the entire group of six species.

One of the last steps in systematic studies is one that we will not complete. Systematists ask the following question: Is this cladogram the simplest cladogram that can be constructed using the data available? This question employs the principle of **parsimony.** As with many natural occurrences, systematists have found that the simplest, most economical explanation for a set of observations is usually the most accurate explanation. This determination usually involves computerized statistical analysis. ▲

6.3 LEARNING OUTCOMES REVIEW—WORKSHEET 6.2

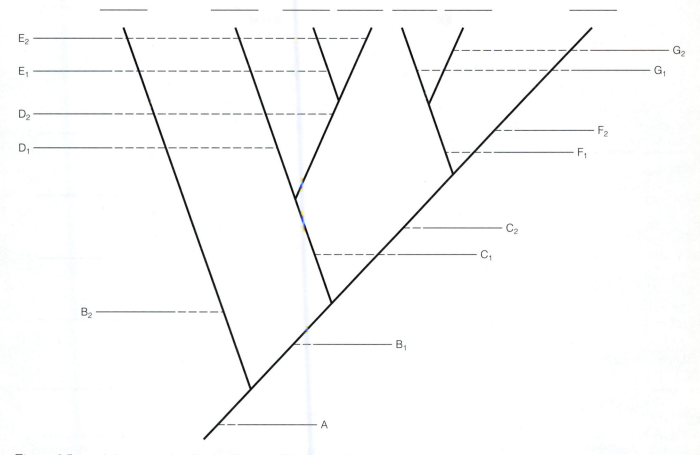

Figure 6.5   A cladogram used in the classification of block animals.

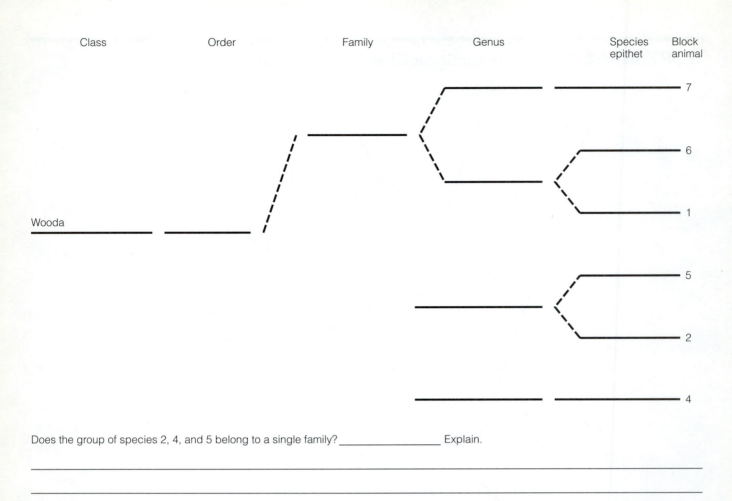

Class	Order	Family	Genus	Species epithet	Block animal

Wooda

Does the group of species 2, 4, and 5 belong to a single family? _____ Explain.

_____

_____

Figure 6.6 A taxonomic hierarchy of block animals. You must decide whether animals 2, 4, and 5 belong to one family or separate families. Complete the hierarchy by drawing in the blank(s) for the family name(s) and the dashed lines that connect the family name(s) with order and genus levels. Explain your decision.

WORKSHEET 6.2   The Classification of Organisms

6.1   A Taxonomic Hierarchy—Learning Outcomes Review

1. Two individuals are members of the same family. They could belong to different
   a. genera.
   b. orders.
   c. classes.
   d. phyla.
   e. b, c, and d could all be correct.

2. Characteristics used in classifying organisms are evolutionarily significant. This means that
   a. shared characteristics arose in a common ancestor.
   b. more recent common ancestry usually means that two organisms share more characteristics than animals with more distant ancestry.
   c. some characteristics that might seem similar in two animal groups may be useless in deriving evolutionary relationships.
   d. All of the above are true.

3. Scientific names are more useful than common names for all of the following reasons except one. Select the exception.
   a. scientific names reflect evolutionary relationships
   b. common names vary from region to region
   c. common names often refer to more than one species
   d. scientific names are always descriptive of the species

6.2   Evolutionary Tree Diagrams—Learning Outcomes Review

4. The goal of systematic studies is to arrange animals into
   a. monophyletic groups.
   b. polyphyletic groups.
   c. paraphyletic groups.

5. Two closely related groups of animals would share
   a. one or more derived characters.
   b. symplesiomorphic characters with each other and with other more distantly related animal groups.
   c. Both a and b are correct.

6. The closely related groups in question 5 would form a subset of a larger lineage called a(n)
   a. outgroup.
   b. Domain.
   c. clade.
   d. paraphyletic group.

6.3   Constructing a Cladogram—Learning Outcomes Review

7. A cladogram is a hypothesis that depicts a monophyletic grouping.
   a. true
   b. false

8. The "-idae" ending is always associated with a genus-level taxonomic grouping.
   a. true
   b. false

9. The family, genus, and species epithet of names in a clade should all reflect the same singular gender.
   a. true
   b. false

10. Scientific names are always descriptive of a character unique to that species.
    a. true
    b. false

11. Two trees are constructed for a set of organisms. Both are consistent with the data available. The researcher selects the tree with fewer branch points to publish. In making this decision, the researcher was applying the principle of parsimony.
    a. true
    b. false

6.1   A Taxonomic Hierarchy—Analytical Thinking

1. Modern systematists are often not concerned about assigning the name of intermediate taxonomic levels between phylum and species (i.e., family, order, and class) to hierarchical names associated with a group of animals. Why do you think this is so?

6.2   Evolutionary Tree Diagrams—Analytical Thinking

2. Explain why polyphyletic groupings and paraphyletic groupings represent errors in taxonomy.

6.3   Constructing a Cladogram—Analytical Thinking

3. The block animal exercise helps us understand the taxonomic hierarchy and principles associated with naming animals. It is not truly adequate to illustrate the concepts used by phylogenetic systematists in grouping animals. Explain the inadequacies in this exercise.

Exercise 7

Animal-like Protists

Prelaboratory Quiz

Study this week's laboratory exercise and then complete the following quiz to assess your preparation for the laboratory.

1. The protozoans perform all functions within the confines of a single plasma membrane. Thus, they are said to display
   a. tissue-level organization.
   b. unicellular organization.
   c. organelle-level organization.
   d. organ-level organization.

2. Members of the super group *Excavata* may have all of the following characteristics EXCEPT
   a. flagella.
   b. pseudopodia.
   c. both symbiotic and free-living species.
   d. photosynthetic and heterotrophic species.

3. Members of the super group *Amoebozoa* include
   a. euglenoids.
   b. *Paramecium*.
   c. amoebas.
   d. all intracellular parasitic protists.

4. All of the following are true of at least some members of the super group *Chromalveolata* EXCEPT
   a. intracellular parasites.
   b. apical complex used in penetrating host cells.
   c. flagella present in all stages.
   d. form spores in their life cycles.

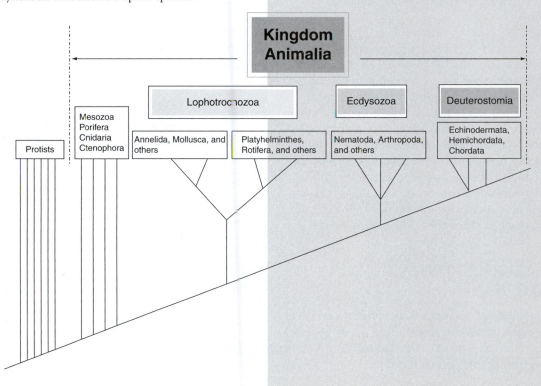

Kingdom Animalia

Lophotrochozoa — Ecdysozoa — Deuterostomia

Protists

Mesozoa
Porifera
Cnidaria
Ctenophora

Annelida, Mollusca, and others

Platyhelminthes, Rotifera, and others

Nematoda, Arthropoda, and others

Echinodermata, Hemichordata, Chordata

5. Vertebrates acquire malaria as a result of the parasite being introduced into the vertebrate through
   a. the bite of a mosquito.
   b. the bite of a fly.
   c. drinking contaminated water.
   d. eating contaminated meat.

6. Members of which of the following possess both a macronucleus and a micronucleus?
   a. Foraminifera
   b. Apicomplexa
   c. Ciliophora

7. True/False The protozoans constitute a monophyletic grouping.

8. True/False Members of the super group *Amoebozoa* possess cilia.

9. True/False The ectoplasm is a gel-like layer of cytoplasm next to the plasma membrane of an amoeba and other protists.

10. True/False Contractile vacuoles are used in osmoregulation (water regulation).

7.1 EVOLUTIONARY PERSPECTIVE

LEARNING OUTCOMES

1. Describe the place of the animal-like protists in the evolution of life on earth.
2. Explain the unicellular grade of organization.
3. Discuss current ideas on the classification of the protozoans.

Ancient members of the Archaea were the first living organisms on this planet. The Archaea and Eubacteria gave rise to the protists during the Precambrian period about 1.5 billion years ago. Most biologists agree that the protists probably arose by way of more than one ancestral group. Various groups of protists, therefore, represent separate evolutionary lineages. The protists are **polyphyletic.** Some protists are plantlike because they are primarily autotrophic. (They produce their own food.) Others are animal-like because they are primarily heterotrophic. (They feed upon other organisms.) In this exercise, you will study three phyla of animal-like protists.

ANIMAL-LIKE PROTISTS: THE PROTOZOA

To the animal biologist, the animal-like protists have been known as the protozoans for many years. Their coverage in zoology courses stems from early classification systems that included them in the animal kingdom. Protozoans are not animals. They are descended from those forms of life that gave rise to multicellular organisms. Protozoans present today have had a long evolutionary history and may be considerably different from the ancestral stock. As a group, however, they display many of the features that probably characterized the earliest protists.

In the protozoans, all functions are performed within the limits of a single plasma membrane. Protozoans, therefore, display **unicellular (cytoplasmic) organization.** Even though there are no tissues or organs in protozoans, there is a division of labor within the organism through specialized **organelles.** Individual protozoans may also associate together to form colonies, and there may be division of labor between members of a colony.

CLASSIFICATION

The classification of the protozoa has undergone substantial revisions in recent years. Older classification systems based on means of locomotion are being abandoned. Newer, evolutionarily significant groups based on morphological and molecular data have resulted in the designation of the following super groups.* We will study representatives of these groups that are animal-like protists (the protozoans).

> Super group *Excavata*
> Super group *Amoebozoa*
> Super group *Rhizaria*
> Super group *Chromalveolata*

7.1 LEARNING OUTCOMES REVIEW—WORKSHEET 7.1

7.2 SUPER GROUP *EXCAVATA*

LEARNING OUTCOMES

1. Describe variations in the structure of common *Excavata*, including two or more of the following: *Euglena*, *Trichonympha*, *Trichomonas*, *Volvox*, and *Trypanosoma*.
2. Describe the life cycle of a *Volvox* colony.

Characteristics: cytosome characterized by a suspension-feeding grove with a posterior-directed flagellum that is used to generate a feeding current.

Euglena

▼ *Euglena* is a common freshwater phytoflagellated protozoan found in ponds and slow-moving streams. Live protozoans can be slowed for study using methylcellulose or a

*Group names are italicized according to the convention established by the International Society of Protozoologists.

similar product. Place a drop of methylcellulose on a clean microscope slide. Spread it over the middle third of the slide. Next, obtain live specimens of *Euglena* from a culture dish, add them to the slide, and cover with a coverslip. Observe under high power, and compare what you see to figure 7.1. Note the green color, the red eyespot, and movements of the flagellum. After viewing your specimens alive, remove the slide from the microscope and add a drop of iodine-potassium iodide (IKI) to the edge of the coverslip. This will kill the euglenoids but will make the flagellum easier to see. Repeat these observations using a prepared *Euglena* slide.

Knowledge about the details of the *Euglena* structure has increased due to the electron microscope. Note the following structures in figure 7.1a, but do not expect to see these details in the light microscope. ▲

flagella	locomotion
eyespot and photoreceptor	photoreception
contractile vacuole	osmoregulation
basal body	base of the flagellum; development and control of the flagellum
chloroplasts	photosynthesis
pyrenoid	starch storage

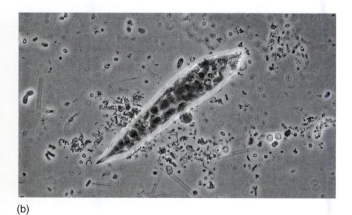

(a)

Figure 7.1 labels: Locomotor flagellum, Photoreceptor, Eyespot, Rudimentary flagellum, Flagellar pocket, Contractile vacuole, Basal body (kinetosome), Pyrenoid, Chloroplast, Mitochondrion, Nucleus, Golgi complex

(b)

Figure 7.1 *Euglena.* (*a*) The ultrastructure of *Euglena* as would be observed in electron micrographs. (*b*) A light photomicrograph of *Euglena* (×250).

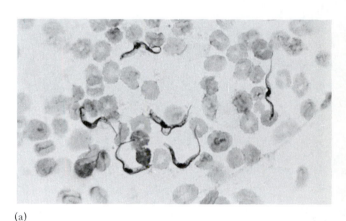

(a)

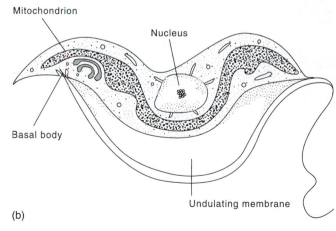

Figure 7.2 labels: Mitochondrion, Nucleus, Basal body, Undulating membrane

(b)

Figure 7.2 *Trypanosoma.* (*a*) A photomicrograph of *Trypanosoma* in vertebrate blood. (*b*) The structure of *Trypanosoma.* The flagellum of *Trypanosomes* often trails behind the protist and is connected to the protist by an undulating membrane.

Other *Excavata*

▼ Examine other *Excavata* as available. Note that most Zoomastigophora are symbionts, including *Trypanosoma* (fig. 7.2), *Trichonympha* (fig. 7.3), and *Trichomonas*.

If termites are available, make a preparation of their hindgut, mastigophoran symbionts. Place a drop of insect saline on a slide. Working in saline and using fine-tipped forceps, grasp the terminal segments of the abdomen of the termite and pull the hindgut free of the remainder of the body. Discard the head, thorax, and remaining abdomen. Cover the hindgut with a coverslip, and gently squash with the eraser end of a pencil. View under high power.

Volvox is a phytomastigophoran that shows colonial organization. If live organisms are available, place a sample from the culture dish in the concavity of a depression slide. Fill the concavity of the slide with culture medium and cover with a coverslip. Be sure there is no air trapped

below the coverslip. Note the graceful rolling of the colony through the water, which results from the activity of the pair of flagella present on each individual (zooid) in the colony (fig. 7.4).

Although most zooids in the colony are unspecialized, reproduction is dependent upon certain individuals being specialized for that function. Asexual reproduction occurs when certain cells withdraw to the watery interior of the colony and form daughter colonies. When the parental colony dies and ruptures, daughter colonies are released. Sexual reproduction occurs in autumn when specialized cells differentiate into macrogametes or microgametes. Macrogametes are large, filled with nutrient reserves, and nonmotile. Microgametes form as a packet of flagellated cells that leave the parental colony and swim to a colony containing macrogametes. Microgametes then break apart and syngamy (fertilization) occurs. The zygote develops into a zygospore by secreting a resistant wall around itself. It is an overwintering stage

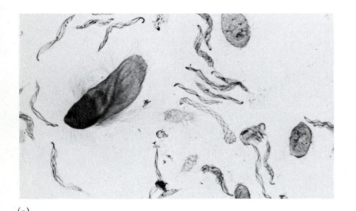

(a)

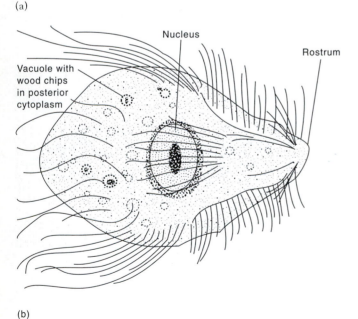

(b)

Figure 7.3 *Trichonympha* is a very complex mastigophoran found in the gut of termites. It possesses thousands of flagella. (a) Photomicrograph (×400). (b) Drawing based on an electromicrograph.

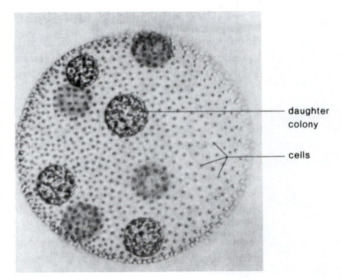

(a)

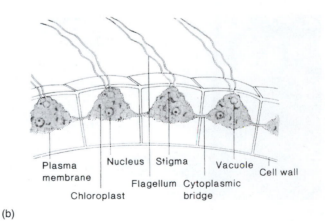

(b)

Figure 7.4 *Volvox*, a colonial flagellate. (a) A *Volvox* colony showing asexually produced daughter colonies. (b) An enlargement of a portion of a body wall.

that is released when the parental colony dies. A small colony is released from the zygospore in the spring. Observe any preparations available that demonstrate these features. ▲

7.2 LEARNING OUTCOMES REVIEW—WORKSHEET 7.1

7.3 SUPER GROUP AMOEBOZOA

LEARNING OUTCOMES

1. Describe the structure of a common amoebozoan, *Amoeba*.
2. Describe the formation of pseudopodia during amoeboid movement.

Characteristics: Amoeboid motility with broad pseudopodia (lobopodia); naked or with tests; cysts common.

AMOEBA

▼ *Amoeba* is a common freshwater protist that lives on the bottom of ponds and slow-moving streams. Examine living specimens. Obtain a sample of water from the bottom of a culture dish, place one or two drops of the sample on a microscope slide, and cover with a coverslip. The amoeba will appear transparent, granular, and irregular in outline. If you have trouble finding a specimen, ask your instructor for help, as they are often difficult to locate. Compare what you see to figure 7.5 and locate as many of the following as possible on your specimen:

ectoplasm	nongranular, gel-like layer of cytoplasm next to the plasma membrane
endoplasm	granular, fluid layer of cytoplasm in the interior of the amoeba
pseudopodia	lobe-like extensions of the amoeba used in locomotion and feeding
food vacuole	food storage and digestion
contractile vacuole	osmoregulation

Save this wet mount for the next observation, in which you will watch the streaming of the cytoplasm and the formation of pseudopodia. ▲

Amoeboid Movement

▼ Sarcomastigophorans as well as other cells (e.g., white blood cells) display a type of movement known as **amoeboid movement.** As you watch your living amoeba, note that the fluid interior (endoplasm) flows forward, forming a pseudopodium. As the endoplasm reaches the tip of the pseudopod, it flows to the outside and changes state to form ectoplasm. At the posterior of the amoeba, ectoplasm is converted into endoplasm and begins flowing toward the advancing pseudopod. Microfilaments, similar to the actin of muscle cells, are believed to be responsible for this movement. (See cell structure, exercise 2.)

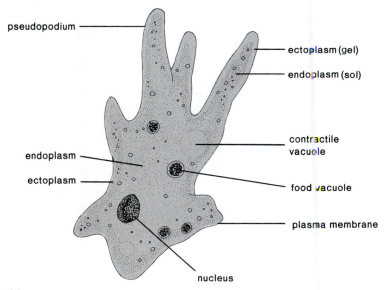

(a)

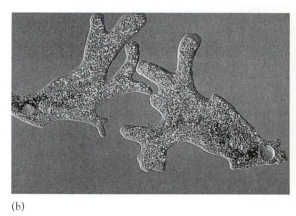

(b)

Figure 7.5 *Amoeba proteus.* (a) A diagrammatic representation of this protist's structure. (b) A light photomicrograph (×100).

Examine a prepared slide of *Amoeba* and compare it to the living specimen. ▲

7.3 LEARNING OUTCOMES REVIEW—WORKSHEET 7.1

7.4 SUPER GROUP RHIZARIA

LEARNING OUTCOME

1. Describe and give examples of members of the super group *Rhizaria*.

Characteristics: Possess thin pseudopodia (filopodia).

▼ Examine rhizarians as available. Rhizarians incude the radiolarians such as the freshwater *Actinosphaerium* and numerous marine species. They secrete a siliceous test through which filopodia extend. Foraminiferans (forams) are primarily marine and secrete a calcium carbonate test. Foraminiferans are known to have existed for millions of years because of the fossilized tests found in limestone and chalk beds. ▲

7.4 LEARNING OUTCOMES REVIEW—WORKSHEET 7.1

7.5 SUPER GROUP CHROMALVEOLATA

LEARNING OUTCOMES

1. Describe the life cycle of the apicomplexan, *Plasmodium*. Identify red blood cells infected with this parasite.
2. Describe the body form of the ciliated protozoan, *Paramecium*.
3. Describe locomotion and feeding by *Paramecium*.

Characteristics: Plastid from secondary endosymbiosis with an ancestral archaeplastid. Plastid then lost in some.
The *Chromoaveolata* is a diverse group of protists united by the common origin of their plastids. One subgroup containing dinoflagellates, apicomplexans, and ciltates is commonly considered protozoan. We will study one important apicomplexan and a ciliate.

APICOMPLEXA

Members of the *Apicomplexa* are all intracellular parasites of animals. They possess an apical complex, a cone-like structure that is used for penetrating host cells. *Plasmodium*, the apicomplexan that causes malaria, is a parasite of red blood cells and liver cells of vertebrates. Multiple fission (schizogony) occurs in liver and red blood cells. Some of the products of this fission enter blood or liver cells for more asexual reproduction. Others enter red blood cells and form gametocytes (gametogeny), which are taken into a mosquito during a blood meal and fuse. The zygote penetrates the gut of the mosquito and is transformed into an oocyst. Meiotic division and multiple fission (sporogeny) produce haploid sporozoites that enter a vertebrate host when the mosquito bites (fig. 7.6). ▼ Examine a demonstration of *Plasmodium* from vertebrate blood. ▲

CILIOPHORA

Ciliates include some of the most complex protozoa. They are widely distributed in freshwater and marine environments. Their group name comes from the presence of cilia that are used for locomotion and for the generation of feeding currents. They also possess a macronucleus and one or more micronuclei. Micronuclei are used in a form of sexual reproduction called conjugation.

Paramecium

▼ *Paramecium* is a common freshwater ciliate. Obtain a drop of culture containing *Paramecium* and place it on a clean microscope slide along with a drop of methylcellulose. Note its "slipper" shape. As it swims, it rotates through the water. You will be able to see the oral groove (figs. 7.7 and 7.8) as it swims. As you watch your specimens, you may be able to see contractile vacuoles forming and discharging near the ends of *Paramecium*. Add a small drop of freshly prepared yeast/congo red solution to the side of the coverslip. (If your slide is starting to dry, or if your specimens are immobilized, make a new preparation.) As stained yeast particles diffuse under the coverslip, *Paramecia* should begin feeding. Observe cilia of the oral groove draw yeast into the cytostome. After a few minutes, you should see food vacuoles forming in the region of the cytostome. Observe a prepared slide of *Paramecium* and compare it to figure 7.7. Which of the following structures can you see on your slide? ▲

pellicle	protective outer covering and maintains cell shape
cilia	locomotion and feeding
trichocysts	protection
macronucleus	controls metabolic processes
micronucleus	hereditary units that participate in sexual reproduction
oral groove, cytostome, and **cytopharynx**	feeding complex
food vacuoles	food storage and digestion
contractile vacuoles	osmoregulation

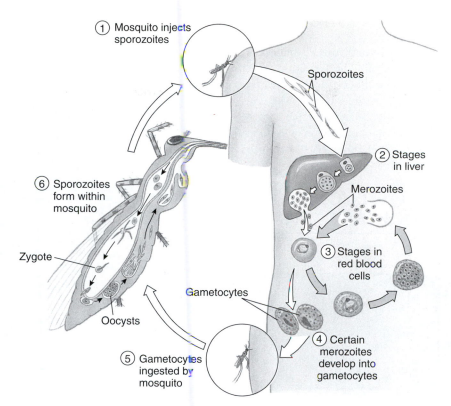

① Mosquito injects sporozoites

Sporozoites

⑥ Sporozoites form within mosquito

② Stages in liver

Merozoites

Zygote

③ Stages in red blood cells

Oocysts

Gametocytes

④ Certain merozoites develop into gametocytes

⑤ Gametocytes ingested by mosquito

Figure 7.6 The *Plasmodium* life cycle involves three phases. Schizogony is multiple fission of the asexual stage and results in the formation of merozoites. Schizogony occurs in liver cells and, later, in the red blood cells of humans. Gametogony occurs in red blood cells and results in the formation of gametocytes. During a blood meal, a mosquito takes in micro- and macrogametes, which fuse to form zygotes. Zygotes penetrate the gut of the mosquito and form oocysts. Meiotic division and sporogony form many haploid sporozoites that may enter a new host when the mosquito bites the host.

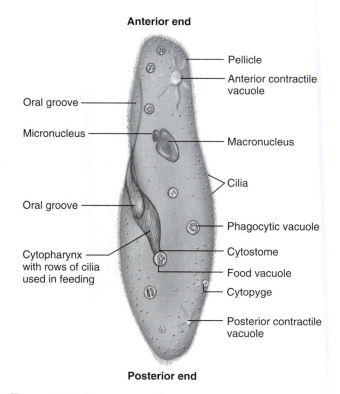

Anterior end

Pellicle

Anterior contractile vacuole

Oral groove

Micronucleus

Macronucleus

Cilia

Oral groove

Phagocytic vacuole

Cytostome

Cytopharynx with rows of cilia used in feeding

Food vacuole

Cytopyge

Posterior contractile vacuole

Posterior end

Figure 7.7 *Paramecium caudatum.*

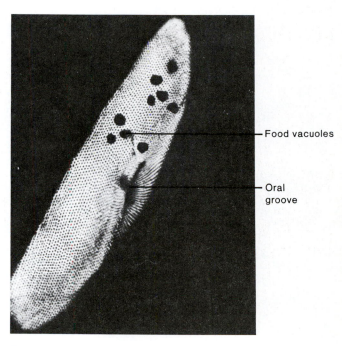

Food vacuoles

Oral groove

Figure 7.8 *Paramecium* (nigrosin stain shows sculpturing of pellicle) (×450).

Other Ciliophora

The presence of at least one macronucleus and micronucleus is characteristic of ciliates. Stained preparations of a small ciliate, called *Colpidium*, show these structures clearly.

▼ Examine a slide of *Colpidium* under high power. Examine any other specimens available for an appreciation of the diversity present in this phylum. They may include *Vorticella* (fig. 7.9), *Stentor* (fig. 7.10), and ciliates from the rumen of cattle. ▲

7.5 LEARNING OUTCOMES REVIEW—WORKSHEET 7.1

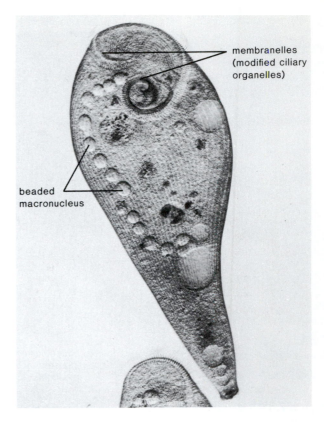

Figure 7.10   *Stentor*.

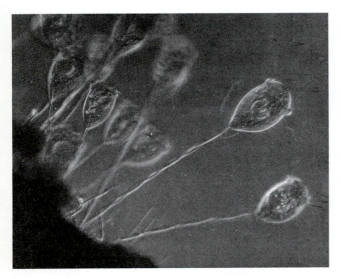

Figure 7.9   *Vorticella*, differential interference optical photomicrograph (×200).

WORKSHEET 7.1   Animal-like Protists

7.1   Evolutionary Perspective—Learning Outcomes Review

1. "Protozoa" is a
   a. phylum of animal-like protists.
   b. nontaxonomic term designating a polyphyletic group of protists.
   c. kingdom of animal-like protists.
   d. group of Archaea from which all animals arose.

2. Protozoans display
   a. unicellular organization.
   b. diploblastic organization.
   c. triploblastic organization.
   d. coelomate organization.

3. Which of the following statements accurately reflects the current state of protozoan taxonomy?
   a. Protozoans belong to 6 clearly-defined monophyletic lineages.
   b. Classification is unsettled, thus formal hierarchical rank designations (class, order, family, etc.) are not used in this laboratory manual.
   c. Protozoans belong to the kingdom Animalia, and their major groups have phylum status.
   d. Protozoans are not animals, but are traditionally covered in zoology courses.
   e. Both b and d are correct.

7.2   Super Group *Excavata*—Learning Outcomes Review

4. Which of the following structures were you able to observe in the *Euglena* using the compound microscope?
   a. chloroplasts
   b. eyespot
   c. Golgi complex
   d. mitochondrion
   e. Both a and b were observed.

5. Members of the *Excavata*, including *Euglena*, *Trichonympha*, and *Trypanosoma* are all characterized by the presence of
   a. cilia.
   b. one or more flagella.
   c. pseudopodia.
   d. exclusively symbiotic life cycles.

6. Observations of *Volvox* demonstrated that
   a. protists are sometimes colonial.
   b. some protists can carry on photosynthesis.
   c. protists reproduce exclusively by fission.
   d. many protists are parasites.
   e. Both a and b are correct.

7. Observations of *Trichonympha* and *Trypanosoma* demonstrated that
   a. protists are sometimes colonial.
   b. some protists can carry on photosynthesis.
   c. protists reproduce exclusively by fission.
   d. many protists are symbionts.
   e. Both a and b are correct.

7.3   Super Group *Amoebozoa*—Learning Outcomes Review

8. The ectoplasm of an amoeba is
   a. fluid (sol) and within the interior of the organism.
   b. gel-like (gel) and near the plasma membrane of the organism.
   c. converted into gel-like endoplasm during amoeboid movement.
   d. Both a and c are correct.

9. Vacuoles found in the *Amoebozoa* may be involved with
   a. food distribution and digestion.
   b. photosynthesis.
   c. osmoregulation.
   d. reproduction.
   e. Both a and c are correct.

10. During amoeboid movement
   a. endoplasm flows forward during pseudopod formation.
   b. ectoplasm is converted into endoplasm posteriorly.
   c. endoplasm is converted into ectoplasm anteriorly.
   d. All of the above are involved with amoeboid movement.

7.4 Super Group *Rhizaria*—Learning Outcomes Review

11. Rhizarians include
   a. amoebas.
   b. *Paramecium*.
   c. *Euglena*.
   d. foraminifera and radiolarians.

12. We know that the super group *Rhizaria* is a very ancient group because
   a. they are all extinct.
   b. fossil rhizarian tests are found in ancient limestone and chalk beds.
   c. Both a and b are correct .

7.5 Super Group *Chromalveolata*—Learning Outcomes Review

You examined a demonstration slide of *Plasmodium*. Select "true" or "false" for each of the following statements in questions 13–16.

13. The stage that you observed was an intracellular parasite found within red blood cells.
   a. true
   b. false

14. This parasite also spends a portion of its life cycle in human liver cells.
   a. true
   b. false

15. The stage that you observed is the oocyst stage.
   a. true
   b. false

16. Gametocytes are ingested by a mosquito and fuse to form a zygote in the mosquito.
   a. true
   b. false

17. During feeding, *Paramecium* forms food vacuoles. Immediately before the food vacuole forms, food is in the
   a. contractile vacuole.
   b. oral groove.
   c. macronucleus.
   d. cytopharynx.

18. Which of the following is responsible for maintaining the "slipper-shape" that you observed in *Parmecium*?
   a. macronucleus
   b. oral groove
   c. pellicle
   d. trichocysts

7.1 Evolutionary Perspective—Analytical Thinking

1. Protozoaons are not animals. They are included in general zoology courses partly because of a tradition that originated when they were considered to be animals. Do you think that zoology courses should omit coverage of these organisms because they are not animals? Explain.

2. Protozoans are single cells. They are also entire organisms. Explain the similarities between animal cells and protozoans that you observed. What differences did you observe between protozoans and animal cells that help us understand how they function as organisms.

7.3 Super Group *Amoebozoa*—Analytical Thinking

3. One amoeba, *Entamoeba histolytica*, is a serious parasite of the human small intestine. It causes debilitating dysentery (severe, bloody diarrhea) that plagues humans living in crowded, unsanitary conditions. Explain why this amoeba depends on the formation of resistant cysts that are released during a diarrhea episode to be transmitted between hosts. Relate your answer to your laboratory observations of *Amoeba*.

7.5 Super Group *Chromalveolata*—Analytical Thinking

4. *Plasmodium* is one of the greatest killers of humanity. It is estimated that 3,000 children die every day from malaria. Examine the life cycle of *Plasmodium*. Suggest the most effective strategies for controlling the spread of this disease. Research the problem using outside references. Compare your ideas to those actually being used.

Exercise 8

Porifera

Prelaboratory Quiz

Study this week's laboratory exercise and then complete the following quiz to assess your preparation for the laboratory.

Name the three cell types that make up the body wall of a sponge and list their function(s).

Name	Function
1. _____	_____
2. _____	_____
3. _____	_____

4. True/False Asconoid sponges have a spongocoel. Water enters the spongocoel through radial canals.

5. True/False Asconoid sponges have the simplest body organization.

6. True/False Incurrent canals of a leucon sponge are lined by choanocytes.

7. True/False The pathway for water leaving a sponge is through the osculum.

8. True/False The largest sponges have the leuconoid body form.

9. True/False In the leuconoid sponges, incurrent canals lead to flagellated chambers.

10. True/False Sponges have a tissue-level organization.

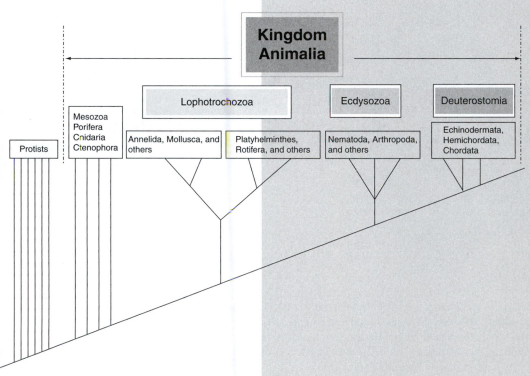

8.1 EVOLUTIONARY PERSPECTIVE

LEARNING OUTCOMES

1. Discuss the kind of evidence that leads zoologists to believe that Animalia is a monophyletic group.
2. Describe the poriferan cellular level of organization, including their three major cell types.

The sponges are members of the phylum Porifera and, like all animals, they are multicellular. The origins of multicellularity are shrouded in mystery. Many zoologists believe that multicellularity could have arisen as cells divided but remained together as a colony of protists. (Recall the structure of *Volvox*.) Although there are numerous variations of this hypothesis, they are all treated here as the **colonial hypothesis.** Ancient flagellated protists are often described as possible ancestral stocks because flagellated protists form colonies. A second proposed hypothesis is the **syncytial hypothesis.** A syncytial cell is a large, multinucleate cell. The formation of plasma membranes within the cytoplasm of a syncytial protist could have formed a small, multicellular organism. This hypothesis is supported by the fact that syncytial organization occurs in some protist phyla.

There is little doubt that the Animalia is monophyletic-derived from a common ancestor. This conclusion is supported by impressive similarities within the Animalia as regards common aspects of cellular structure and molecular data. The most likely candidate group of ancestral protists is the choanoflagellates, a group of protists possessing a basket-like feeding collar surrounding the base of a flagellum.

THE CELLULAR LEVEL OF ORGANIZATION

The Porifera consist of loosely organized cells with no well-defined tissues. They are asymmetrical or radially symmetrical, sessile, and mostly marine. (See the discussion of symmetry on page xi.) Even though sponges do not have tissues, their bodies are more than colonies of independent cells. Sponge cells are specialized for specific functions. This is often referred to as division of labor. The major cell types and their functions are as follows (fig. 8.1):

pinacocytes Thin, flat cells covering the outer surface and some of the inner surface. May be mildly contractile, thus having the ability to change the shape of the sponge. May be specialized into porocytes that surround pores and can contract to regulate circulation of water to the interior of the sponge.

choanocytes Flagellated cells lining sponge chambers. Create water currents and filter suspended food particles from the water. Sometimes called collar cells.

mesenchyme cells Amoeboid cells embedded in the noncellular matrix. May store, digest, and distribute food, or be specialized to form reproductive cells, secrete spicules, secrete spongin, or secrete fibrils. Spicules and spongin give structural support to the sponge

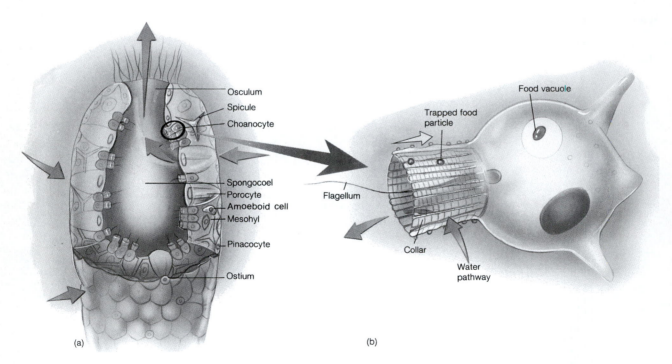

Figure 8.1 Morphology of a simple sponge, including body-wall organization. (*a*) In this example, pinacocytes form the outer body wall, and amoeboid cells and the spicules are found in the mesohyl. Ostia are formed by porocytes. (*b*) Choanocytes are cells in which the flagellum is surrounded by a collar of microvilli that traps food particles. Food is moved toward the base of the cell, where it is incorporated into a food vacuole and passed to amoeboid cells, where digestion takes place. Solid arrows show the water flow pattern. The open arrow shows the direction of movement of trapped food particles.

colony. Cells are loosely arranged in jellylike material called the **mesohyl.** Amoeboid cells, spicules or spongin, and various fibrils are embedded in the mesohyl.

8.1 LEARNING OUTCOMES REVIEW—WORKSHEET 8.1

8.2 THREE PORIFERAN BODY FORMS

LEARNING OUTCOMES

1. Observe an asconoid sponge. Describe the pattern of water movement through the sponge, including the function and location of choanocytes.
2. Observe a syconoid sponge. Describe the pattern of water movement through the sponge, including the function and location of choanocytes.
3. Observe demonstrations of leuconoid sponges. Explain why this body form is characteristic of the largest sponges.

Members of the phylum Porifera are the sponges. Sponges are divided into three classes. Rather than trying to characterize each poriferan class, we will examine the three body forms characteristic of sponges.

CLASSIFICATION

Class Calcarea
    Calcium carbonate spicules in the form of needles or three- or four-rayed stars. Asconoid, syconoid, or leuconoid. Marine.
Class Hexactinellida
    Siliceous spicules in the form of six-rayed stars. Syconoid or leuconoid. Marine, mostly deep water.
Class Demospongiae
    Spicules siliceous (but not six-rayed) and/or spongin. Leuconoid. Freshwater (one family) and marine.

Poriferan body forms are described based upon the arrangement of canals. Recall from exercise 1 that when observing preserved specimens under a dissecting microscope, it is best to immerse them completely in water to prevent the refraction of light off moist surfaces. Failure to do so will result in a distorted image.

THE ASCONOID SPONGE— *LEUCOSELENIA* (CALCISPONGIAE)

▼ **Asconoid sponges** have the simplest organization. Choanocytes line the spongocoel, drawing water through small openings called **ostia** and expelling it through the osculum (fig. 8.2a). Examine *Leucoselenia* slides using the low power of a compound microscope. *Leucoselenia* is very small, and a small sponge cluster will be mounted under the coverslip of your slide. Look at the tip of one of these clusters. Focus up and down carefully. You should see the **osculum** as the edges of this opening come into and go out of focus. Three-rayed **spicules** should be obvious protruding from the body wall. The **spongocoel** can be detected by focusing up and down on the body wall. The central portion of the sponge should be lighter in color, as the large central cavity allows light to easily pass through this region of the sponge. ▲

THE SYCONOID SPONGE—*SCYPHA* (CALCISPONGIAE)

▼ **Syconoid sponges** have a tubular design similar to the ascon sponge, but the body wall is folded. The "folds" form **radial canals** and **incurrent canals** (fig. 8.2b). **Choanocytes** line the radial canals rather than the **spongocoel.** Examine preserved specimens of *Scypha* under a dissecting microscope. Externally locate the **osculum,** which opens at the top of the sponge, and the many small **dermal pores** along the sides. Dermal pores open into the canal system of the sponge. Choanocytes create water currents that draw water through dermal pores to the interior of the sponge. Eventually water is discharged through the osculum.

With a sharp scalpel or razor blade make a longitudinal section through the specimen. Locate internal openings to the radial canals, the **apopyles.** The large cavity into which the incision was made is the spongocoel.

Examine a cross section of *Scypha* using a compound microscope. Examine first under lower power (fig. 8.3). Note the central spongocoel. Radial canals radiate from the spongocoel. They connect through small openings (prosopyles) to incurrent canals that open to the outside of the sponge. Recall that the dermal pore is the external opening to the incurrent canal and the apopyle is the internal opening of the radial canal. Return to the radial canal and change to high power. Examine the inner cell layer of the radial canal. Choanocytes make up this layer (figs. 8.4 and 8.5). Some slides are prepared with the calcium carbonate **spicules** intact; in other slides, spicules are dissolved before mounting. If your slide shows spicules, note their triradiate structure. If your slide does not show them, obtain and examine a separate preparation of spicules (fig. 8.6). ▲

THE LEUCONOID SPONGE

Leuconoid sponges represent the most complex body form. The canal system is extensively branched (fig. 8.2c). Small incurrent canals lead to **flagellated chambers** lined by choanocytes. Flagellated chambers discharge water into excurrent canals that eventually lead to an osculum. Usually there are many oscula in each sponge. The "bath sponge" is an example of a leuconoid sponge. The skeleton of this sponge is made of a soft protein, called **spongin,** rather than calcium carbonate or silica.

▼ Examine demonstration materials showing the leuconoid body form. ▲

SPONGE EVOLUTION

The ascon-to-sycon-to-leucon progression probably does not reflect an exact sequence in the evolution of sponge body forms. (Most sponges, however, are leuconoid.) What seems clear is that natural selection resulted in increased body size and greater surface area for choanocytes.

▼ Look again at demonstrations of the three sponge body forms. Answer analytical thinking question 3 in worksheet 8.1 now. ▲

8.2 LEARNING OUTCOMES REVIEW—WORKSHEET 8.1

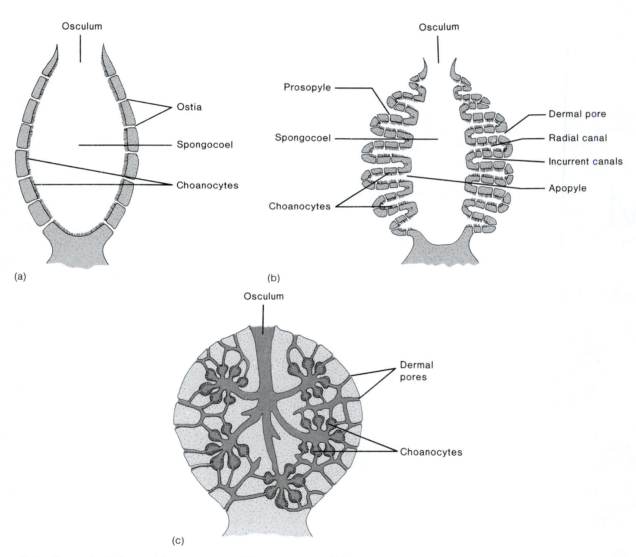

Figure 8.2 Sponge body forms: (*a*) ascon sponge; (*b*) sycon sponge; (*c*) leucon sponge.

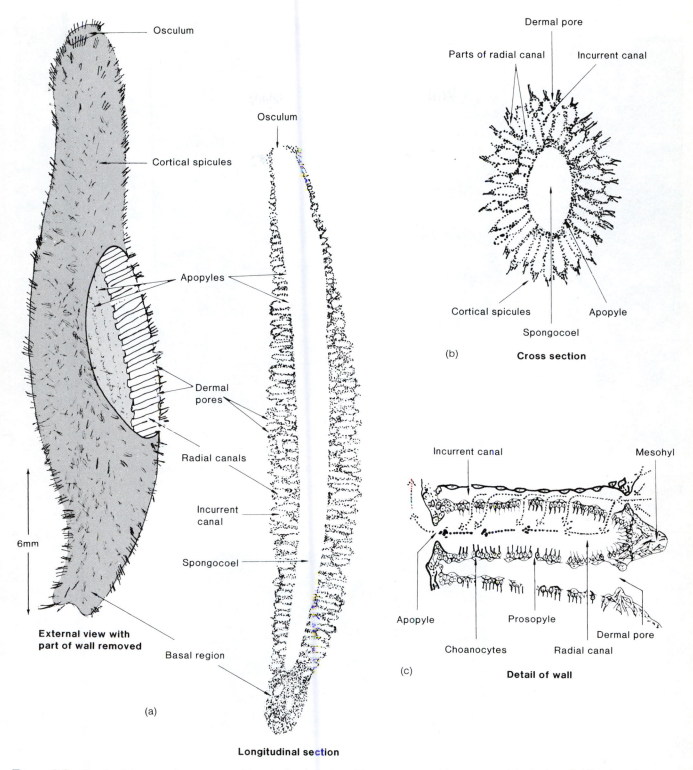

Figure 8.3 *Scypha*: (*a*) external structure and longitudinal section; (*b*) cross section; (*c*) structure of the body wall. The dotted arrows in (*c*) show the path of water movement.

flagellum   collar

cell body

Figure 8.4 Choanocyte (scanning electron micrograph) (×9,400).

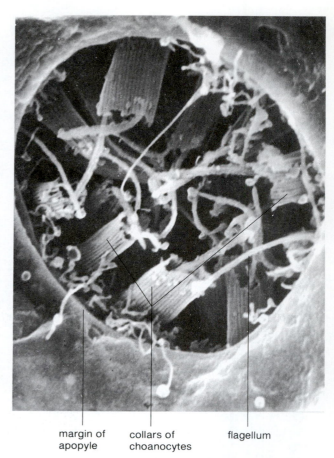

margin of   collars of   flagellum
apopyle   choanocytes

Figure 8.5 Choanocytes as viewed through an apopyle (scanning electron micrograph) (×6,200).

Figure 8.6 Sponge spicules (×150)

WORKSHEET 8.1 Porifera

8.1 Evolutionary Perspective—Learning Outcomes Review

1. Multicellularity may have arisen as flagellated cells remained together forming a colony and then an organism. This idea is referred to as the
   a. colonial hypothesis.
   b. syncytial hypothesis.
   c. choanocyte hypothesis.

2. Virtually all biologists agree that the
   a. poriferan lineage is unique and, therefore, Animalia is polyphyletic.
   b. poriferan lineage is within the Animalia and, therefore, Animalia is monophyletic.
   c. poriferan lineage is paraphyletic.

3. This cell type can be specialized to serve a variety of functions in a sponge, for example spicule formation, reproduction, and food distribution.
   a. pinacocytes
   b. mesenchyme cells
   c. choanocytes.

8.2 Three Poriferan Body Forms—Learning Outcomes Review

4. In the ascon sponge *Leucoselenia*, choanocytes
   a. line radial canals.
   b. line the spongocoel.
   c. line incurrent canals.
   d. line the osculum.

5. Which of the following is the sequence of structures water flows through in a sycon sponge?
   a. incurrent canal, prosopyle, apopyle, radial canal, spongocoel
   b. radial canal, apopyle, incurrent canal, spongocoel, prosopyle
   c. prosopyle, radial canal, apopyle, spongocoel, incurrent canal
   d. incurrent canal, prosopyle, radial canal, apopyle, spongocoel

6. Dermal pores would be found
   a. on the outside of a sponge body wall.
   b. lining the spongocoel of a scycon sponge.
   c. lining a radial canal of a sycon sponge.
   d. within flagellated chambers of a leucon sponge.

7. Which of the following statements explains why leuconoid sponges are usually larger than either asconoid or syconoid sponges?
   a. leuconoid sponges are always more highly evolved.
   b. leuconoid sponges have greater surface area for choanocytes.
   c. leuconoid sponges have a simpler canal system.
   d. leuconoid sponges always use a spongin skeleton.

8.1 Evolutionary Perspective—Analytical Thinking

1. Why is it more accurate to describe a sponge as having cellular level of multicellular organization rather than colonial organization?

8.2 Three Poriferan Body Forms.—Analytical Thinking

2. Describe the filtering surface of a syconoid sponge. How would you distinguish between a dermal pore and an apopyle, and between the radial canal and the incurrent canal in a slide cross section of this sponge?

3. You studied three sponge body forms in this exercise.
   a. Which of these body forms is most common?

   b. Which of these three body forms is largest in overall size?

   c. Which of these three body forms has the most complex canal systems?

   d. Describe the evolutionary connection between the previous three observations.

Exercise 9

Cnidaria

Prelaboratory Quiz

Study this week's laboratory exercise and then complete the following quiz to assess your preparation for the laboratory.

1. Members of the phyla Cnidaria and Ctenophora possess
   a. radial or biradial symmetry.
   b. diploblastic, tissue-level organization.
   c. oral and aboral ends.
   d. All of the above are correct.

2. The life cycle of many cnidarians involves an alternation between _____ and _____ stages.
   a. aquatic, terrestrial
   b. polyp, medusa
   c. bilateral, radial
   d. diploblastic, triploblastic

3. Members of the class Hydrozoa include
   a. sea anemones.
   b. true jellyfish.
   c. corals.
   d. colonial hydroids and *Hydra*.

4. Members of the class Anthozoa include
   a. *Hydra*.
   b. corals and sea anemones.
   c. true jellyfish.
   d. freshwater cnidarians.

5. The larval stage of cnidarians is called the
   a. planula.
   b. hydranth.
   c. polyp.
   d. medusa.

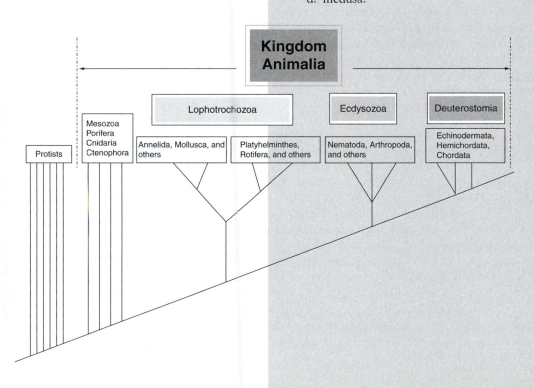

6. True/False The epidermis of cnidarians contains epithelio-muscular cells and cnidocytes.

7. True/False The phylum name Cnidaria is derived from the presence of specialized cells called cnidocytes, which are used in reproduction.

8. True/False With a few exceptions, members of the phylum Cnidaria are monoecious.

9. True/False Members of the class Anthozoa do not display alternation of generations.

10. True/False The predominate life-history stage in members of the class Scyphozoa is the asexually reproducing polyp.

9.1 EVOLUTIONARY PERSPECTIVE

LEARNING OUTCOMES

1. Describe the relationship of the Cnidaria and Ctenophora to other animal phyla.
2. Explain the diploblastic radiate organization of the Cnidaria and Ctenophora.

Members of the phylum Cnidaria are radially symmetrical. (See the discussion of symmetry on page xi.) The Cnidaria, along with another lesser known phylum (Ctenophora), are called the radiate phyla. The radiate phyla are closely related to the ancestors of all other animal groups. Like the poriferans, they are probably derived independently of other phyla from animal ancestors (see p. 131). As in all discussions of the evolutionary relationships among higher taxonomic groups, one must be careful to avoid picturing modern cnidarians, or any modern animals, as being ancestral. Modern species are the result of hundreds of millions of years of evolution and may be very different from ancestral forms.

THE RADIATE ANIMALS— TISSUE-LEVEL ORGANIZATION

Radial animals have no anterior and posterior (see terms of direction on page x). As a result, terms of direction are defined based on the position of the mouth opening. The end of the organism that contains the mouth is the **oral end,** and the opposite end is the **aboral end.** The radial symmetry of some cnidarians is modified into **biradial symmetry.** In biradial symmetry, a single plane, passing through a central (oral/aboral) axis, divides the organism into mirror images. Biradial symmetry occurs when a single, or a paired, structure (e.g., feeding or sensory tentacles) occurs in a basically radial animal. It differs from bilateral symmetry in that there is no distinction between dorsal and ventral surfaces. Radial symmetry

is advantageous for sedentary animals because sensory receptors are evenly distributed around the body. These organisms can respond to stimuli that come from all directions.

Cnidarians also possess **diploblastic, tissue-level organization.** This means that similar cells are organized into tissues, and all cells are derived from two embryological layers. Ectoderm of the embryo gives rise to the outer layer of the body wall, called the **epidermis,** and the endoderm of the embryo gives rise to the inner layer of the body wall, called the **gastrodermis.** Cells of the epidermis and gastrodermis are specialized to perform functions such as protection, food gathering, coordination, movement, digestion, and absorption.

9.1 LEARNING OUTCOMES REVIEW—WORKSHEET 9.1

9.2 PHYLUM CNIDARIA

LEARNING OUTCOMES

1. Describe a generalized cnidarian life cycle and how that life cycle is represented by members of the Hydrozoa (*Obelia*), Scyphozoa (*Aurelia*), and Anthozoa.
2. Explain diploblastic, tissue-level organization using your observations of a cross section of *Hydra*.

The phylum Cnidaria includes five classes of animals, including jellyfish, sea anemones, corals, and their relatives. The phylum name comes from the presence of specialized cells used in defense, feeding, and attachment. The **cnidocytes** contain stinging organelles called **nematocysts** (fig. 9.1).

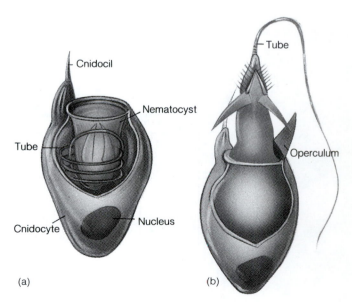

(a)   (b)

Figure 9.1 Cnidarian cnidocytes: (*a*) undischarged; (*b*) discharged.

ALTERNATION OF GENERATIONS

The life cycle of a typical cnidarian displays **alternation of generations**—it alternates between an often sessile **polyp** (hydroid) stage and a swimming **medusa** (jellyfish) stage (fig. 9.2). The polyp stage is often an asexual stage, and the medusa is often the sexual stage. In some cnidarian classes, either the polyp or medusa is reduced or missing.

With a few exceptions, the Cnidaria are **dioecious**—the sexually reproducing stage has individuals that are either male or female. In the Cnidaria, it is usually impossible to distinguish sexes by superficial examination.

GASTROVASCULAR CAVITY

The gastrovascular cavity is a large central cavity that serves to receive and digest food. The single opening on the oral surface serves as both mouth and anus. Tentacles surrounding the oral opening aid in feeding. The gastrovascular cavity of cnidarians also acts as a hydrostatic compartment. This cavity, along with contractile cells associated with the body wall, acts as a **hydrostatic skeleton** to provide support and accomplish movement.

THE BODY WALL

The body wall of cnidarians consists of two cellular layers. The epidermis consists of **epithelio-muscular cells** and cnidocytes. Epithelio-muscular cells are contractile and aid in movement. The inner gastrodermis lines the gastrovascular cavity and contains **gland cells** for the production and release of enzymes. The gastrodermis also has flagellated, **nutritive-muscular cells** that contain food vacuoles produced from phagocytosis of partly digested food particles. Between the epidermis and gastrodermis is the **mesoglea.** It is a noncellular gel and is abundant in the medusa.

CLASSIFICATION

Class Hydrozoa
Class Scyphozoa
Class Staurozoa
Class Cubozoa
Class Anthozoa

CLASS HYDROZOA

Characteristics: Polyp and medusa stages present, though one may be reduced in importance; cnidocytes present in the epidermis; gametes produced epidermally; no wandering mesenchyme cells in the mesoglea; medusa with velum; freshwater and marine. Examples: *Obelia, Hydra, Gonionemus,* and *Physalia.*

Obelia

We will first examine the colonial hydroid *Obelia* because it clearly illustrates alternation of generations (best represented in the Hydrozoa). *Obelia* is a marine hydrozoan whose colonial hydroid grows attached to some substrate. Because it is colonial, the hydroid will have multiple polyps. *Obelia* has two types of polyps. Other species have up to five.

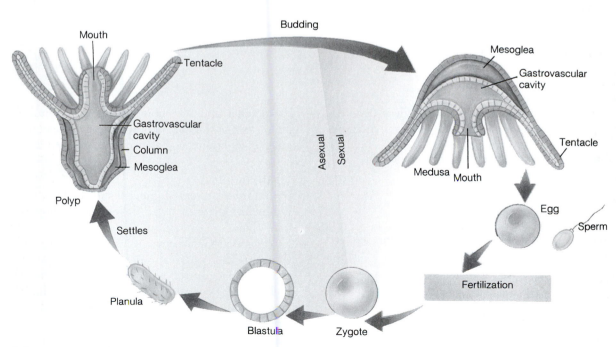

Figure 9.2 A generalized cnidarian life cycle showing alternation between medusoid and polypoid forms.

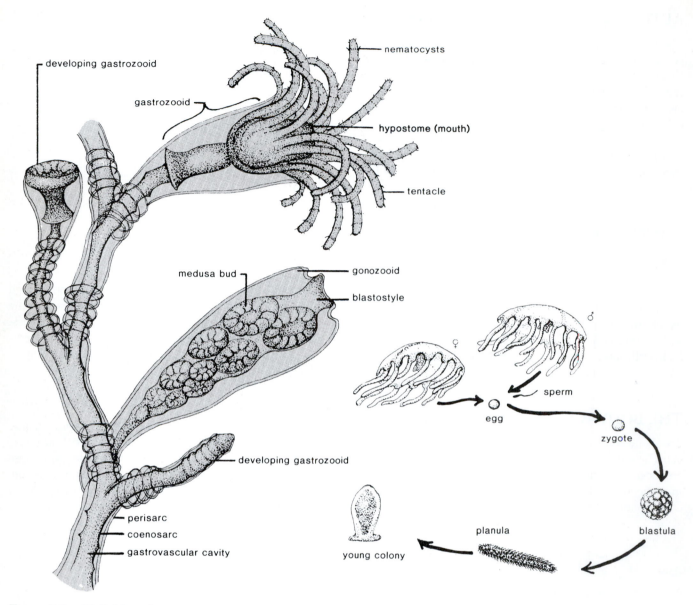

Labels in figure (clockwise/reading order):

developing gastrozooid

gastrozooid

nematocysts

hypostome (mouth)

tentacle

medusa bud

gonozooid

blastostyle

developing gastrozooid

perisarc

coenosarc

gastrovascular cavity

sperm

egg

zygote

blastula

planula

young colony

Figure 9.3 *Obelia* life cycle.

▼ Examine a slide of the *Obelia* hydroid under a dissecting microscope. Using figure 9.3, locate the following structures:

coenosarc	cellular body wall surrounding the gastrovascular cavity
perisarc	nonliving, protective, chitinous sheath surrounding coenosarc
gastrozooid (hydranth)	feeding polyp with mouth, tentacles, and hydrotheca
gonozoid (gonangium)	reproductive polyp, observe: blastostyle with medusa buds, gonotheca.

Next, obtain a slide of the *Obelia* medusa. The *Obelia* medusa is very small and should be observed with a compound microscope. You should be able to locate the exumbrellar and subumbrellar surfaces, manubrium, tentacles, and, at the margin of the medusa, clear, round statocysts (equilibrium and balance receptors). ▲

Gonionemus

▼ Obtain a preserved *Gonionemus* medusa. *Gonionemus* is a hydrozoan with a well-developed medusa and a minute polyp.

Examine *Gonionemus* using a dissecting microscope. Compare your observations to figures 9.4 and 9.5. The convex surface of the medusa is referred to as the **exumbrella** and the concave surface the **subumbrella.** Place your *Gonionemus*, subumbrellar surface up, in a dish of water. Note the consistency of the medusa. This is due to the large quantity of mesoglea present. The mouth opening is contained on the **manubrium** that hangs from the middle of the

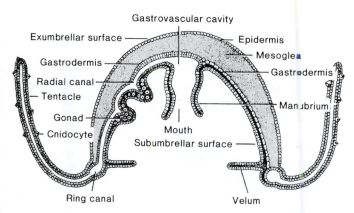

Figure 9.4 The structure of *Gonionemus* (oral/aboral section).

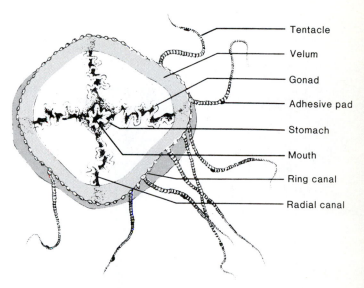

Figure 9.5 *Gonionemus*, oral view.

oral surface. Four sets of **gonads** radiate from the manubrium to the margin of the medusa. Note that a membranous shelf extends inward from the margin of the umbrella. This is the **velum** and is diagnostic of hydrozoan medusae. It is believed to aid in swimming. At the margin of the medusa, **tentacles** with nematocysts and adhesive pads can be seen. At the base of each tentacle are photoreceptors and equilibrium and balance receptors. We will not attempt to locate these.

Internally, the gastrovascular cavity divides into four **radial canals** that connect at the margin of the medusa with a **ring canal.** Radial canals are located just aboral to the gonads and can be seen through the exumbrellar surface. These canals distribute food throughout the medusa. ▲

Hydra

Hydra is a freshwater hydrozoan that lacks the medusa stage (fig. 9.6a). (Its sexual and asexual reproductive cycles will not be described here. Consult your text if instructed to do so.)

▼ Obtain a prepared slide mounted with a cross section of *Hydra*. Note the diploblastic organization of the body wall. Locate epidermis, gastrodermis, mesoglea, and gastrovascular cavity (fig. 9.6b). Turn to worksheet 9.1, Analytical Thinking question 1. Diagram the cross section of *Hydra* that you are observing. Draw what you see in the slide; do not simply try to reproduce the artist's rendering in figure 9.6b. Label the epidermis, gastrodermis, mesoglea, and gastrovascular cavity. Examine living *Hydra* by placing a specimen in a dish and observing the *Hydra* under a dissecting microscope. Allow the *Hydra* to recover from the shock of handling, then gently touch it with a needle. Observe the response. Transfer it to a depression slide, cover with a coverslip, and examine the tentacles closely. The cnidocytes can be seen as small "bumps." ▲

Other Hydrozoans

Another freshwater cnidarian belongs to the genus *Craspedacusta*. This hydrozoan has a small polyp stage that lives in temperate freshwater lakes during the spring and summer. In the fall, the medusa is produced. Zygotes overwinter and develop into a polyp the following spring.

▼ Examine any specimens available. ▲

Physalia is called the Portuguese man-of-war. Although not evident from looking at a specimen, *Physalia* represents a colony containing both polyps and medusae. Different types of polyps are specialized for feeding, protection, and asexual reproduction. Medusae are specialized for swimming, forming gas-filled floats, forming oil-filled floats, defense, or sexual reproduction. *Physalia* and closely related hydrozoans normally float or swim in deep-water areas of oceans.

▼ Examine a demonstration specimen. ▲

CLASS SCYPHOZOA

Characteristics: Polyp reduced or absent. Medusa prominent without velum; gametes produced gastrodermally; cnidocytes present in gastrodermis and epidermis; mesoglea with wandering mesenchyme cells of epidermal origin. Marine. Examples: *Aurelia, Rhizostoma.*

Aurelia

▼ The Scyphozoa are referred to as the "true jellyfish." Obtain a specimen of *Aurelia* and place it in a water-filled bowl. Spread it aboral-side up. Note the eight notches at the margin of the jellyfish. These are the location of **rhopalia,** sense organs for photoreception (ocellus) and equilibrium and balance (statocysts). Note also the **radial canals, ring canal,** and **tentacles.** Next, turn the specimen oral-side up (fig. 9.7). Find the mouth that is surrounded by four **oral lobes** (modifications of the manubrium). These are armed with nematocysts and are used in food capture. Sexes are separate, and **gonads** lie within modifications of the gastrovascular cavity called **gastric pouches.** Gonads are the horseshoe-shaped structures lining the margin of each gastric pouch. ▲

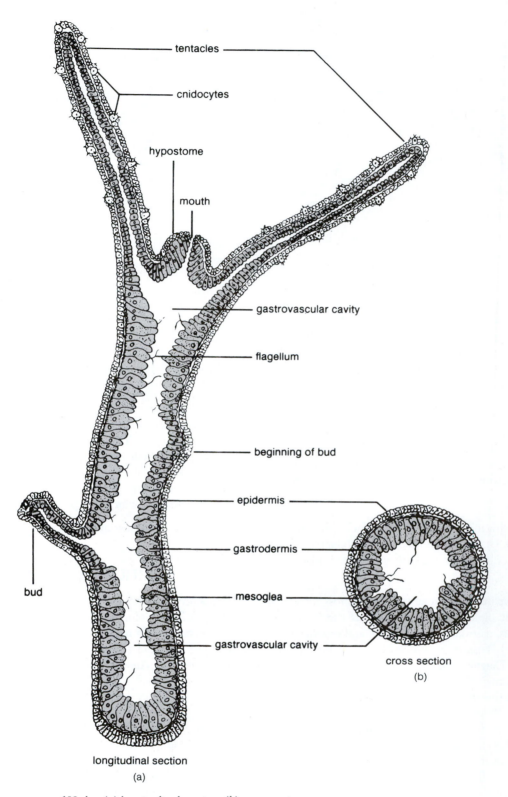

Figure 9.6 The structure of *Hydra*: (*a*) longitudinal section; (*b*) cross section.

The zygote develops into a planula that settles and forms a small polyp. The polyp is first called a **scyphistoma,** and then a **strobila.** The strobila produces miniature medusae, called **ephyrae,** by budding. The ephyrae are stacked one on top of the other on the strobila, the top one being closest to maturity. As ephyrae mature, they break loose from the strobila and take up a free-swimming life-style. They then grow to sexually mature adults.

▼ Examine any materials available that demonstrate the life cycle of *Aurelia.* ▲

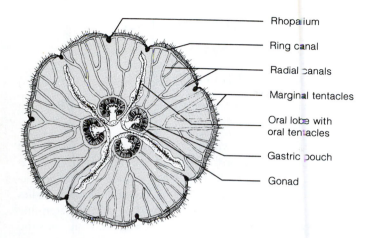

Figure 9.7 *Aurelia* medusa, oral view.

Labels: Rhopalium, Ring canal, Radial canals, Marginal tentacles, Oral lobe with oral tentacles, Gastric pouch, Gonad

CLASS ANTHOZOA

Characteristics: Polyp only, no medusa; solitary or colonial; gastrovascular cavity divided by mesenteries; cnidocytes present in the gastrodermis. Sea anemones and corals.

Metridium

▼ Examine the sea anemone *Metridium*. The body is cylindrical with the mouth and tentacles located on the **oral disk.** The **pedal disk** is the point of attachment to the substrate. Between the pedal and oral disks is the **column.** Inside the mouth note numerous ridges and one, or occasionally more, smooth grooves called **siphonoglyphs.** This ciliated groove directs water into the **pharynx.**

If your instructor directs you to do so, find the following internal structures by making a longitudinal incision through the body wall (fig. 9.8). ▲

pharynx	tubular passageway from the oral opening to the gastrovascular cavity
gastrovascular cavity	cavity below and around the gullet
mesenteries	extensions of the body wall into the gastrovascular cavity; increase surface area for absorption and secretion
gonads	on the margins of incomplete septa (those not connecting to the pharynx)
acontia	filaments bearing nematocysts at the aboral end of the gastrovascular cavity

The Anthozoa lack a medusa stage. Gonads in the dioecious polyp produce gametes that are released externally, where fertilization takes place. After settling to the substrate, the planula grows into the adult polyp. Reproduction also occurs asexually by **fragmentation** (regeneration of individuals from pieces of the polyp).

Other Anthozoa

▼ Examine demonstrations of other Anthozoa, particularly the stony and soft corals. Stony corals are responsible for the

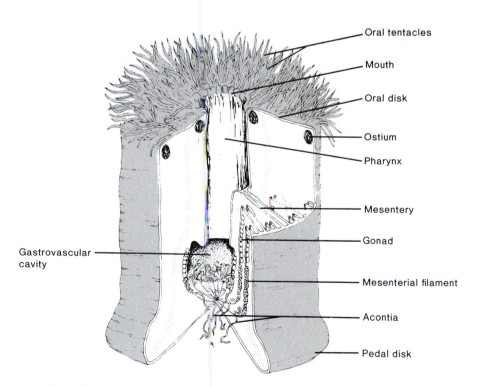

Labels: Oral tentacles, Mouth, Oral disk, Ostium, Pharynx, Mesentery, Gonad, Mesenterial filament, Acontia, Pedal disk, Gastrovascular cavity

Figure 9.8 The structure of *Metridium*.

formation of coral reefs. They secrete a calcium carbonate exoskeleton around their pedal disk and column. Note the small openings in a piece of hard coral where an anthozoan once lived. Soft corals are common in warm waters and are often very colorful. They secrete an internal skeleton of calcium carbonate or protein. They include sea fans, sea whips, red corals, and pipe organ corals. ▲

Classes Staurozoa and Cubozoa

Cubozoans are found in the warm tropical waters off the coast of Australia They have cuboidal medusae and tentacles that hang from the corners of the medusa. *Chironex fleckeri*, the sea wasp, has extremely potent nematocysts and has caused more human suffering and death than any other cnidarian.

Staurozoans are small marine cnidarians found in higher latitudes of the Atlantic Ocean and northwestern Pacific coast of North America. Others have been described from the Antarctic waters. They are in the form of a goblet and attach to substrate by the stem of the "goblet." A series of tentacles attach to the margin of the "goblet." We will not examine any specimens from these classes.

9.2 LEARNING OUTCOMES REVIEW—WORKSHEET 9.1

9.3 EVOLUTIONARY RELATIONSHIPS

LEARNING OUTCOME

1. Explain the evolutionary relationships among the Cnidarian classes.

▼ A cladogram showing a traditional interpretation of the evolutionary relationships among the cnidarian classes is shown in worksheet 9.1 (fig. 9.9). Examine the synapomorphies depicted on the cladogram. Based upon your studies in this week's laboratory, you should be able to determine the position of each of the classes studied. Complete the cladogram and answer the questions that follow it. ▲

9.3 LEARNING OUTCOMES REVIEW—WORKSHEET 9.1

WORKSHEET 9.1   Cnidaria

9.1   Evolutionary Perspective—Learning Outcomes Review

1. Members of the phyla Cnidaria and Ctenophora are
   a. asymmetrical.
   b. radially symmetrical.
   c. bilaterally symmetrical.
   d. triploblastic and radially symmetrical.

2. Which of the following statements accurately describes the relationships of the Cnidaria to other animal phyla?
   a. They are the stock from which all other animal phyla arose.
   b. They diverged from the ancestral animals very early in evolution; they share a common ancestry with the Ctenophora.
   c. They diverged from the ancestral animals very early in evolution; they do not share common ancestry with any other animal phylum.
   d. None of the above is correct.

3. Cnidarians possess
   a. cellular-level organization.
   b. diploblastic organization.
   c. triploblastic acoelomate organization.
   d. triploblastic coelomate organization.

9.2   Phylum Cnidaria—Learning Outcomes Review

4. Most cnidarians are
   a. monoecious.
   b. dioecious.

5. Alternation of generations is very clearly reflected in the life stages of
   a. *Hydra*, a hydrozoan.
   b. *Obelia*, a hydrozoan.
   c. *Metridium*, an anthozoan.
   d. *Aurelia*, a scyphozoan.
   e. Both b and d are correct.

6. Which of the following structures is a protective, nonliving chitinous sheath around the *Obelia* hydranth?
   a. coenosarcs
   b. perisarc
   c. gonozooid
   d. gastrozooid

7. You are examining a small cnidarian polyp. You observe a structure with a series of medusa-like buds stacked on top of one another. This polyp would be called a(n)
   a. scyphistoma.
   b. strobila.
   c. ephyra.
   d. gonozood.

8. One of the medusa-like buds released from the structure described in the previous question would be a(n)
   a. scyphistoma.
   b. strobila.
   c. ephyra.
   d. gonozood.

9. The following structures except one are found in many members of the class Anthozoa. Select the exception.
   a. pedal disk
   b. siphonoglyph
   c. pharynx
   d. gastric pouches

10. In exploring a tide pool, you touch a sea anemone that is standing erect and tall. After touching the anemone, it contracts its body, withdraws its tentacles, and becomes an amorphous mass of tissue. As you watch the anemone, you notice it slowly extending itself to its previous erect position. This extension is accomplished by taking water into its gastrovascular cavity using a ciliated groove called the
    a. radial canal.
    b. oral lobe.
    c. pedal disk.
    d. gastric pouch.
    e. siphonoglyph.

11. When you examined *Hydra*, you touched the animal and it contracted and then extended its body in a fashion similar to that described in question 10. Which of the following statements describe(s) what you saw?
    a. Flagellated cells in the gastrovascular cavity filled the cavity with water.
    b. The gastrovascular cavity acted as a hydrostatic skeleton.
    c. Muscle cells of the body wall caused the contraction but not the extension.
    d. All of the above are correct.

9.3 Evolutionary Relationships—Learning Outcomes Review

12. Most zoologists now believe that members of the class _____ are most closely related to the ancestral cnidarians.
    a. Staurozoa
    b. Scyphozoa
    c. Anthozoa
    d. Hydrozoa
    e. Cubozoa

9.1 Evolutionary Perspective—Analytical Thinking

1. Diagram a cross section of *Hydra*. Which adult tissue layer is derived from ectoderm? Which adult tissue layer is derived from endoderm? Compare the diploblastic, tissue-level organization observed in this exercise to the cellular-level organization of the Porifera. Are there functions carried out by diploblastic animals that are not possible for the Porifera that result from this diploblastic structure?

9.2 Phylum Cnidaria—Analytical Thinking

2. Describe the functions of the hydrostatic skeleton of cnidarians. What evidence of the hydrostatic skeleton did you observe for in this laboratory exercise?

9.3 Evolutionary Relationships—Analytical Thinking

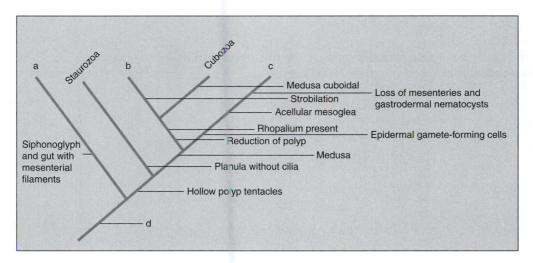

Figure 9.9 Evolutionary relationships among the Cnidaria. Fill in the class names.

Answer the following based on your laboratory observations and figure 9.9.

3. Fill in the following blanks to complete the cladogram.
   a. _____
   b. _____
   c. _____
   d. _____ (symmetry), _____(common larval stage), _____ (unique cell type)

4. What synapomorphies distinguish the Anthozoa from other cnidarians?

5. Staurozoa, Scyphozoa, Cubozoa, and Hydrozoa form a clade distinguished by the presence of _____.

6. What distinguishes the scyphozoan/cubozoan clade from the Hydrozoa?

7. Look at the characters described in questions 3–6. Which one of these characters were you unable to actually observe in the laboratory preparations?

8. Based on this cladogram, which of the following statements is probably TRUE?
   a. The ancestral cnidarians were probably polyp-like animals.
   b. The ancestral cnidarians were probably medusa-like animals.
   c. The ancestral cnidarians probably had both polyp and medusa stages.
   d. Explain the reasons for your choice.

Exercise 10

Platyhelminthes

Prelaboratory Quiz

Study this week's laboratory exercise and then complete the following quiz to assess your preparation for the laboratory.

1. Members of the phylum Platyhelminthes are characterized by all of the following EXCEPT
   a. organ-level organization.
   b. triploblastic body plan.
   c. coelomate body plan.
   d. bilateral symmetry.

2. In members of the phylum Platyhelminthes, muscles and excretory structures arise from the embryonic germ layer called the
   a. ectoderm.
   b. mesoderm.
   c. endoderm.

3. Animals that are bilaterally symmetrical usually have a head where nervous tissue and sensory structures are found. This development of a head in a bilaterally symmetrical animal is called
   a. organ organization.
   b. cephalization.
   c. diploblastic organization.
   d. centralization.

4. Members of the class Turbellaria include the
   a. free-living flatworms.
   b. monogenetic flukes.
   c. tapeworms.
   d. roundworms.

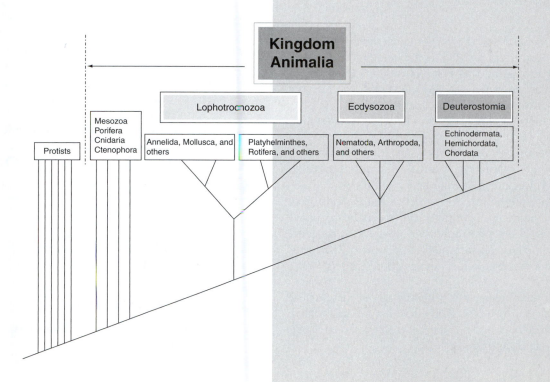

143

5. Members of the class Cestoidea include the
   a. free-living flatworms.
   b. monogenetic flukes.
   c. tapeworms.
   d. roundworms.

6. True/False   A host that harbors the immature stage of a parasite is called the definitive host.

7. True/False   Snails are the usual intermediate host of members of the class Trematoda.

8. True/False   Members of the class Cestoidea have a blind-ending digestive tract.

9. True/False   The definitive host of a tapeworm may be infected by eating flesh of an intermediate host containing a cysticercus larva.

10. True/False   Mature or gravid proglottids are found near the scolex of a tapeworm.

10.1 EVOLUTIONARY PERSPECTIVE

LEARNING OUTCOMES

1. Describe the relationship of members of the Platyhelminthes to other animal phyla.
2. Discuss the implications of bilateral symmetry as reflected in body structure and the organization of the nervous system.
3. Explain how the addition of a third germ layer (mesoderm) influences the structure and function of an animal.

The members of the phylum Platyhelminthes, the flatworms, exhibit bilateral symmetry and triploblastic organization. Three of the four classes within this phylum are parasitic, and the fourth class contains the free-living flatworms.

The evolutionary relationships between the Platyhelminthes and other phyla are controversial. As described below, members of this phylum are the first animals studied in this laboratory manual that are bilaterally symmetrical. Recent molecular and mophological analyses place the Platyhelminthes within a major division of bilaterally symmetrical animals called Lophotrochozoa (see p. 143). The lophotrochozoan phyla and the ecdysozoan phyla share protostomate development patterns. Consult your textbook for further discussion of these relationships.

Within the phylum, it is generally agreed that ancient free-living flatworms (class Turbellaria) were ancestral to modern turbellarians and members of the two classes of parasites (Trematoda and Cestoidea).

THE TRIPLOBLASTIC ACOELOMATE ANIMALS—THE ORGAN LEVEL OF ORGANIZATION

The Platyhelminthes include free-living flatworms, like the planarians, and the parasitic tapeworms and flukes. The term "flatworm" refers to the fact that the body is dorsoventrally flattened. Flatworms are the first organisms we are studying to have tissues organized into organs and the first to demonstrate bilateral symmetry. **Bilateral symmetry** means that one plane passing through the longitudinal axis of an organism divides it into right and left halves that are mirror images. It is characteristic of active, crawling, or swimming organisms and usually results in the formation of a distinct head **(cephalization)** where nervous tissue and sensory structures accumulate. This reflects the importance to the organism of monitoring the environment it is meeting—rather than that through which it has just passed—and results in the presence of definite anterior and posterior ends. The Platyhelminthes and all phyla above them on the evolutionary tree are bilaterally symmetrical or have evolved from bilaterally symmetrical ancestors.

Although there was some interdependence of tissues in the Cnidaria, interdependence is more highly developed in the Platyhelminthes. Thus, the flatworms display the **organ level of organization.** Three major sets of organs characterize the phylum. The excretory system consists of flame cells and their associated ducts. The nervous system consists of a pair of anterior ganglia, usually with two ventral nerve cords running the length of the organism. Nerve cords are interconnected by transverse nerves to form a ladderlike structure. The digestive tract is incomplete. A single opening serves for ingestion of food and elimination of wastes.

The Platyhelminthes are **triploblastic** and **acoelomate.** There are three embryonic germ layers: ectoderm, endoderm, and mesoderm. As with the Cnidaria, ectoderm gives rise to the outer epithelium, and the endoderm gives rise to the lining of the gut tract. The third germ layer, **mesoderm,** gives rise to the tissue between ectoderm and endoderm, including muscle, excretory structures, and undifferentiated cells referred to as parenchyma. The term "acoelomate" refers to the fact that there is no body cavity (fluid-filled space) between or within any of the embryonic germ layers.

10.1 LEARNING OUTCOMES REVIEW—WORKSHEET 10.1

10.2 PHYLUM PLATYHELMINTHES

LEARNING OUTCOMES

1. Describe the triploblastic acoelomate organization as reflected in the structure of *Dugesia*.

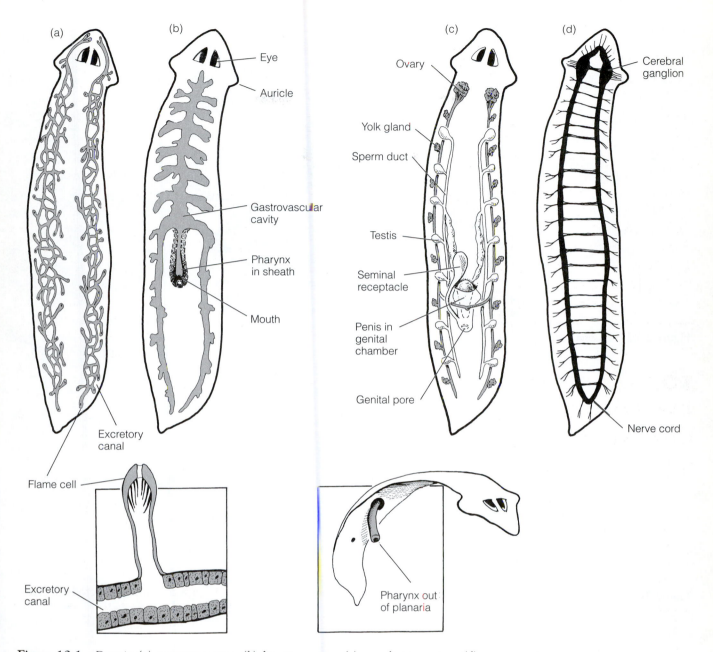

Figure 10.1 *Dugesia*: (a) excretory system; (b) digestive system; (c) reproductive system; (d) nervous system.

2. Discuss the classification of the Platyhelminthes and identify and describe class representatives.
3. Explain the life cycles of the trematode *Clonorchis sinensis* and the cestode *Taenia pisiformis*. Identify life-cycle representatives in slide preparations.

CLASSIFICATION

Class Turbellaria
Class Monogenea
Class Trematoda
Class Cestoidea

CLASS TURBELLARIA

Characteristics: Free-living, ciliated epidermis containing rodlike rhabdites, ventral mouth opening, monoecious, both sexual reproduction and asexual fission common. Example: *Dugesia* (planaria).

▼ Examine a prepared whole mount of *Dugesia* using a dissecting microscope (fig. 10.1). Note the presence of a definite anterior end. Two sets of specialized sensory receptors should be obvious. Photoreceptors can be seen dorsally. **Auricles** can be seen as lateral extensions of the head. Chemoreceptors are located on the auricles.

Most turbellarians are carnivores and feed on live invertebrates or scavenge on larger dead animals. If your

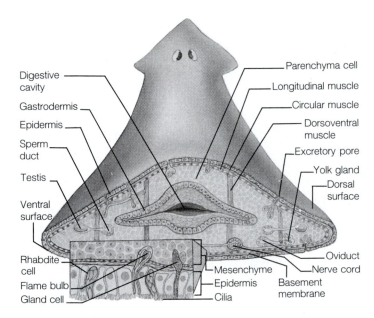

Figure 10.2 *Dugesia.* Cross section through anterior third.

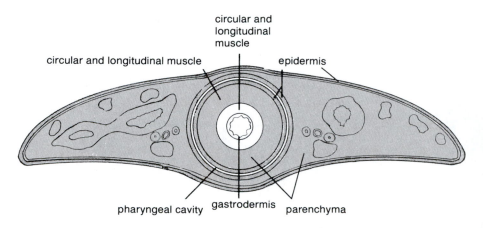

Figure 10.3 *Dugesia.* Cross section through the pharyngeal region.

specimen has the digestive cavity (**gastrovascular cavity**) stained, note that it is in the form of a single tube anteriorly and then divides posteriorly. The mouth opening is located midventrally. In feeding, a protrusible **pharynx** is extended through the mouth opening. Digestion is both extracellular and intracellular.

Examine a living specimen in a water-filled dish. Observe the gliding movement. Epidermal mucous glands lay down a sheet of mucus as cilia and body-wall musculature propel the animal. Examine a demonstration of *Dugesia* feeding on fresh beef liver.

Planarians have remarkable powers of regeneration. This is easily demonstrated. Place a planarian on a cube of ice. Allow it to become immobilized and then cut it in half (or some other section) with a razor blade. Rinse the pieces into a dish containing pond or spring water. Cover to prevent evaporation. Change the water in the dish every second day until regeneration is completed.

The excretory system of a freshwater planarian functions primarily in osmoregulation. Planarians live in a hypotonic environment and, therefore, are continually taking on water by osmosis. Protonephridia consist of networks of fine tubules that originate as tiny enlargements called **flame cells.** Flame cells remove this excess water from the tissues. Flickering cilia within the flame cells create currents that drive excess fluids through the tubule system to the outside.

Obtain a prepared slide showing a cross section of *Dugesia* (figs. 10.2 and 10.3). Examine it under low power of a compound microscope. A section through the middle of the planarian will show the pharynx; sections through the anterior and posterior thirds will not show the pharynx. In an anterior or posterior section, note the acoelomate organization and the derivatives of the three embryonic germ layers. The **epidermis** is ectodermal in origin, the undifferentiated **parenchyma (mesenchyme)** that fills the interior is mesodermal in origin, and **gastrodermis** that lines the gastrovascular cavity is endodermal in origin. In locating the latter, look for the space of the gastrovascular cavity and then look for the cells lining that cavity. The digestive tract is branched, so expect to see more than one section through the gastrovascular cavity. In the ventral epidermis, locate the small, rodlike **rhabdites.** Rhabdites are characteristic of the class and can be discharged to form a protective mucous coat around the planarian. Muscle cells can also be seen in this slide. Just inside the epidermis is a layer of **circular muscle,** and just inside that is a layer of longitudinal muscle. The **longitudinal muscle** will be seen as small dots because cells are cut in cross section. Strands of **dorsoventral muscle** run between the dorsal and ventral surfaces. Draw the cross section of *Dugesia* studied here in worksheet 10.1, analytical thinking question 1. Avoid the temptation to reproduce the drawing in figure 10.2. Draw what you see in the microscope. Answer the questions related to your illustration. Examine a section through the pharynx. Compare what you see to figure 10.3. ▲

Reproductive structures will not be studied because they are difficult to see in most preparations. The Turbellaria are, however, **monoecious.** Recall that monoecious species have both ovaries and testes in one individual. In the Turbellaria, as in most monoecious taxa, cross-fertilization is the rule. After copulation, fertilized eggs are deposited in the substrate of freshwater ponds and streams. Development may be direct (young resemble the adults), or in some marine turbellarians, larval tages may hatch from the eggs. As mentioned previously, the regenerative powers of turbellarians are well developed, and sexual reproduction is common.

▼ Exercise 1 of this laboratory manual had an activity investigating phototaxis of planarians. If you did not complete that exercise earlier, your instructor may want you to turn there at this time. In addition to investigating phototaxis, experiments can be designed to study the response of planaria to temperature and the direction of the pull of gravity. ▲

CLASS MONOGENEA

Characteristics: Monogenetic flukes; mostly ectoparasites on vertebrates (usually on fishes, occasionally on turtles, frogs, copepods, squids); bear opisthaptor; body wall with tegument. Examples: *Gyrodactylus* and *Polystoma.*

▼ Members of the Monogenea, Trematoda, and Cestoidea possess an outer epidermis modified as a tegument. It is nonciliated and syncytial (a continuous layer of fused cells). It is adaptive for parasites in these classes because it aids in the transport of nutrients, gases, and wastes across the body wall and protects against enzymes and the host's immune system. Monogenetic flukes have a single generation in their life cycle, that is, one adult develops from one egg. Embryonic development leads to a larval stage called an oncomiracidium. Most monogenetic flukes are external parasites of fishes, where they attach to gill filaments and feed

on epithelial cells, mucus, or blood. Observe any demonstration slides that are available. Note the large attachment structure called an **opisthaptor.** ▲

CLASS TREMATODA

Characteristics: Flukes; body wall with tegument; leaflike or cylindrical; usually with oral and ventral suckers; adults, parasites of vertebrates. Examples: *Fasciola, Clonorchis (Opisthorchis), Schistosoma.*

▼ Obtain a prepared slide of the human liver fluke *Clonorchis sinensis.* It illustrates trematode structure clearly (fig. 10.4). Using a dissecting microscope, note the tapered anterior and more rounded posterior ends. At the anterior tip is the **oral sucker.** It serves as a holdfast and surrounds the mouth opening. Posterior to the oral sucker is the **ventral sucker,** or **acetabulum.** It also serves as a holdfast. Extending posteriorly from the mouth is the blind-ending digestive tract. The muscular **pharynx** is just posterior to the mouth. The pharynx is followed by a short **esophagus.** The digestive tract divides to become Y-shaped. Each arm of the "Y" is called a **cecum.** The remaining internal structures are components of the reproductive complex. Most trematodes are monoecious, with cross-fertilization the rule. ▲

Trematodes typically have a two- or three-host life cycle. The host that harbors the adult stage of the parasite is the **definitive host.** The host that harbors the immature stage is the **intermediate host.** The life cycle of trematodes can be extremely complex. The life cycle of *Clonorchis sinensis* (fig. 10.5) represents one of the many variations found in the class.

Clonorchis sinensis adults live in the bile passageways of the human liver, where they feed on human cells and cell fragments. Mature eggs pass through the host via the bile duct into the intestine and out of the host with the feces. **Miracidia** hatch after eggs are eaten by the first intermediate host, the snail. Miracidia develop into **sporocysts.** Sporocysts develop rediae asexually, and **rediae,** in turn, develop **cercariae.** Hundreds of cercariae leave for each miracidium that infected the snail. Cercariae penetrate the second intermediate host, the fish, and develop into **metacercariae,** which encyst in muscle. Metacercariae have the adult body form but are sexually immature. When metacercariae are ingested by humans eating incompletely cooked fish, metacercariae develop into adults.

▼ Examine any slide of preserved or living materials that illustrate the life cycle of *Clonorchis sinensis.* Examine other trematodes available for study. ▲

CLASS CESTOIDEA

Characteristics: Tapeworms; body wall with tegument bearing microvilli; body consisting of scolex with hooks and/or suckers and a long tapelike strobila divided into

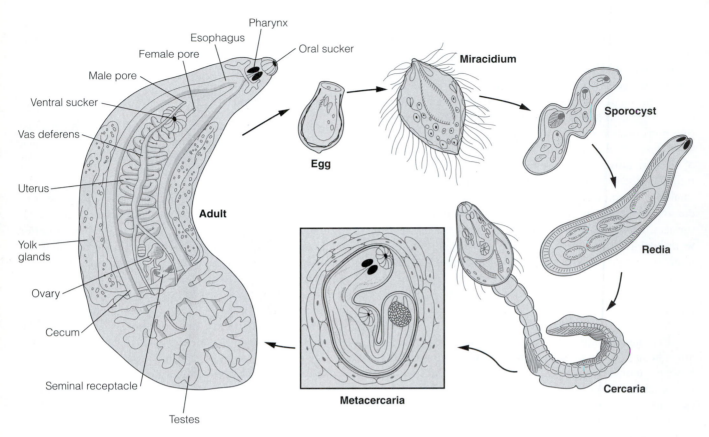

Figure 10.4 *Clonorchis sinensis*, adult and larval stages.

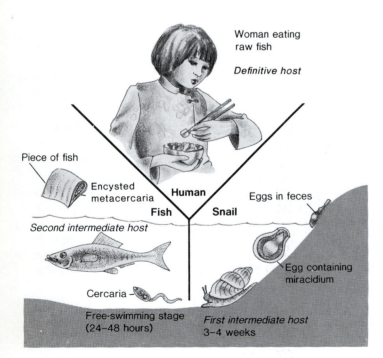

Figure 10.5 The life cycle of *Clonorchis sinensis*. Humans become infected by eating uncooked or poorly cooked fish containing metacercariae. Eggs passed with feces contain miracidia, which are eaten by snails. Miracidia develop into sporocysts within the snail. Sporocysts develop rediae, which develop cercariae. Hundreds of cercariae leave the snail and encyst in fish muscle as metacercariae.

proglottids; no digestive tract; monoecious; parasites of vertebrates. Examples: *Taenia, Hymenolepis*.

Cestodes, with some exceptions, have two hosts in their life cycles. Figure 10.6 shows stages in the life cycle of a typical cestode, the dog tapeworm *Taenia pisiformis*. Adults generally live in the digestive tract of their vertebrate host, where they absorb nutrients from the host's intestine. The **scolex** is a holdfast structure. Anterior proglottids contain **immature** reproductive structures, middle proglottids contain **mature** reproductive structures, and posterior proglottids contain eggs and are referred to as **gravid** proglottids. As proglottids are produced behind the scolex, they move posteriorly and begin accumulating fertilized eggs. Cestodes are monoecious. Cross-fertilization with another individual or proglottid is the rule. When a proglottid is gravid, reproductive organs degenerate, and the proglottid breaks off the end of the **strobila.** These "bags" of eggs then pass out of the host with the feces.

Fertilized eggs develop into an encapsulated, six-hooked larva called an **oncosphere** or hexacanth. The onchosphere is ingested by the intermediate host and uses its hooks to penetrate the wall of the digestive tract. In the intermediate host, the onchosphere encysts as a second larval stage. The second larval stage is either a **cysticercus** or **cysticercoid** larva. When the intermediate host is eaten, the second larval stage enters the definitive host.

▼ Obtain a slide of the dog tapeworm *Taenia pisiformis* (fig. 10.6). (These slides usually have representative regions

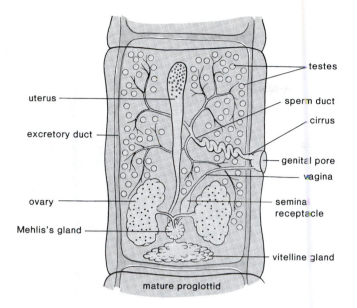

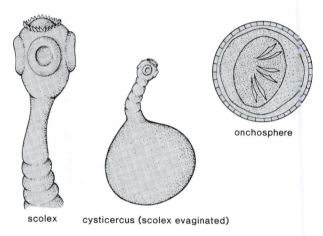

Figure 10.6 *Taenia pisiformis*, adult and larval stages.

Figure 10.7 *Taenia pisiformis*, scolex with hooks.

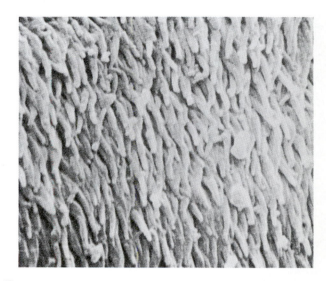

Figure 10.8 The tegument of *Hymenolepsis diminuta* showing microvilli.

of the tapeworm mounted under one coverslip.) Examine the scolex with its suckers (fig. 10.7). Just behind the scolex note a number of immature proglottids. Mature proglottids are indicated by the presence of reproductive structures and some fertilized eggs. Gravid proglottids are full of fertilized eggs. These segments are ready to break free and pass out of the host. Note the absence of a digestive tract. Nutrients are absorbed from the host's intestine across the tapeworm body wall. Note the microvilli that increase the surface area for absorption (fig. 10.8). Examine any materials available that illustrate aspects of the life cycle described here. In *T. pisiformis*, the intermediate host is the rabbit, and the larval stage in that host is a cysticercus.

Human tapeworms include the beef tapeworm *Taeniarhynchus saginata*, the pork tapeworm *Taenia solium*, and others. Examine any demonstration materials available. ▲

10.2 LEARNING OUTCOMES REVIEW—WORKSHEET 10.1

10.3 EVOLUTIONARY RELATIONSHIPS

LEARNING OUTCOMES

1. Explain the evolutionary relationships among the platyhelminth classes.

▼ A cladogram showing a traditional interpretation of the evolutionary relationships among the platyhelminth classes is shown in worksheet 10.1 (fig. 10.9). Examine the synapomorphies depicted on the cladogram. Based upon your studies in this week's laboratory, you should be able to determine the position of each of the classes studied. Complete the cladogram and answer the questions that follow it. ▼

10.3 LEARNING OUTCOMES REVIEW—WORKSHEET 10.1

WORKSHEET 10.1 Platyhelminthes

10.1 Evolutionary Perspective—Learning Outcomes Review

1. Members of the phylum Platyhelminthes are
   a. asymmetrical.
   b. radially symmetrical.
   c. bilaterally symmetrical.
   d. triploblastic and radially symmetrical.

2. Which of the following statements accurately describes the relationships of the Platyhelminthes to other animal phyla?
   a. They are the stock from which all other animal phyla arose.
   b. They are Lophotrochozoans and closely related to the Arthropoda.
   c. They are Lophotrochozoans and closely related to the Annelida.
   d. They are Ecdysozoans.

3. Members of the Platyhelminthes possess
   a. cellular-level organization.
   b. diploblastic organization.
   c. triploblastic acoelomate organization.
   d. triploblastic coelomate organization.

10.2 Phylum Platyhelminthes—Learning Outcomes Review

4. Most flatworms are
   a. monoecious.
   b. dioecious.

5. Which of the following structures observed in *Dugesia* is mesodermal in origin?
   a. epidermis
   b. gastrodermis
   c. parenchyma
   d. muscle
   e. c and d are both mesodermal in origin

6. Compare the cross sections you drew for *Hydra* (Exercise 9) and *Dugesia* (Exercise 10). Which of the following was present in *Dugesia* but not *Hydra*?
   a. epidermis (derived from ectoderm)
   b. gastrodermis (derived from endoderm)
   c. mesoglea
   d. parenchyma (derived from mesoderm)

7. The stage in the life cycle of *Clonorchis sinensis* infective to humans is
   a. a cercaria found in a snail.
   b. a metacercaria found in fish flesh.
   c. redia found in a snail.
   d. miracidium found in the water.

8. Which of the following structures of *Clonorchis sinensis* is "Y-shaped?"
   a. the digestive tract
   b. the female reproductive structures
   c. the male reproductive structures

9. The anterior proglottids of a tapeworm strobila are
   a. reproductively mature.
   b. reproductively immature.
   c. gravid (full of fertilized eggs).
   d. are the oldest proglottids.

10. Tapeworms have all of the following EXCEPT
    a. male reproductive organs.
    b. larval stages that infect an intermediate host.
    c. a digestive tract.
    d. female reproductive organs.

11. An immature stage of a tapeworm that is infective to the definitive host could be a
    a. sporocyst.
    b. cysticercus.
    c. oncosphere.
    d. miracidium.

12. When walking your dog, you observed small, white proglottids in some fecal material. Which of the following is TRUE about your dog?
    a. Your dog is harboring the adult stage of a tapeworm. It picked up the infection by eating a fish containing sporocysts.
    b. Your dog is harboring the cysticercus stage of a tapeworm.
    c. Your dog is harboring the adult stage of a tapeworm. It picked up the infection by eating rabbit containing metacercaria.
    d. Your dog is harboring the adult stage of a tapeworm. It picked up the infection by eating rabbit containing the cysticercus stage of the tapeworm.

13. Which of the following statements help explain a planarian's response to light?
    a. Bilateral symmetry is usually accompanied by cephalization.
    b. Organ-level organization is present.
    c. Cephalization results in the formation of an anterior end.
    d. Platyhelminths have a nervous system consisting of cerebral ganglia and longitudinal nerve cords.
    e. All of the above help explain a planarian's response to light.

10.3   Evolutionary Relationships—Learning Outcomes Review

14. Most zoologists now believe that members of the class _____ are most closely related to the ancestral platyhelminths.
    a. Turbellaria
    b. Trematoda
    c. Cestoidea
    d. Monogenea

10.1   Evolutionary Perspective—Analytical Thinking

1. Diagram a cross-section of *Dugesia*. Label the following: epidermis, gastrodermis, parenchyma, dorsoventral muscle, circular and longitudinal muscles, and rhabdites. Which adult tissue is derived from ectoderm? Which adult tissue is derived from endoderm? Which adult structures are derived from mesoderm? Compare the triploblastic organ-level organization observed in this exercise to the diploblastic tissue-level organization observed in exercise 9. Are there functions carried out by triploblastic animals that are not possible for diploblastic animals?

10.2   Phylum Platyhelminthes—Analytical Thinking

2. Imagine that you are a health official in an oriental country. You discover an infestation of *Clonorchis sinensis* in a small rural community. What recommendations would you make regarding the control of this parasite?

10.3   Evolutionary Relationships—Analytical Thinking

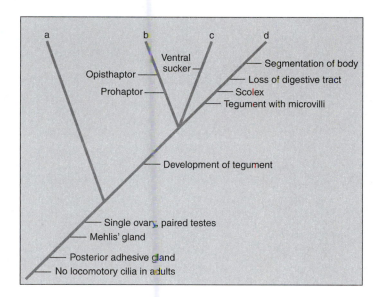

Figure 10.9   Evolutionary relationships among the Platyhelminthes.

Answer the following based on your laboratory observations and figure 10.9.

3. Fill in the blanks to complete the cladogram.
    a. _____
    b. _____
    c. _____
    d. _____

4. What single derived character distinguishes members of the classes Cestoidea, Monogenea, and Trematoda from all other members of the phylum Platyhelminthes? _____

5. Name at least one derived character for each of the following flatworm classes:
    a. Monogenea _____
    b. Trematoda _____
    c. Cestoidea _____

6. Name one class of the phylum Platyhelminthes for which there are no known synapomorphies.

    _____

7. According to figure 10.9, which of the four classes of Platyhelminthes is most closely related to ancestral members of this phylum? _____

8. Explain why an absence of synapomorphies in a lineage leads systematists to the conclusion that members of that lineage are closely related to the ancestral members of the taxon.

Exercise 11

Mollusca

Prelaboratory Quiz

Study this week's laboratory exercise and then complete the following quiz to assess your preparation for the laboratory.

1. Members of all of the following phyla are lophotrochozoans EXCEPT
   a. Annelida.
   b. Arthropoda.
   c. Platyhelminthes.
   d. Mollusca.

2. One function of the mantle of most molluscs is to
   a. filter feed.
   b. secrete the shell.
   c. house visceral organs.
   d. prevent desiccation.

3. A rasping, tonguelike structure found in many molluscs is called the
   a. radula.
   b. pen.
   c. beak.
   d. siphon.

4. Snails and slugs are members of the class
   a. Monoplacophora.
   b. Bivalvia.
   c. Gastropoda.
   d. Polyplacophora.
   e. Scaphopoda.

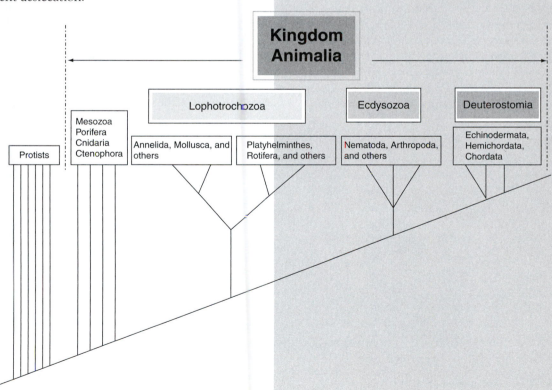

5. Chitons are members of the class
   a. Monoplacophora.
   b. Bivalvia.
   c. Gastropoda.
   d. Polyplacophora.
   e. Scaphopoda.

6. True/False   A glochidium is a larval stage of most marine bivalves.

7. True/False   In most members of the class Cephalopoda, the mantle is modified to promote jetlike propulsion.

8. True/False   In this week's laboratory, we will be carrying out a dissection of a freshwater mussel.

9. True/False   All molluscs possess an open circulatory system.

10. True/False   Blood, in molluscs, serves as a hydraulic skeleton in addition to carrying nutrients.

11.1 EVOLUTIONARY PERSPECTIVE

LEARNING OUTCOME

1. Describe the relationship of the Mollusca to other animal phyla.

Molluscs, along with Platyhelminthes, Annelida, and others are Lophotrochozoans (see p. 155). The molluscs are a very diverse group. The fossil record indicates that representatives of all modern classes were present in prehistoric seas 500 million years ago. Since then they have undergone adaptive radiation into freshwater and terrestrial habitats, including tropical rain forests and deserts. Members of the class Cephalopoda are often cited as the most highly evolved invertebrate group. Their complexity of structure and function rivals that found in many chordates.

11.2 PHYLUM MOLLUSCA

LEARNING OUTCOMES

1. Describe characteristics and give examples of members of the phylum Mollusca.
2. Characterize each of the classes studied, including habitat, feeding and reproductive habits, and distinctive features of body organization.
3. Recognize representatives of the molluscan classes.

The molluscs make up a very large group (second only to the arthropods in number of species) containing the snails, bivalves, chitons, squid, octopuses, and others. The molluscs are triploblastic and coelomate. The coelom, however, is secondarily reduced in size and importance in molluscs, forming only the pericardial cavity surrounding the heart and cavities of the gonads and nephridia.

The molluscan body consists of three regions: the **head-foot**, the **visceral mass**, and the **mantle** (fig. 11.1). The head-foot functions largely in locomotion and retracting the body into a shell. It is muscular and relatively fast in responding to stimuli. The visceral mass contains organ systems devoted to digestion, reproduction, excretion, and other visceral functions. In a generalized mollusc, the visceral mass is carried dorsal to the head-foot. The mantle is a fleshy tissue that is attached to the visceral mass and secretes a **shell.** The shell is made up of calcium carbonate and protein. With a few exceptions, the shell can be considered characteristic of the phylum. Between the mantle and the visceral mass is the **mantle cavity.** The mantle cavity contains gills and openings to excretory, digestive, and reproductive systems. The mantle and mantle cavity may be variously modified for sensory reception and locomotion.

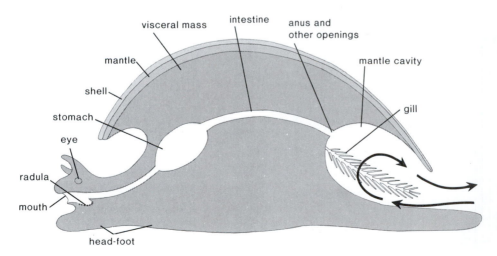

Figure 11.1   The generalized mollusc.

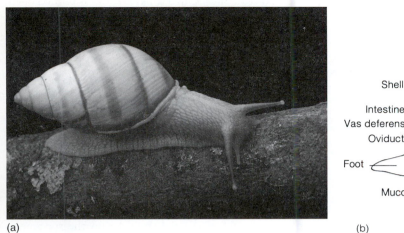

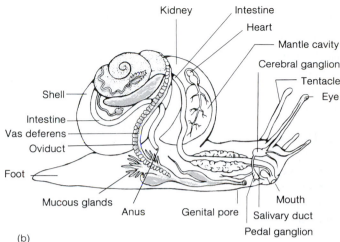

Figure 11.2 Class Gastropoda: (*a*) A land (pulmonate) gastropod. The banded tree snail (*Orthalicus*) is shown here. (*b*) Internal structure of a pulmonate gastropod.

A rasping tonguelike structure, the **radula,** is present in all mollusc classes except the bivalves. It consists of a chitinous belt of rasping teeth. This belt overlies a fleshy, tonguelike structure supported by a cartilaginous odontophore. As the radula is moved back and forth over the odontophore, food is rasped into small particles and passed back to the digestive tract.

Most molluscs have an **open circulatory system.** The heart pumps blood through vessels to large tissue spaces called blood sinuses, where the blood bathes the tissues. Blood is then returned to the gills (or lungs) and the heart. The cephalopods are the sole exception to this generalization. Their circulatory system is closed; blood is confined to vessels. Blood in molluscs performs two functions: carrying nutrients and oxygen and serving as a **hydraulic skeleton.** A hydraulic skeleton serves much the same purpose as the hydrostatic skeleton of cnidarians. In the hydraulic skeleton, fluids are moved through tissue spaces rather than a body cavity. The muscles associated with these movements are often some distance from the body part being moved. Oxygen is carried by the respiratory pigment, hemocyanin. Rather than being confined to blood cells, hemocyanin is carried in solution in the blood. Additionally, molluscs have more complex digestive, excretory, nervous, and sensory systems than previous phyla. Considerable specialization of systems for individual lifestyles is seen in all the above systems. Molluscs are usually dioecious, although some are monoecious. Many aquatic molluscs have larval stages that will be described later.

CLASSIFICATION*

Class Solenogastres
Class Caudofoveata
Class Polyplacophora

*This taxonomic listing reflects a phylogenetic arrangement. The observations that follow, however, begin with molluscs that are familiar to most students.

Class Monoplacophora
Class Scaphopoda
Class Bivalvia
Class Gastropoda
Class Cephalopoda

CLASS GASTROPODA—SNAILS, SLUGS, LIMPETS

Characteristics: Shell, when present, usually coiled; body symmetry distorted; some monoecious. Examples: *Physa, Helix, Busycon.*

▼ Obtain a living snail. If the specimen is aquatic, be sure to keep the snail submerged for most of the following set of observations. Allow your snail to recover from being handled, then compare it to figure 11.2. Identify the head-foot, the mantle, and the shell. The mantle can usually be observed around the lower edge of the shell. The visceral mass is covered by the shell. It will be observed later on a preserved snail that has had the shell removed.

Hold the shell apex up with the opening **(aperture)** toward you. Is the aperture on the right or left?_____ If on the right, the shell is **dextral;** if on the left, the shell is **sinistral.** This is an important taxonomic characteristic. Note the whorls (complete turns) of the shell and fine lines of growth. The apex is the oldest part of the shell.

If the specimen is aquatic, the following must be done with the snail submerged in a dish of water. In an undisturbed living snail, note the head with its two pairs of **tentacles. Eyes** may be located at the tip or base of the more dorsal tentacles. Touch one of the tentacles of your living snail with a needle. Notice the quick response that draws the tentacle back. This response is the result of fast-acting retractor muscles. Watch as the tentacle is reextended. It is a much slower process because it is accomplished by a hydraulic mechanism. Blood is shifted from other parts of the body by the contraction of distant muscles. Blood gradually fills the tissue spaces of the tentacle, thus extending the tentacle.

Allow your snail to attach itself firmly to the bottom of a glass dish. Hold the dish up and examine the foot through the bottom of the dish using a hand lens. Note the mouth opening and the method of locomotion. The gliding movement is accomplished by waves of muscular contraction passing over the foot. Observe a prepared slide showing a gastropod radula. Details of radular structure are taxonomically important.

Study a specimen of the large marine snail *Busycon*, that has been removed from its shell. Locate the head-foot, mantle, and visceral mass. Note the calcareous **operculum** on the dorsal aspect of the posterior half of the foot. This protective plate closes over the aperture of the shell when the snail withdraws into the shell. On the head find the tentacles and eyes. On the left side, the mantle is modified into a **siphon** that allows water to circulate over the fan-shaped gill. Most of the spire of the visceral mass is liver (bluish gray) and gonad (yellow, on the surface of the liver near the tip of the spire). Observe an empty *Busycon* shell and visualize how the structures fit into the shell. Look at demonstrations of the egg cases of *Busycon*. ▲

Other Gastropoda
▼ Examine any available demonstration materials that show subclass representatives. ▲

Pulmonates
    Garden slugs
    Other terrestrial snails
Prosobranchs
    Limpets
    Abalone
    Conchs
Opisthobranchs
    Sea slugs
    Sea hares

DISSECTION METHODS

In the exercise that follows, you will be performing the first of a number of dissections called for in this manual. It is therefore appropriate to give some instructions for dissection. Contrary to what might be one's first impression, dissection involves only limited cutting. The purpose of dissection is to expose body parts for viewing in as natural a state as possible. This usually involves the separation of body parts along natural parting lines. For this you need at least a scalpel, scissors, forceps, dissecting needle, and dissecting pins. The scalpel is used for entry into the body of an organism. The initial cuts are shallow so the thickness and consistency of the tissue can be determined. Once the body cavity is opened, scissors can be used to continue cutting. This is usually done on larger specimens where one blade of the scissors can easily fit into the body cavity. It is important to keep the lower blade of the scissors parallel to and next to the inner surface of the body wall. This avoids cutting into the internal structures. Once the body wall is pinned open, use forceps for cleaning fat or other adhering tissue from the surface of the structure. The dissecting needle will be one of your most useful instruments. It is used for gently lifting, moving, and separating tissues and organs. A heavy metal probe, or the handle of your scalpel, can be used for separating and lifting larger, heavier structures.

Before starting your dissection, you must read through the entire description, looking at any figures available. Instructions for dissection are always given based on the anatomical terminology noted on pages x–xi. Review that terminology before beginning. All terms are in reference to the animal, not the dissector. For example, if the dissection is carried out on the ventral aspect of the organism, a structure described as being on the right will be on the specimen's right, but your left. If you are dissecting a small specimen with delicate internal structures, it is advisable to carry out the dissection with the specimen completely submerged in water. Water makes delicate structures easier to move around because they have a tendency to float. Remember also that if you want to observe a dissected specimen under a dissecting microscope, the specimen must be pinned near the edge of the dissecting pan to fit into the field of view. Finally, figures and photographs are for purposes of orienting you to the location of specific structures. Visual aids cannot be used in guiding the dissection, and they are not a substitute for the dissection experience. Follow the text closely in dissecting. All instructions and precautions are given there.

CLASS BIVALVIA—CLAMS, OYSTERS, MUSSELS

Characteristics: Body enclosed in a shell consisting of two valves, hinged dorsally; no head or radula; wedge-shaped foot: Examples: *Anodonta*, *Mytilus*, and *Venus*.

External Structure
▼ Examine a preserved freshwater mussel. Note the two **valves** and their point of attachment on the dorsal surface. The two valves are held together by an elastic **hinge ligament.** A swollen area **(umbo)** is near the anterior end of the hinge and is the oldest part of the shell. Hold the bivalve with the hinge oriented up and the umbo oriented away from you. You are now looking at the bivalve as if you were looking at your dog or cat from a posterior view. The end away from you is anterior, the end toward you is posterior, the hinge is dorsal, and the lower margin of the bivalve is ventral. The left valve is on your left, and the right valve is on your right. Note the **lines of growth** that diverge from the umbo toward the edge of the shell.

Internal Structure
Bivalves usually come from suppliers with the valves "pegged" open. Locate the position of **anterior** and **posterior adductor muscles** that are holding the valves closed (fig. 11.3). Slip a scalpel between the mantle and the left valve. Cut each adductor muscle at its point of attachment to the shell.

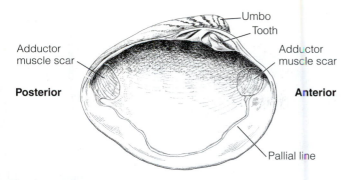

Figure 11.3 Internal structure of a bivalve shell.

When both muscles are cut, loosen the mantle over the entire area of the left valve and open the valves.

Examine the inner surface of the empty valve and the **mantle** covering the body of the bivalve (see fig. 11.3). Just ventral to the hinge, the mantle covers the **pericardial sac** and the **heart.** These structures are very delicate: be careful not to disturb them until directed to do so (fig. 11.4). Examine the adductor muscles that you cut on opening the bivalve and their points of attachment on the empty shell. The adductor muscles have a "catch" mechanism that allows them to remain contracted for long periods with little energy expenditure and without fatigue. When the adductor muscles relax, elasticity of the **hinge ligament** forces the valves open. Next to each adductor muscle is a smaller **retractor muscle** that brings the foot into the shell. Find the retractor muscles and their points of attachment on the shell. Below the anterior adductor muscle is the **anterior protractor muscle** that helps extend the foot.

A thickening at the margin of the mantle, the **pallial muscle,** attaches the mantle to the shell. The **pallial line** on the shell marks a line of attachment for this muscle. The pallial line arches between the points of attachment of the adductor muscles. The mantle of the left and right valves comes together posteriorly to form **incurrent** and **excurrent apertures** that allow water to enter and leave the mantle cavity (see fig. 11.4). Water enters through the ventral incurrent aperture and leaves through the dorsal excurrent aperture. In many bivalves these apertures are modified into long **siphons.** After a bivalve has burrowed into the substrate, the siphons can be extended to the water/substrate interface for gas exchange and filter feeding.

Return to the empty valve. Just below the hinge are tongue-and-groove modifications of the shell, called **teeth,** that prevent twisting. Detach the empty valve and break it with a hammer. Examine a cross section of the shell. There are three layers in all mollusc shells. Outside is a layer of polymerized protein called **periostracum.** The middle **prismatic layer** is the thickest and is made of calcium carbonate. The inner layer is also calcium carbonate and is called the **nacreous layer.**

The space between the mantle and the body is the mantle cavity. Lift and carefully remove the mantle to expose the visceral mass, foot, and associated structures (see fig. 11.4). The muscular, wedge-shaped **foot** is at the ventral aspect of the body. The **gills** cover much of the soft **visceral mass** that makes up the bulk of the body. At the anterior margin of the visceral mass, note the flaplike **labial palps.** Labial palps surround, and direct food toward, the mouth. Water coming in from the incurrent aperture reaches the ventral aspect of the gills and passes dorsally through the gills into a space above the gills formed by the attachment of the gills to the mantle on one side and the visceral mass on the other. Use your scalpel to make a shallow slice into this **suprabranchial chamber** (fig. 11.5). Once in the suprabranchial chamber water, is directed posteriorly and out of the mantle cavity through the excurrent aperture (see fig. 11.5 and fig 11.6). In the process, suspended food particles are filtered and gas exchange occurs. Food particles are transported by cilia to **food grooves** along

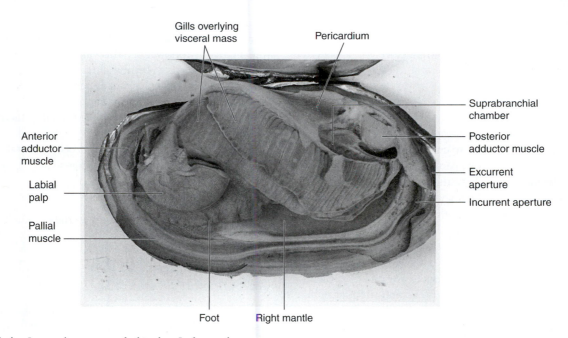

Figure 11.4 Internal structure of a bivalve. Left mantle cut away.

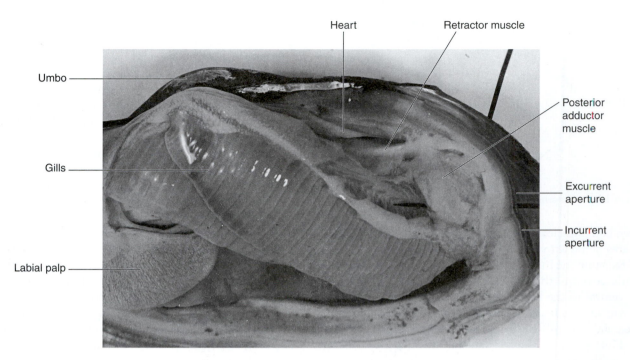

Umbo

Gills

Labial palp

Heart

Retractor muscle

Posterior adductor muscle

Excurrent aperture

Incurrent aperture

Figure 11.5 Internal structure of a bivalve. The mantle covering the posterior portion of the suprabranchial chamber has been removed. Note the pin inserted through the excurrent aperture with its point resting in the suprabranchial chamber. Water circulates from the incurrent aperture, through the gills, into the suprabranchial chamber, and exits the bivalve through the excurrent aperture.

the ventral margin of the gills. Cilia in the food grooves transport food to the labial palps.

Digestion begins as food is entangled in a mucoid string, which enters the stomach. A consolidated mucoid mass, the crystalline style, projects into the stomach from a diverticulum, called the style sac. Rotation of the crystalline style against a chitinized gastric shield dislodges enzymes from the style. Undigestible materials are sent on to the intestine, and partially digested food enters a digestive gland where intracellular digestion occurs. The intestine and digestive gland will be observed later.

The circulatory system of bivalves is an open system. A heart is enclosed by the **pericardium** (homologous to a reduced coelomic peritoneum) and located dorsal to the visceral mass. The pericardium is a thin membrane covering the top of the visceral mass (see fig. 11.4). Carefully remove the pericardium to see the heart. The **heart** wraps around the intestine where the intestine emerges from the visceral mass (see fig. 11.5). The **intestine** is running posteriorly to empty at the excurrent aperture. The heart consists of two parts, a thick-walled **ventricle** surrounding the intestine and two thin-walled **auricles** attached at either side of the ventricle. If you were careful in removing the pericardium, you should see both.

Blood leaves the ventricle through the anterior and posterior aortae and goes to tissue sinuses for exchange of nutrients, gases, and wastes. Blood in the posterior aorta goes to the mantle and then returns to the auricles. Blood in the anterior aorta goes to the viscera and the foot. This blood returns to the auricles via the nephridia and the gills.

The excretory system consists of a pair of dark **nephridia** lying below the pericardial sac. Nephridia remove metabolic

waste products from the blood and release the waste into the mantle cavity near the excurrent aperture.

Use your scalpel to make a sagittal section of the foot and visceral mass (fig. 11.7). Within the visceral mass, note cut sections of intestine. The yellowish tissue surrounding the intestine is gonad. Anteriorly you will have cut through a greenish digestive gland (see fig. 11.7 and fig. 11.8).

Sexes in bivalves are separate, but difficult to distinguish. Gonads are within the visceral mass, and gametes are released in the vicinity of the excurrent aperture. Fertilization is usually external, with swimming larval stages. In marine bivalves, the first larval stage, the **trochophore** (fig. 11.9a), is followed by the **veliger** (fig. 11.9b) and the adult. In freshwater bivalves, fertilization occurs within the mantle cavity of the female, and zygotes develop into **glochidia** (fig. 11.9c) within the gill chambers. Glochidia have tiny valves with teeth that are used in attaching to the gill filaments, fins, or skin of a fish. Each glochidium lives as a parasite for several weeks, then drops off to take up adult life. Examine slides of any larval stage available. ▲

Other Bivalves
▼ Examine any demonstration materials available. ▲

CLASS CEPHALOPODA—SQUID, OCTOPUSES, NAUTILI

Characteristics: Foot modified into a circle of tentacles and a siphon; shell reduced or absent; head in line with the elongate visceral mass.

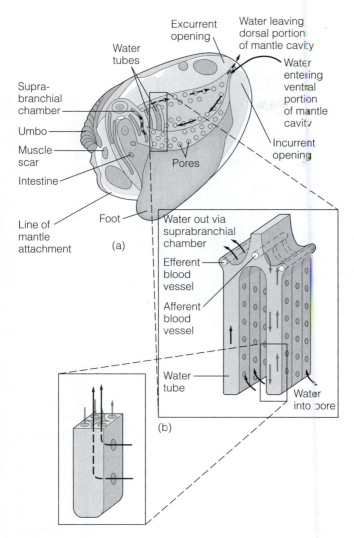

Figure 11.6 Circulation of water and blood through the gills of a bivalve. Arrows show path of water. (*a*) Incurrent (solid arrows) and excurrent (dashed arrows) water currents circulate through the bivalve gills. Food is filtered as water enters water tubes through pores in the gills. (*b*) Cross section through a portion of a gill. Water (light arrows) passing through a water tube passes in close proximity to blood (dark arrows). Gas exchange occurs between water and blood in the water tubes.

The cephalopods are the most highly evolved molluscs and in many ways the most advanced of all nonchordates. They are all marine, dioecious, and active predators. Experiments have shown them to be capable of degrees of learning unparalleled by other nonvertebrates. Cephalopods use jet propulsion in locomotion and are, therefore, fast swimmers. Ecologically, cephalopods are less important. They make up less than 0.5% of all molluscan genera. The reasons for this are not completely known.

Loligo—The Squid, External Structure

▼ Examine the external structure of *Loligo* (fig. 11.10). Note the head, which bears eight **arms** and two **tentacles.** Eyes on the head are structurally much like the eyes of vertebrates. They can apparently form clear images and discriminate some colors. Note that the head is lined up with the elongate **mantle.** The mantle bears **lateral fins** and encloses the **visceral mass.** Near the region where the mantle and the head meet, note the conical **siphon** protruding from below the mantle. Water drawn into the mantle cavity can be forcefully expelled through the siphon when muscles of the mantle contract. Jetlike propulsion results. Muscles attached to the siphon can direct the jet of water in different directions.

Look closely at the eight arms and two longer tentacles that encircle the mouth (fig. 11.11). In males, one arm (the **hectocotylus**) is modified for sperm transfer. The suckers on it are smaller and attached to the arm by a longer stalk. Using the hectocotylus, the male reaches into his mantle cavity, removes a spermatophore (package of sperm), and transfers it to the mantle cavity of the female. At the base of the arms and tentacles is the mouth opening. Spread the tentacles and note the chitinous **beak** used in tearing prey. Below the beak is the radula (not seen without dissection). The shell of *Loligo* is embedded in the mantle and is referred to as the **pen.** The tip of the chitinous pen can be seen as a point in the mantle opposite the siphon. The shell of cephalopods varies from being well developed in the nautili to being completely gone in the octopuses. ▲

Other Cephalopods

▼ Examine any demonstrations available, including

octopuses
cuttlefish
Nautilus shells
cuttlebone of *Sepia*
pen of *Loligo* ▲

CLASS POLYPLACOPHORA—CHITONS

Characteristics: Elongate, dorsoventrally flattened; head reduced in size; shell consisting of eight dorsal plates. Example: *Chiton.*

▼ Examine demonstrations of chitons. Note the dorsal plates and the broad, flat, muscular foot. This foot allows chitons to attach to the rocky substrate common to intertidal regions. The mantle cavity is reduced to a groove running on either side of the body between the foot and the margin of the animal. ▲

CLASS SCAPHOPODA—TOOTH SHELLS OR ELEPHANT'S-TUSK SHELLS

Characteristics: Body enclosed in a tubular shell that is open at both ends; tentacles; no head. Example: *Dentalium.*

▼ Examine demonstration specimens. Note especially the structure of the shell. The foot protrudes from the larger end of the shell. The small posterior end is open to facilitate the circulation of water. As the scaphopod burrows through sand and mud, the posterior end protrudes from the substrate to allow water to circulate through the mantle cavity. ▲

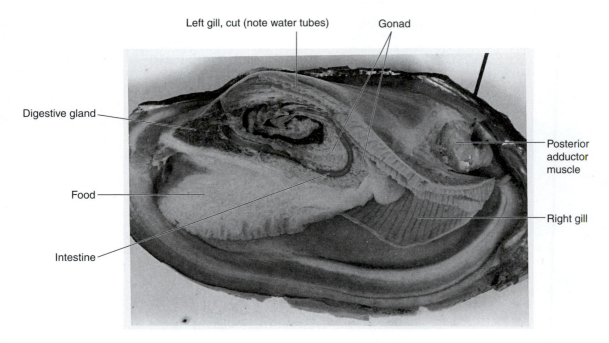

Figure 11.7   Internal structure of a bivalve. Left gill cut away. Visceral mass cut midsagitally to expose internal structures.

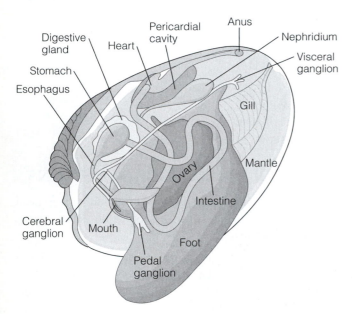

Figure 11.8   Internal structure of a bivalve. Sagittal section of the head-foot and visceral mass.

CLASS MONOPLACOPHORA

Characteristics: Single-arched shell; foot broad and flat; certain structures serially repeated. Example: *Neopilina*.

This is a group of molluscs that, prior to 1952, was known only from fossils. In 1952, this limpetlike animal was dredged up from a depth of 3,520 meters off the Pacific coast of Costa Rica. For many years *Neopilina* was thought to represent the ancestral form of the molluscs and, because of a suggestion of metamerism (serial repetition of body parts), might have linked the molluscs closely with the arthropods

and annelids. Most biologists have now discarded this idea; they view the Monoplacophora as an interesting group, but not representative of ancestral molluscs. The suggested metamerism is secondarily derived.

CLASS SOLENOGASTRES

Characteristics: Shell, mantle, and foot lacking; worm-like; head poorly developed; grooved ventral surface. Example: *Neomenia*.

This is a small group of moderately deep-water marine molluscs.

CLASS CAUDOFOVEATA

Characteristics: Worm-like with scalelike spicules on the body wall; lack shell, foot, and nephridia.

Members of the class Caudofoveata range in size from 2 mm to 14 cm and live in vertical burrows on the deep-sea floor. Approximately 70 species have been described.

11.2 LEARNING OUTCOMES REVIEW—WORKSHEET 11.1

11.3 EVOLUTIONARY RELATIONSHIPS

LEARNING OUTCOMES

1. Describe the relationships among members of the molluscan classes.

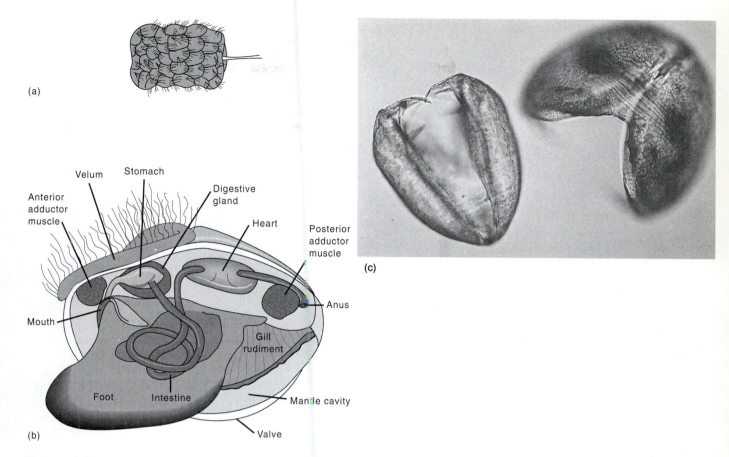

(a)

Anterior
adductor
muscle

Velum   Stomach

Digestive
gland

Heart

Posterior
adductor
muscle

Mouth

Anus

Gill
rudiment

(b)

Foot   Intestine

Mantle cavity

Valve

(c)

Figure 11.9 Larval stages of molluscs: (*a*) trochophore; (*b*) veliger; (*c*) photomicrograph of glochidia (×50).

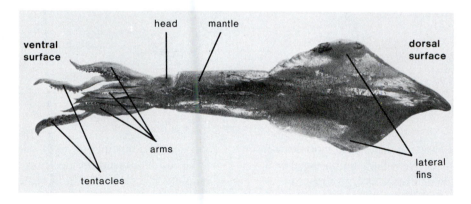

**ventral
surface**

head   mantle

**dorsal
surface**

arms

lateral
fins

tentacles

Figure 11.10 *Loligo*, anterior view.

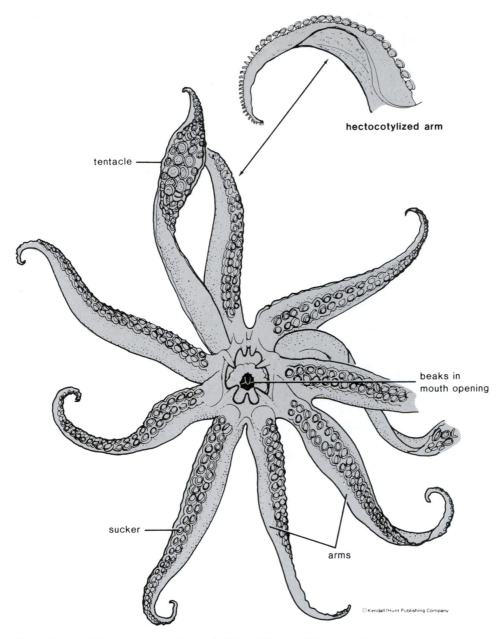

hectocotylized arm

tentacle —

beaks in
mouth opening

sucker —

arms

© Kendall/Hunt Publishing Company

Figure 11.11 Ventral view of the mouth area of a squid. © Kendall/Hunt Publishing Company.

2. List three characters used to distinguish molluscs from other phyla and one derived character for members of each of the following classes: Polyplacophora, Gastropoda, Cephalopoda, Bivalvia, and Scaphopoda.

▼ A cladogram showing a traditional interpretation of the evolutionary relationships among the molluscan classes is shown in worksheet 11.1 (fig. 11.12). Examine the synapomorphies depicted on the cladogram. Based

upon your studies in this week's laboratory, you should be able to determine the position of the classes studied. Complete the cladogram and answer the questions that follow it. ▲

11.3 LEARNING OUTCOMES REVIEW—WORKSHEET 11.1

WORKSHEET 11.1   Mollusca

11.1   Evolutionary Perspective—Learning Outcomes Review

1. Molluscs are members of the subkingdom
   a. Lophotrochozoa.
   b. Deuterostomia.
   c. Ecdysozoa.

2. Members of which of the following phyla are most closely related to the molluscs?
   a. Cnidaria
   b. Annelida
   c. Arthropoda
   d. Chordata

11.2   Phylum Mollusca—Learning Outcomes Review

3. All of the following are unique molluscan features except one. Select the exception.
   a. mantle
   b. mantle cavity
   c. radula
   d. shell
   e. visceral mass

4. Most molluscs have a(n) _____ circulatory system. The _____ are an exception.
   a. open; gastropods
   b. closed; bivalves
   c. open; polyplacophora
   d. open; cephalopods
   e. closed; bivalves

5. Imagine you are holding a snail shell with its apex up and the aperture facing you. You observe that the aperture is on the left side of the shell. This shell has the _____ morphology.
   a. sinistral
   b. dextral

6. A calcareous plate that closes the aperture of a snail shell when the snail is withdrawn is the
   a. mantle.
   b. foot.
   c. operculum.
   d. siphon.

7. The umbo of a bivalve valve is
   a. dorsal and posterior.
   b. dorsal and anterior.
   c. ventral and posterior.
   d. ventral and anterior.

8. After passing through the incurrent aperture of a bivalve mantle, water is
   a. in the mantle cavity.
   b. about to pass through the water tubes of the gills.
   c. in the suprabranchial chamber.
   d. immediately diverted to the excurrent aperture.
   e. Both a and b are correct.

9. Food grooves of bivalves are found
   a. along the ventral margin of the gills.
   b. on the labial palps.
   c. on the ventral margin of the foot.
   d. on the pallial muscle.

10. Which of the following is true of the bivalve heart?
    a. It consists of two ventricles and one auricle.
    b. It is located within the visceral mass.
    c. It wraps a portion of the intestine.
    d. It is located ventral to the gills.
    e. Both a and d are correct.

11. The siphon of a squid aids in
    a. circulating blood to the gills.
    b. tearing prey for ingestion.
    c. jet propulsion.
    d. structural support; it replaces the shell.

12. In trying to distinguish between male and female squid, one should examine the structure of the
    a. arms.
    b. sipon.
    c. fins.
    d. tentacles.

13. While walking along a rocky shoreline you observed a mollusc attached firmly to a submerged rock. The mollusc had a broad foot and eight dorsal plates. This mollusc is a member of the class
    a. Solenogastres.
    b. Gastropoda.
    c. Caudofoveata.
    d. Polyplacophora.

11.3   Evolutionary Relationships—Learning Outcomes Review

The following are important derived characters that you observed or read about in this laboratory exercise. Match the class name that is associated with this derived character. There may be more than one character for some classes. Some classes may not be used.

14. _____ torsion

15. _____ loss of radula

16. _____ ink sac

17. _____ shell with eight plates

18. _____ closed circulatory system

19. _____ bivalve shell

20. _____ tusk-shaped shell
    a. Bivalvia
    b. Cephalopoda
    c. Gastropoda
    d. Polyplacophora
    e. Monoplacophora
    f. Scaphopoda

11.1 Evolutionary Perspective—Analytical Thinking

1. The cephalopods were described in this section as the most highly evolved invertebrate group. Do you agree with this assessment? Consider your reading, observations in this lab exercise, and your study of other phyla up to this point in your general zoology class. Explain your answer.

11.2 Phylum Mollusca—Analytical Thinking

2. Compare and contrast the structure and function of the hydraulic skeleton found in most molluscs to the hydrostatic skeleton found in cnidarians. Members of one molluscan class lack a hydraulic skeleton. Name the class and explain the reason for the absence of the hydraulic skeleton in these animals.

11.3 Evolutionary Relationships—Analytical Thinking

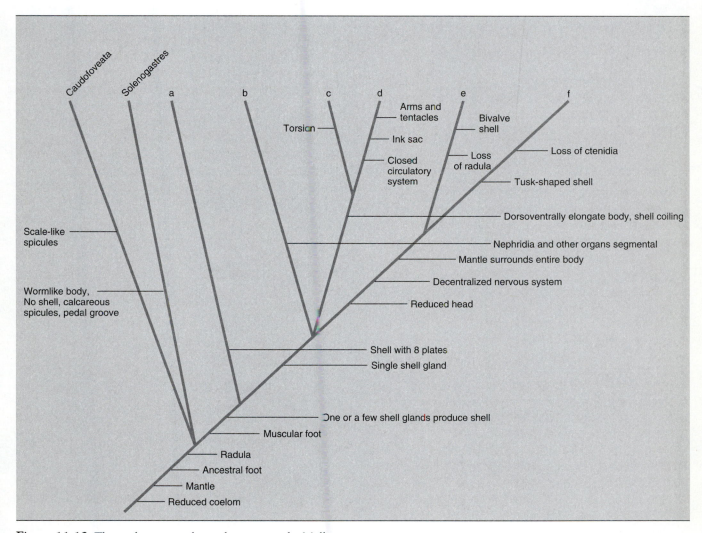

Figure 11.12 The evolutionary relationships among the Mollusca.

Answer the following based on your laboratory observations and figure 11.12.

3. Fill in the blanks to complete the cladogram.
   a. _____
   b. _____
   c. _____
   d. _____
   e. _____
   f. _____

4. Members of what class of molluscs were once thought to display segmentation similar to that seen in annelids?

5. Is the segmentation seen in the above group of molluscs a characteristic that was present in molluscan ancestors of both of these groups, or is it more recently derived?

6. What is the function of the coelom of molluscs?

7. Name three symplesiomorphic characters of the phylum Mollusca. _____ , _____ , and _____

8. Is the presence of a shell an ancestral character for all molluscs? Explain your answer.

9. What derived characters distinguish the classes Gastropoda and Cephalopoda from other molluscs? _____ _____

   What group of cephalopods retains both characters? _____

10. The presence of a bivalve shell and the absence of a radula are synapomorphies for members of the class _____. These characters reflect important aspects of the life-style of these animals. Explain why this statement is true.

11. Is the protective structure of a bivalve a single shell of two pieces or two shells that work together? Explain your answer based on your examination of figure 11.12.

Exercise 12

Annelida

Prelaboratory Quiz

Study this week's laboratory exercise and then complete the following quiz to assess your preparation for the laboratory.

1. The annelids are most closely related to members of which of the following phyla?
   a. Arthropoda
   b. Mollusca
   c. Nematoda
   d. Echinodermata

2. The serial repetition of body parts is called
   a. tagmatization.
   b. metamorphosis.
   c. metamerism.
   d. paedomorphosis.

3. All of the following are consequences of metamerism EXCEPT
   a. lessened impact of bodily injury.
   b. the ability to specialize certain regions of the body to specific functions.
   c. efficient locomotion using localized changes in body shape and hydrostatic pressure.
   d. All of the above are consequences of metamerism.

4. The largest of the two annelid classes is
   a. Polychaeta.
   b. Clitellata.

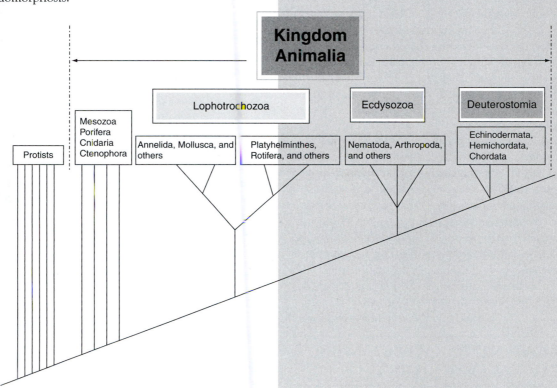

5. All of the following are annelid features EXCEPT
   a. an open circulatory system.
   b. nephridia used in excretion.
   c. a complete digestive tract.
   d. a ventral nervous system.
   e. a body wall with circular and longitudinal muscle layers.

6. True/False   Polychaetes have paired, lateral extensions of the body wall in each segment called parapodia.

7. True/False   A clitellum is a swollen area in the anterior portion of oligochaetes and polychaetes that secretes mucus during sperm exchange.

8. True/False   Members of the class Clitellata are monoecious. Most members of the class Polychaeta are dioecious.

9. True/False   Members of the subclass Oligochaeta have free-swimming trochophore larval stages.

10. True/False   Seminal receptacles of the earthworm store sperm prior to sperm exchange with another worm.

12.1 EVOLUTIONARY PERSPECTIVE

LEARNING OUTCOME

1. Describe the relationships of the annelids to other animal phyla.

Members of the phylum Annelida are lophotrochozoans and have a complete digestive tract, a solid ventral nerve cord, and segmentally arranged body parts. The segmentation seen in the Annelida has been interpreted as reflecting shared common ancestry with the Arthropoda (insects, crustaceans, and their relatives). Molecular evidence that places annelids within the Lophotrochozoa and arthropods within the Ecdysozoa is compelling. The segmental arrangement of body parts in these two groups must have arisen independently.

The Annelida is traditionally divided into two classes, Polychaeta and Clitellata. The latter is divided into two subclasses, Oligochaeta (earthworms and their relatives) and Hirudinea (leeches). A growing body of morphological and molecular evidence suggests that the Polychaeta is a polyphyletic grouping and that other groups of worms previously considered separate phyla (Sipuncula, Echiura, and Siboglinidae) are actually annelids. This recent research is changing our view of evolutionary relationships within the Annelida and between Annelida and other lophotrochozoans. Until the taxonomic relationships are settled, the traditional classification is retained in this laboratory manual.

12.1 LEARNING OUTCOMES REVIEW—WORKSHEET 12.1

12.2 PHYLUM ANNELIDA

LEARNING OUTCOMES

1. Describe characteristics and give examples of members of the phylum Annelida.
2. Characterize each of the classes studied, including habitat, feeding and reproductive habits, and distinctive features of body organization.
3. Recognize representatives of the annelid classes.

The Annelida, or segmented worms, is a relatively large phylum containing the earthworms, leeches, and marine worms. The most striking and significant evolutionary feature of the annelids is metamerism. **Metamerism** refers to the serial repetition of body parts. In a generalized form, metameric animals consist of identical repeating segments, each capable of operating relatively independently (i.e., they each contain their own excretory, nervous, reproductive, and circulatory structures). No animal shows this degree of metameric development, but the annelids come the closest. Advantages that result from metamerism include the following:

1. more efficient locomotion, utilizing localized changes in body shape and hydrostatic pressure (used in efficient crawling, burrowing, and swimming);
2. lessened impact of bodily injury; and
3. **tagmatization,** the ability to devote certain segments of the body to specialized functions (e.g., specialization for locomotion, sensory function, and feeding).

Annelids are triploblastic and bilaterally symmetrical. They are also coelomate. Recall that a coelom is a body cavity formed within mesoderm. The inner body wall of an annelid is lined by a thin mesodermal sheet, the peritoneum. Visceral organs are lined on the outside by a mesodermal sheet, the serosa. The coelom is especially important in the Annelida in its role as a hydrostatic skeleton. The metameric organization of the body allows hydrostatic compartments to be used independently in locomotion. For example, in anterior segments, circular muscles might be contracting, causing elongation and forward extension during crawling. At the same time, posterior segments can have longitudinal muscles contracted to increase the cross-sectional area of the worm and thus maintain contact with the substrate.

In addition, annelids have

1. a body wall with circular and longitudinal muscle layers,
2. a complete digestive tract,
3. a ventral nervous system showing some cephalization,
4. a closed circulatory system,
5. nephridia used in excretion, and
6. both monoecious and dioecious taxa.

CLASSIFICATION

Class Polychaeta
Class Clitellata
    Subclass Oligochaeta
    Subclass Hirudinea

CLASS POLYCHAETA

Characteristics: Mostly marine; head with eyes and tentacles; parapodia bearing numerous setae; trochophore larvae. Examples: *Nereis, Arenicola, Sabella.*

The class Polychaeta is the largest of the three annelid classes. Polychaetes are mostly marine, living in muddy and sandy offshore areas. The prototype might be visualized as having many undifferentiated segments, each bearing lateral extensions (parapodia) containing many setae (hence the name). Evolution has resulted in variations on this theme. Many modern polychaetes show various degrees of tagmatization and cephalization. They may be free-swimming, burrowing, or tube-dwelling polychaetes. They may feed as predators or on detritus or suspended organic particles. In the latter instances, they may develop intricate filtering devices. Some polychaetes obtain a substantial portion of their nutrients from organic matter dissolved in seawater. While the uptake of dissolved organic matter has been suggested for many aquatic organisms, it has been demonstrated to be significant in relatively few.

Nereis—The Clamworm

▼ Examine a preserved specimen with the aid of a hand lens or dissecting microscope. Examine the head (fig. 12.1). Note the **prostomium,** which bears four dark eyes, a pair of small

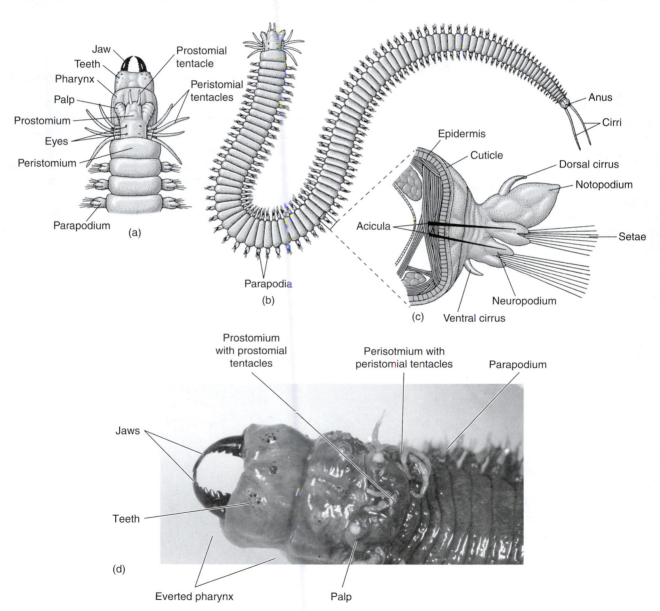

Figure 12.1 *Nereis* structure. (*a*) Dorsal view of the head with the pharynx extended. (*b*) Dorsal view of the entire worm. (*c*) Structure of a parapodium. (*d*) Photograph of dorsal view of head with pharynx extended.

tentacles, and fleshy **palps.** In back of the prostomium is the **peristomium** bearing four pairs of **peristomial tentacles.** The pharynx, bearing pincerlike **jaws,** is probably inverted into the head. In feeding, the pharynx is everted, and the jaws are used in capturing prey. Insert fine-tipped forceps or a teasing needle into the pharyngeal opening and evert the pharynx to see the jaws. Posteriorly, each segment bears a pair of lateral **parapodia.** Each parapodium is divided into a dorsal lobe (notopodium) and a ventral lobe (neuropodium). The internal structure of *Nereis* will not be studied, since it is similar to the earthworm to be studied next. *Nereis,* like most polychaetes, is dioecious. Fertilization and development are external. A ciliated larval stage called a **trochophore** develops, which will eventually metamorphose into the adult. Examine a slide showing a trochophore larva (fig. 12.2). Examine any available demonstrations of other polychaetes. ▲

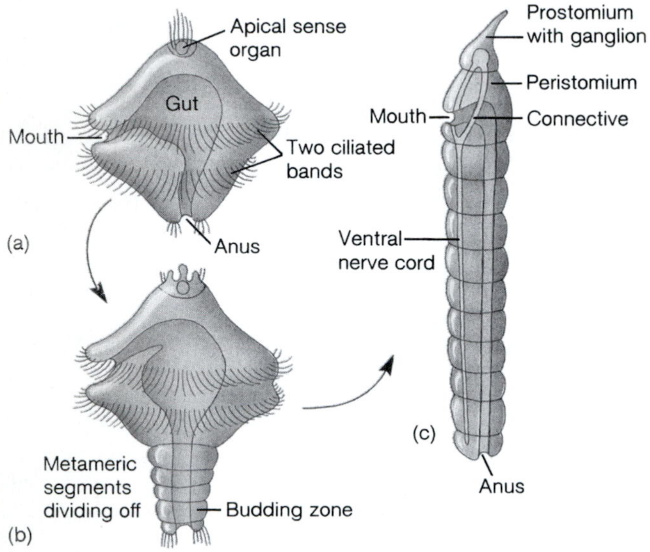

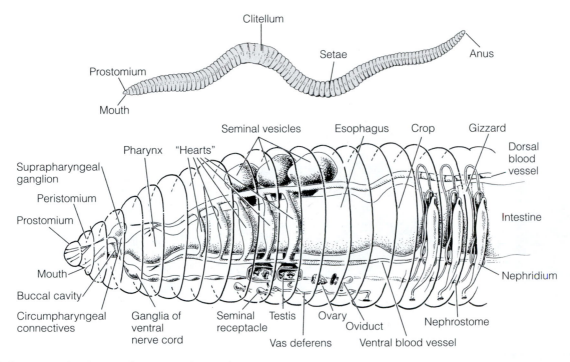

Figure 12.2   The development of a polychaete trochophore larva. *(a)* A trochopore buds segments *(b)* near the anus and *(c)* settles to the substrate as an immature worm. From *A Life of Invertebrates* by W. D. Russell-Hunter. Copyright © 1979 W. D. Russell-Hunter. Reprinted by permission.

CLASS CLITELLATA

Characteristics: Clitellum used in cocoon formation; monoecious direct development; few or no setae. Examples: *Lumbricus* and *Hirudo.*

The class Clitellata includes the earthworms and their relatives and the leeches. The two subclasses that are studied in the following sections formerly had class-level status, but current research provides very strong support for monophyly of the class.

Subclass Oligochaeta—*Lumbricus terrestris*

▼ Examine a preserved earthworm using a hand lens or dissecting microscope (fig. 12.3). Note the obvious metamerism. Distinguish anterior and posterior and dorsal and ventral on your worm as you carry out the following observations. The slightly swollen area in the anterior third of the worm is the **clitellum.** This region secretes mucus that holds two worms together for sperm exchange. The clitellum also forms a cocoon around developing embryos. Along the lateral and ventrolateral margins of the worm are small **setae** (four pair per metamere). Run your fingers along the

Figure 12.3   External and internal structure of an earthworm.

sides of the worm and feel the rough texture that the setae give the body wall. Can you distinguish dorsal and ventral on the worm? Setae provide secure contact with the substrate while burrowing and crawling. Setae are manipulated by muscles attached at the base of each seta. The anterior end of the worm has a mouth opening. The lobe over the mouth is the **prostomium.** The first segment, which has the mouth and is just posterior to the prostomium, is the **peristomium.** The posterior segment bears the anus. Other external features of interest are openings from various internal systems. Most are difficult to see. The openings of the sperm ducts are on the ventral surface of metamere 15 and are conspicuous. Find them. Smaller openings for ova are on the ventral surface of segment 14 but are difficult to see in preserved worms. Other openings include openings from seminal receptacles (between 9 and 10, and 10 and 11) and nephridial openings (on the ventrolateral surface of each segment except 1–3 and the last). ▲

Lumbricus—Internal Structure

▼ Recall that setae are ventral and lateral. Run your fingers along the surface of the worm. Locate the setae and distinguish dorsal and ventral. Pin your worm dorsal side up near one edge of a dissecting pan. Using a sharp scalpel, make a shallow longitudinal cut along the dorsal surface of the worm midway between anterior and posterior ends of the worm. Your incision should be very shallow so you do not cut into the dorsal vessel and intestinal tract. If you do cut too deep you will release intestinal contents into the body cavity. Cutting too deeply at this point is not serious. As you extend the cut anteriorly, compensate with shallower incisions. Note that the body wall is held in place by **septa** (internal divisions between metameres). Using a sharp scalpel, cut the septa along the length of the worm on both sides of the intestine. These septa divide the coelom into separate cavities. Pin the body wall to the bottom of the pan, inclining the pins away from the worm. Continue your incision anteriorly until you reach the prostomium. Remove any pins at the very anterior end of the worm. Be very careful as you cut through the anterior five segments. The suprapharyngeal ganglion is located very near the body wall in this region. It will be studied later. It is not necessary to extend the cut posteriorly to the anus. Keep your preparation moist at all times.

Digestive System Observe the digestive tract running from mouth to anus. Anteriorly there are a number of modifications of the tract. Just posterior to the mouth opening is the muscular **pharynx** (fig. 12.4). The torn muscles associated with the pharynx were attached to the body wall. When these muscles contract, food particles are sucked into the mouth. The **esophagus** is posterior to the pharynx and is surrounded by cream-colored bodies that will be studied later. The esophagus expands into a thin-walled storage structure, the **crop.** Probe the wall of the crop gently and note its texture. Just posterior to the crop is the muscular **gizzard.** This is a grinding structure with thick, muscular walls (obvious on gentle probing). Food is passed from the gizzard to the **intestine,** where further digestion and absorption occur. The intestine ends at the anus.

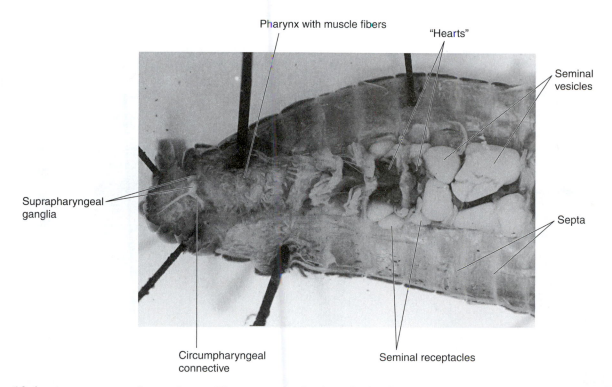

Figure 12.4 Anterior region of an earthworm. The peristomium has been displaced laterally to expose a circumpharyngeal connective.

Circulatory System Circulation in the earthworm is through a series of closed vessels. The two main vessels that can be seen in your dissection are the **dorsal** and **ventral blood vessels.** These vessels are the main pumping structures. In the dorsal vessel, blood moves anteriorly. The dorsal vessel is the dark line running along the dorsal surface of the digestive tract. In the middle of your worm, carefully cut through and remove about 3 cm of the digestive tract. The ventral blood vessel can usually be seen adhering to the segment of intestine removed or to the ventral body wall (fig. 12.5). In the ventral vessel, blood moves posteriorly (an exception is noted below). Segmental branches off the ventral vessel supply the intestine and body wall with blood (polychaetes have a third set of branches going to the parapodia). These branches eventually break into capillary beds to pick up or release nutrients and oxygen. Gas exchange occurs between the capillary beds of the body surface and the environment. Oxygen is carried by the respiratory pigment hemoglobin, which is dissolved in the fluid portion of the blood. From these capillary beds, blood is collected into larger vessels that eventually unite with the dorsal vessel. At the level of the esophagus, segmental branches are expanded into five pairs of "hearts" or aortic arches (see fig. 12.4). Although these are contractile, they only function in pumping blood between dorsal and ventral vessels. As noted earlier, dorsal and ventral vessels are the primary pumping structures. Most blood entering the ventral vessel is pumped posteriorly. Some blood moves from the aortic arches anteriorly. Locate the aortic arches. They are dark, expanded structures on either side of the esophagus. They are not present in all annelids. Other vessels associated with the nerve cord will be noted later. In an anesthetized worm, waves of contraction can be observed passing along the dorsal vessel. Observe this demonstration if available.

Reproductive System Earthworms are monoecious, with cross-fertilization the rule. Gonads are too small to see, but associated structures are clearly visible. Note the large, cream-colored structures associated with segments 9–12 (see fig. 12.4). These are **seminal vesicles.** Testes are associated with seminal vesicles. Sperm are passed from the testes to the seminal vesicles for storage prior to copulation. During copulation, sperm exit through a duct system opening at segment 15. The ovaries are located in segment 13, and their duct system opens to the outside in segment 14. Two pairs of small, round, cream-colored structures are present on the ventrolateral body wall in segments 9 and 10. These **seminal receptacles** receive sperm during copulation.

During copulation, two worms line up facing in opposite directions with the ventral surfaces of their anterior ends in contact with each other. This lines up the clitellum of one worm with the genital segments of the other worm. Worms are held in place by a mucous sheath secreted by the clitellum that envelopes the anterior halves of both worms (fig. 12.6a). Sperm are released from the sperm duct and travel along the external, ventral body wall in sperm grooves formed by the contraction of special muscles (fig. 12.6b).

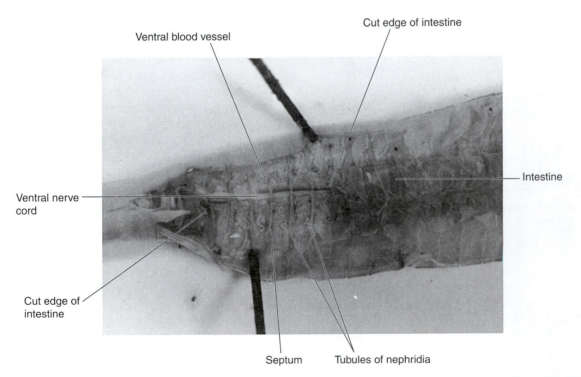

Figure 12.5 Posterior region of an earthwarm. A section of intestine has been removed to expose the ventral nerve cord (white) and ventral blood vessel (dark colored).

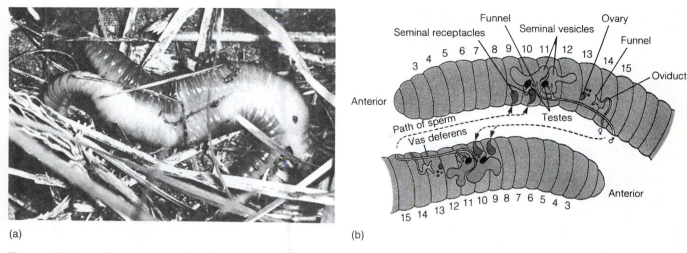

(a)

(b)

Figure 12.6 Earthworm copulation: (a) two worms within a mucous sheath; (b) path of sperm. (b) From *A Life of Invertebrates* by W. D. Russell-Hunter. Copyright © 1979 W. D. Russell-Hunter. Reprinted by permission.

Figure 12.7 Earthworm cocoons.

Muscular contractions along this groove help propel sperm toward the openings of the seminal receptacles.

Following copulation, the clitellum forms a **cocoon** for the deposition of eggs and sperm (fig. 12.7). The cocoon consists of mucoid and chitinous materials that encircle the clitellum. A food reserve, albumin, is secreted into the cocoon by the clitellum, and the worm begins to back out of the cocoon. Eggs are deposited into the cocoon as the cocoon passes the oviductal openings, and sperm are released as the cocoon passes the openings of the seminal receptacles. Fertilization occurs within the cocoon and, as the worm continues backing out, the ends of the cocoon are sealed and the cocoon is deposited in moist soil. Development is direct (without a larval stage) inside the cocoon.

Excretory System **Nephridia** are organs of excretion in the annelids. The form of the nephridium varies depending on the particular annelid. Those found in the earthworm are highly evolved metanephridia. Flood your worm with water. Observe the body wall under a dissecting microscope. Note the coiled tubule attached to the body wall of each segment.

The funnel-shaped **nephrostome** is attached to the septum dividing two segments and opens into the anterior segment. The **tubule** is within the posterior segment and connects to the body wall in that segment (see fig. 12.5). The tubule then opens to the outside. Body fluids, ammonia, and urea are drawn into the nephrostome and excreted.

Filtration of the blood across the tubule wall can occur because of the close association between capillaries and the nephridium. These physiological processes are not clearly understood.

Nervous System The nervous system consists of a **ventral nerve cord** with **segmental ganglia** and a pair of large ganglia anterior and dorsal to the pharynx (see figs. 12.4 and 12.5). Nerve fibers arise from segmental ganglia and innervate structures in each segment. In the region of your worm where the intestine was removed, note the ventral nerve cord with its segmental ganglia (seen as slight swellings). If your middorsal incision is not all the way to the prostomium, extend it at this time and note the **suprapharyngeal ganglion** (loosely called the "brain"). This ganglion is connected by **circumpharyngeal connectives** to the ventral nerve cord. ▲

Cross Section ▼ Using a dissecting microscope or the lowest power of a compound microscope, examine a microscope slide showing a cross section of *Lumbricus*. Locate those structures studied earlier using figure 12.8. In addition, note the noncellular, protective **cuticle** covering the **epidermis** (hypodermis). Just below the epidermis, note the **circular muscle** and below that, note the **longitudinal muscle**. The **coelomic lining,** or **peritoneum,** lies below the longitudinal muscle. Covering the dorsal vessel and the intestine are cells called **chloragogen cells.** These are involved with glycogen and fat synthesis and urea formation. Folding to the interior of the intestine is the **typhlosole,** which increases the surface area for secretion and absorption. Two **lateral neural**

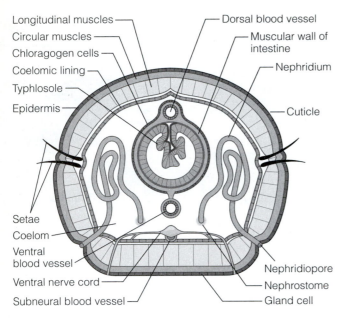

Figure 12.8 Earthworm cross section. Nephridial funnels (nephrostomes) open into the next anterior segment of the worm.

blood vessels and one **subneural blood vessel** are associated with the ventral nerve cord. Depending on the section studied you may also see a nephridium and a setum. Examine any available demonstrations of other oligochaetes. ▲

Class Hirudinea—Leeches

Characteristics: Body with 34 segments and many annuli; direct development; anterior and posterior suckers; setae absent; parapodia absent. Example: *Hirudo*.

Leeches are the most highly specialized annelids. They are predominantly freshwater, but a few are marine and terrestrial. They are predators with mouthparts usually adapted for fluid feeding, preying mostly on other invertebrates such as earthworms, slugs, and insect larvae. Those few leeches that feed on vertebrate blood have salivary secretions that prevent the clotting of blood.

▼ Examine any demonstration specimens available for study. Note that leeches lack parapodia and head appendages. They are dorsoventrally flattened and taper anteriorly. Anteriorly there is an **anterior sucker,** and posteriorly there is a **posterior sucker.** Leeches have 34 segments, but segments are difficult to distinguish externally because they have become secondarily divided. Several secondary divisions, called **annuli,** are in each true segment. Locate annuli on the specimen you are observing (fig. 12.9). ▲

Leeches move by an undulating, swimming motion or by a looping action. These motions are made possible by more complex musculature than is present in other annelids. In addition, the coelom has lost its metameric partitioning; rather than using independent coelomic compartments in locomotion, the leech has a single hydrostatic

(a)

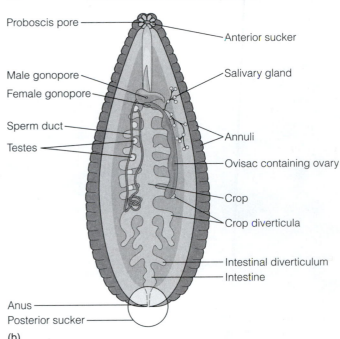

(b)

Figure 12.9 Leech structure. (a) *Hirudo medicinalis,* the medicinal leech, has been used for bloodletting. (b) Internal structure of a leech.

cavity. After attaching the larger, posterior sucker to the substrate, circular muscles of the body wall contract to extend the leech anteriorly. The smaller anterior sucker that surrounds the mouth is attached to the substrate, and the posterior sucker is released. Longitudinal muscles then bring the posterior end of the leech anteriorly to repeat the sequence.

Leeches are monoecious, and sperm transfer and egg deposition usually occur in the same manner as described for *Lumbricus*. They possess a clitellum, which is not obvious during nonbreeding periods. A penis may aid in sperm transfer in some species. Other leeches may transfer sperm by expelling a packet of sperm (a spermatophore) from one leech through the integument and into the body cavity of another. Development is direct.

12.2 LEARNING OUTCOMES REVIEW—WORKSHEET 12.1

12.3 EVOLUTIONARY RELATIONSHIPS

LEARNING OUTCOMES

1. Describe the current status of higher taxonomy of the Annelida.
2. List a combination of characters that distinguish Annelida from other animal phyla.

▼ A cladogram showing a modern interpretation of the evolutionary relationships among the annelids is shown in worksheet 12.1 (fig. 12.10). The phylum Annelida is monophyletic, but "Polychaeta" is probably a paraphyletic grouping. Figure 12.10 shows additional lineages nested within the "polychaete annelids." The Clitellata is a monophyletic group believed to have evolved from a group of polychaetes that invaded freshwater, and then later their descendants invaded terrestrial environments. The phylogeny shown in figure 12.10 is based on evidence from molecular biology, and therefore does not depict snyapomorphic characters. Answer the worksheet questions that follow the cladogram. ▲

12.3 LEARNING OUTCOMES REVIEW—WORKSHEET 12.1

WORKSHEET 12.1 Annelida

12.1 Evolutionary Perspective—Learning Outcomes Review

1. Annelids are members of the subkingdom
   a. Lophotrochozoa.
   b. Deuterostomia.
   c. Ecdysozoa.

2. Members of which of the following phyla are most closely related to the Annelida?
   a. Cnidaria
   b. Mollusca
   c. Arthropoda
   d. Chordata

12.2 Phylum Annelida—Learning Outcomes Review

3. Which of the following are annelid features?
   a. metamerism
   b. wormlike; body wall with circular and longitudinal muscles
   c. ventral nervous system
   d. nephridia used in excretion
   e. closed circulatory system
   f. All of the above are annelid features

4. Members of the class Polychaeta are characterized by all of the following except one. Select the exception.
   a. mostly marine
   b. numerous setae
   c. trochophore larvae
   d. clitellum
   e. All of the above characterize the Polychaeta.

5. The prostomium of the clamworm, *Nereis*, has all of the following except one. Select the exception.
   a. tentacles
   b. jaws
   c. palps
   d. eyes

6. Lateral extensions of the body wall of *Nereis*
   a. are called parapodia.
   b. are divided into dorsal and ventral lobes.
   c. are found on all body segments posterior to the peristomium.
   d. All of the above are true.

7. How can you tell dorsal and ventral on the earthworm?
   a. The clitellum is on the dorsal surface.
   b. The clitellum is on the ventral surface.
   c. The setae are ventral and lateral in position.
   d. The setae are dorsal and lateral in position.

8. Which of the following structures of the earthworm digestive system had a fuzzy appearance due to torn muscles used in sucking food into the digestive tract?
   a. esophagus
   b. pharynx
   c. crop
   d. gizzard

9. Which of the following structures of the earthworm digestive system was firm to the touch because it is muscular and used in grinding food?
   a. esophagus
   b. pharynx
   c. crop
   d. gizzard

10. Large, cream-colored structures surrounding the esophagus of an earthworm are
    a. seminal vesicles and store sperm prior to copulation.
    b. seminal receptacles and store sperm after copulation.
    c. seminal vesicles and store sperm after copulation.
    d. testes.

11. Blood flows _____ in the dorsal vessel of an earthworm.
    a. anterior to posterior
    b. posterior to anterior

12. A nephridium has its tubule in segment 19. The nephrostome of that nephridium will open into segment
    a. 18.
    b. 19.
    c. 20.

13. Where did you look to find a nephridium in your earthworm?
    a. along the lateral body wall.
    b. ventrally under the intestine.
    c. attached to the intestine.
    d. dorsal to the intestine.

14. In an earthworm, the _____ is just above the circular muscle of the body wall and the _____ is just below circular muscle.
    a. cuticle; epidermis
    b. epidermis; coelomic lining
    c. epidermis; longitudinal muscle
    d. longitudinal muscle; coelomic lining

15. The leach body has 34 segments, but this is not easy to determine because
    a. no external divisions can be seen.
    b. segments are subdivided externally into annuli.
    c. the coelom has lost its metameric partitioning to create a single hydrostatic cavity.
    d. Both b and c are correct.

12.3 Evolutionary Relationships

16. Annelida should be abandoned as a phylum name because its members are polyphyletic.
    a. true
    b. false

17. Polychaeta will probably be abandoned as a class name in the future because its members are paraphyletic.
    a. true
    b. false

18. Ancient members of the class Clitellata gave rise to other members of the phylum.
    a. true
    b. false

12.1   Evolutionary Perspective—Analytical Thinking

1. What is meant by the term "paraphyletic grouping"? What is the problem with such a grouping? Explain why the polychaetes are probably paraphyletic.

12.2   Phylum Annelida—Analytical Thinking

2. The presence of separate metameric hydrostatic compartments is cited as allowing efficient locomotion in the Annelida. In the leeches, however, internal metameric compartments are lost in order to accomplish their form of locomotion. Explain why metameric compartments are efficient for locomotion in the Polychaeta and Oligochaetea but not in the Hirudinea.

12.3   Evolutionary Relationships—Analytical Thinking

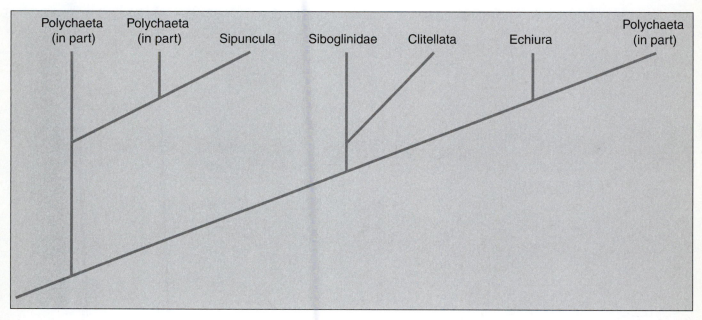

Figure 12.10   Evolutionary relationships among the Annelida. This cladogram is a molecular phylogeny that depicts the "Polychaeta" as a paraphyletic grouping, with Clitellata and other groups nested within the polychaetes.

Answer the following based on the background in the laboratory manual, your textbook, and the cladogram shown above.

3. Explain why the cladogram in figure 12.11 does not depict any synapomorphic characters.

4. Explain how the pattern of branching seen in this cladogram reflects paraphyly.

5. Annelida is monophyletic. If you were to place a set of characters common for the phylum at the base of the tree, what characters would you use? Are any of these characters unique to the phylum, or is it the combination of characters that define phylum members?

6. The Clitellata forms a monophyletic clade. A derived character for members of this clade would be the _____. What might explain the evolution of the character in this lineage?

7. Name two synapomorphies that could distinguish members of the Hirudinea from all other annelids?

Exercise 13

The Pseudocoelomate Body Plan: Aschelminths

Prelaboratory Quiz

Study this week's laboratory exercise and then complete the following quiz to assess your preparation for the laboratory.

1. A body cavity that is located between mesoderm on the inside of the body wall and endoderm of a gut tract is a
   a. coelom.
   b. pseudocoelom.
   c. gut.
   d. cloaca.

2. The pseudocoelomate phyla are
   a. only distantly related; some of these animals are lophotrochozoans and others are ecdysozoans.
   b. all closely related lophotrochozoan phyla.
   c. all closely related ecdysozoan phyla.
   d. all closely related deuterostomate phyla.

3. All of the following are aschelminths EXCEPT
   a. nematodes.
   b. rotifers.
   c. flukes.
   d. nematomorphs.

4. In this week's laboratory, we will perform a dissection of
   a. an earthworm.
   b. a nematode.
   c. a rotifer.
   d. a gastrotrich.

5. The development of an individual from an unfertilized egg is _____. It is encountered in this week's laboratory during observations of the phylum _____.
   a. vitellogenesis; Nematoda
   b. parthenogenesis; Nematoda
   c. vitellogenesis; Rotifera
   d. parthenogenesis; Rotifera
   e. parthenogenesis; Nematomorpha

6. True/False   The aschelminths include a diverse group of closely related phyla.

7. True/False   All nematodes are parasites.

8. True/False   A female *Ascaris* is smaller than the male and has a hooked posterior tip.

9. True/False   Mesenteries are mesodermal sheets that suspend the gut tract of pseudocoelomate animals within their body cavities.

10. True/False   In temperate regions, most rotifers spend the winter as dormant zygotes.

13.1 EVOLUTIONARY PERSPECTIVE

LEARNING OUTCOME

1. Explain why the term "aschelminth" is a term of convenience and is not taxonomically descriptive.

The aschelminths include a group of phyla that share one common feature, the pseudocoelom. They include the phyla Rotifera, Acanthocephala, Nematoda, Kinorhyncha, Gastrotricha, Nematomorpha, and others (fig. 13.1). Except for their one common feature, they are a diverse group of animals, only distantly related. Recent evidence from molecular biological investigations has revealed that some of the phyla just listed, including the Rotifera and the Acanthocephala, are lophotrochozans (see p. 171). The remaining phyla, including Nematoda, Kinorhyncha, Gastrotricha, and Nematomorpha, are ecdysozoans. The Ecdysozoa also include the Arthropoda (insects, crustaceans, and their relatives) and are united by the fact that they molt a secreted cuticle during growth. This molting process is called ecdysis.

The conclusion that is drawn from this information is that the term "aschelminth" is a term of convenience that refers to a very diverse, and very distantly related, assemblage of animals. The common feature that they share, the pseudocoelom, must have arisen independently in two major lineages of bilateral animals.

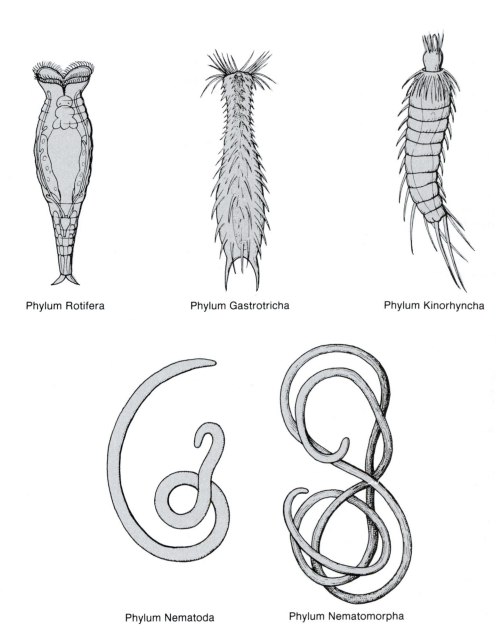

Phylum Rotifera

Phylum Gastrotricha

Phylum Kinorhyncha

Phylum Nematoda

Phylum Nematomorpha

Figure 13.1   Representatives of the pseudocoelomate phyla.

Recall that the pseudocoelom is a body cavity that forms between mesoderm of the body wall and endoderm of the gut. Because mesoderm is restricted to the body wall, there is no muscular or connective tissue associated with the gut tract. No mesodermal sheet covers the inner surface of the body wall, and no membranes suspend organs within the body cavity. An important use of the pseudocoelom in the aschelminths is as a hydrostatic skeleton. In the exercises that follow, you will first examine the lophotrochozoan aschelminths and then the ecdysozoan aschelminths.

13.1 LEARNING OUTCOMES REVIEW—WORKSHEET 13.1

13.2 ASCHELMINTHS THAT DO NOT MOLT

LEARNING OUTCOMES

1. Describe characteristics and give an example of members of the phylum Rotifera.
2. Explain parthenogenesis.
3. Describe characteristics of members of the phylum Acanthocephala.

M embers of the phyla Rotifera and Acanthocephala are lophotrochozoans. In the following exercises, you will examine one representative of each of these two phyla.

PHYLUM ROTIFERA

Rotifers are mostly freshwater pseudocoelomates that derive their name from a ciliated **corona.** When cilia are beating, the impression is of rotating wheels. The body is elongate, consisting of a **head,** a **trunk,** and a **foot.**

We will examine a living specimen of *Philodina* (fig. 13.2). *Philodina* is frequently used in introductory courses because it clearly shows details of internal and external anatomy. *Philodina* is the common bird-bath rotifer. It is transferred between bird baths in the feathers of birds. It, as well as other related rotifers, is well known for its ability to withstand desiccation. Dried scrapings from a bird bath will remain viable on a laboratory shelf for years and spring to life when water is added. This phenomenon is called cryptobiosis and may be the source of the animals your instructor used in preparing for the following observations.

▼ Obtain a drop of water from the bottom of a culture dish and make a wet mount. Observe the cylindrical body with head, trunk, and foot. The foot is used in attaching to the substrate. The ciliated corona can be seen anteriorly. Note the beating cilia. Cilia create water currents used in feeding. Note the periodic retraction of the corona. This is also associated with feeding. Just below the corona, look for a muscular grinding structure

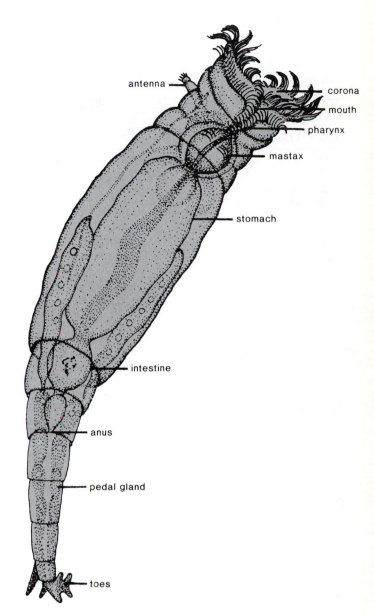

Figure 13.2 The structure of *Philodina*.

called the **mastax.** It is associated with the gut tract and is easily seen in living rotifers because of its rhythmical contractions. ▲

Rotifers are dioecious. Frequently, as in *Philodina*, reproduction occurs by the development of unfertilized diploid eggs. This is called **parthenogenesis.** In fact, males are unknown in *Philodina*. In other species, parthenogenesis occurs in the spring and summer. In autumn, parthenogenic males are produced. Fertilization results in a zygote that overwinters and hatches into a female in the spring. This is an adaptation to withstand the harsh winter conditions of temperate lakes. Other rotifers exhibit normal sexual reproduction.

PHYLUM ACANTHOCEPHALA

▼ Adult acanthocephalans, or spiny-headed worms, are endoparasites in the intestinal tract of vertebrates (especially fishes). Juveniles are parasites of crustaceans and insects.

The body of the adult is elongate and has a proboscis covered with recurved spines that it uses to attach to the intestinal wall of its host. Examine a prepared slide of an adult acanthocephalan. ▲

13.2 LEARNING OUTCOMES REVIEW—WORKSHEET 13.1

13.3 ASCHELMINTHS THAT MOLT

LEARNING OUTCOMES

1. Describe characteristics and give an example of members of the phylum Nematoda.
2. Identify the external and internal structure of dissected *Ascaris*.
3. Identify internal structures in cross sections of male and female *Ascaris*.
4. Recognize members of the phylum Nematomorpha.

Members of the phyla Nematoda and Nematomorpha are ecdysozoans. In the following exercises, you will examine one representative of each of these two phyla.

PHYLUM NEMATODA

Characteristics: Round in cross section, covered by a flexible, nonliving cuticle; lack motile cilia and flagella; longitudinal muscles only. Examples: *Ascaris*, *Trichinella*, *Necator*.

Ascaris—External Structure

▼ Obtain a specimen of *Ascaris*. It will most likely be *Ascaris suum*, the pork ascarid. A species indistinguishable from the pork ascarid occurs in humans, *Ascaris lumbricoides*. Nematodes are dioecious. Males are smaller, and the posterior tip is hooked. Females are larger and lack the hook. Determine the sex of your worm. Examine the external structure of the worm using a dissecting microscope or hand lens. Note the cylindrical appearance of the worm. Distinguish anterior from posterior. The anterior tip of both sexes is less pointed than the posterior. In looking straight down on the anterior tip, note the three lips that surround a triangular **mouth.** Near the posterior tip one can see a slitlike **anus.** The anus is ventral and can be used to distinguish dorsal from ventral. Find lateral lines running down both sides of the worm. Within each is an excretory canal.

Internal Structure

Place your worm dorsal surface up along one side of a dissecting pan (so you can fit the specimen under a dissecting microscope) and place a pin in each end. Internal structures in preserved *Ascaris* are delicate, so it is best to submerge the specimen entirely during dissection and observation. Water will float internal organs and result in less damage. Using a sharp scalpel, make a longitudinal incision along the length of the worm. Cut just through the body wall, being careful not to damage internal structures. Pin the body wall to the bottom of the dissecting pan as you proceed.

Internal structure is shown in figure 13.3. Note the **pseudocoelom.** Running the length of the body cavity is a flat, ribbonlike **intestine.** Anteriorly, the digestive tract is modified into the **pharynx.** This structure is muscular and has a pumping function. The remainder of the body cavity is filled with reproductive structures. In looking at reproductive structures, examine your worm first and then trade with someone who has the opposite sex. You are responsible for both.

Female

The female reproductive tract is Y-shaped. Note the mass of coiled and folded tubules. The duct system is single near its attachment at the genital opening (located within the anterior one-third of the worm) and double at the opposite extreme. **Ovaries** are the smallest diameter tubules at the free ends of the "Y." As you trace the tubules proximally, note that their diameters increase. Find a region where the tubule diameter is intermediate. This region is the **oviduct.** There is no grossly observable transition between either ovary and oviduct or oviduct and **uterus,** the largest diameter tubules. The oviduct carries eggs between the ovary and the uterus. Fertilization occurs in the uterus, and eggs are held here prior to their release. The two uteri unite to form the **vagina.** Copulatory spicules of the male are inserted into the vagina for sperm transfer. The position of the **genital pore** can be located internally where the vagina attaches to the body wall.

Male

The reproductive tract of the male is a single continuous tube. It ends at the posterior tip where it unites with the digestive tract to form a **cloaca** (an opening common to digestive and reproductive tracts). The **testis** is the region at the smaller, free end of the tubule. Sperm formed in the testis travel down the intermediate-diameter **vas deferens** to the **seminal vesicle,** where sperm are stored. The region of the tubule near the cloaca is the **ejaculatory duct.** The ejaculatory duct propels sperm into the female vagina during copulation. Locate these regions. As with the female, there is no grossly observable boundary between them.

Ascaris Cross Section

Obtain a prepared *Ascaris* slide that shows male and female cross sections (figs. 13.4 and 13.5). The larger section is the female. Examine it under low power. The outer body wall is covered by a nonliving, protective cuticle. Below the cuticle, note the body wall, the pseudocoelom, and other internal structures. Recall that the pseudocoelomates are

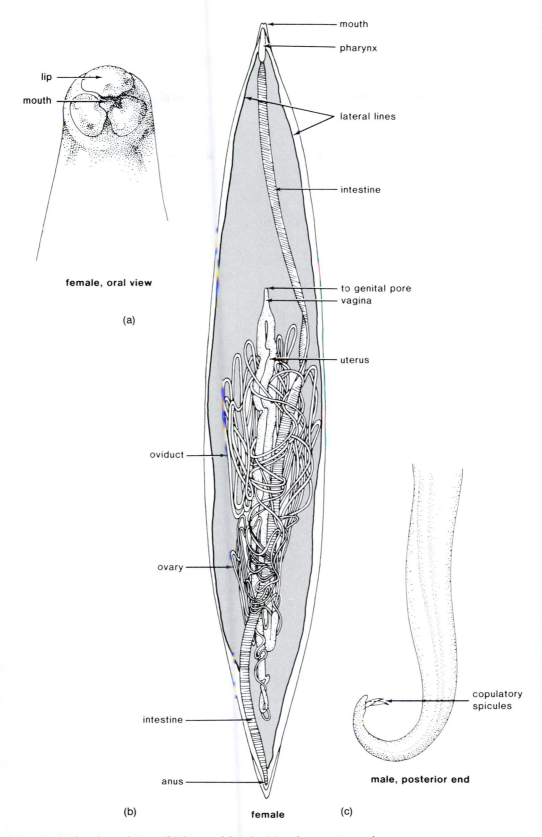

mouth

pharynx

lateral lines

intestine

to genital pore

vagina

uterus

oviduct

ovary

intestine

anus

lip

mouth

female, oral view

(a)

(b)

female

copulatory
spicules

male, posterior end

(c)

Figure 13.3   *Ascaris:* (a) female, oral view; (b) dissected female; (c) male, posterior end.

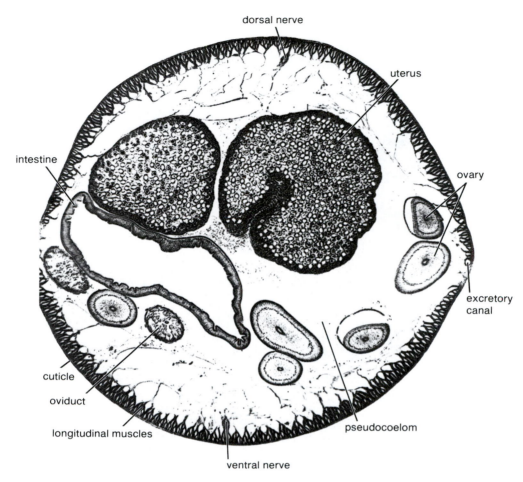

Figure 13.4 Cross section of an *Ascaris* female.

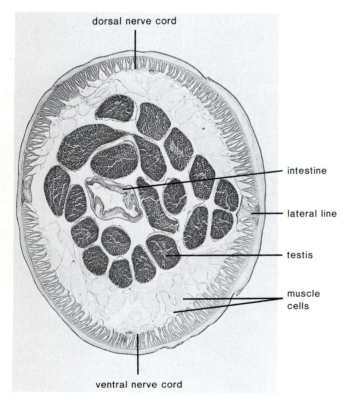

Figure 13.5 Cross section of an *Ascaris* male.

triploblastic. Derivatives of the three germ layers can be seen in your cross section. Just below the cuticle is a layer of cells called the epidermis. It is ectodermal. Locate lateral lines, lateral thickenings of the epidermal layer. Within each lateral line is an excretory canal.

The flattened intestine is endodermal. There is no mesoderm associated with the intestine to help it maintain its shape.

Muscles and reproductive structures are mesodermal. Observe **longitudinal muscles** just inside the epidermis. When muscles contract, they act against the hydrostatic skeleton. Because there are no circular muscles, a thrashing movement results. You will observe this later. Cross sections of females are usually taken through the middle third of the body and thus show two sections of uteri. Each uterus is a large-diameter tubule containing eggs. Oviducts are smaller in diameter with a central cavity (lumen) for the passage of eggs from the ovary to the uterus. Sections through the ovary are nearly solid.

Examine the cross section of a male. In addition to the body wall and intestine, note cross sections through the reproductive tract. The testis will be represented in numerous small-diameter sections. The cells inside are amoeboid sperm cells (recall there are no cilia or flagella in the nematodes). You should also see several sections through the larger vas deferens and one section through the very large seminal vesicle. Sperm should also be present in these sections. ▲

Life Cycles

After fertilization, eggs of nematodes are usually released to the environment, where development takes place. (There are a number of exceptions.) Larvae look similar to adults but are smaller and sexually immature. Larvae molt as they grow to adulthood. In *Ascaris*, eggs are deposited with feces. Development through the initial larval stage occurs within the very resistant egg capsule. When ingested, larvae hatch and migrate through the intestinal wall and then, via the bloodstream, to the lungs. They finally migrate up the respiratory tract and are swallowed, thus making their way back to the small intestine, where they mature (fig. 13.6).

Trichinella spiralis is a nematode that causes trichinosis in nearly all meat-eating animals. Humans usually become involved by eating incompletely cooked pork. Eggs hatch within the female, and larvae are released from the female worm into the intestine of the host. These larvae migrate through the intestinal wall to the lymphatic system and then the circulatory system. They encyst in active striated muscle, such as the diaphragm and intercostal muscles, where they remain viable for many years (fig. 13.7).

Many other life-cycle variations occur in parasitic nematodes. These nematodes include the hookworms like *Ancylostoma* and *Necator*, whose adults live attached to the intestinal wall, where they feed on blood. Eggs hatch after being released into the soil, and the larvae penetrate the skin of the host. Filarial nematodes depend upon arthropod vectors for transmission to new hosts. The adult dog heartworm, for example, lives in the ventricles of the heart.

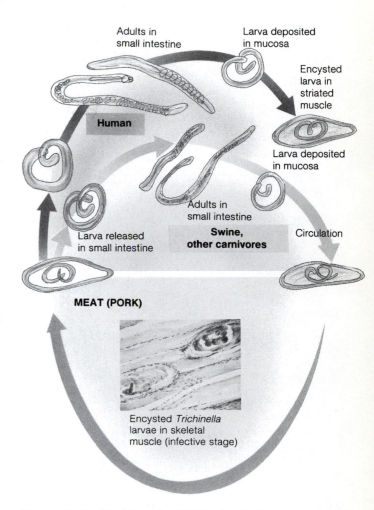

Figure 13.7   The life cycle of *Trichinella spiralis*.
Source: Redrawn from original from Centers for Disease Control and Prevention, Atlanta, GA.

Larvae move into the peripheral circulation, where they can be picked up by a mosquito. When the mosquito bites another dog, they are introduced into a second host.

▼ Examine demonstrations of the following parasites as they are available: ▲

Trichinella spiralis
    Adult stages from the intestine
    Larvae encysted in muscle
Necator americanus and *Ancylostoma duodenale*
    Hookworms
Enterobius vermicularis
    Pinworm

Free-Living Nematodes

In the preceding activities we have emphasized parasitic nematodes. By far most nematodes are free living. They occur in virtually every aquatic and terrestrial habitat. One commonly used in the laboratory is the vinegar eel *Tubatrix aceti* (fig. 13.8). It is relatively transparent, thus showing internal structure in living specimens.

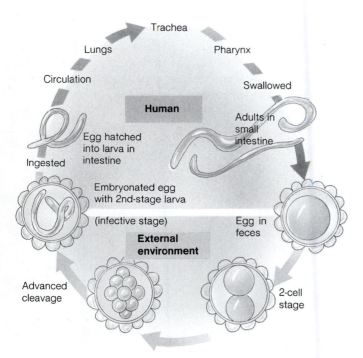

Figure 13.6   The life cycle of *Ascaris lumbricoides*.
Source: Redrawn from original from Centers for Disease Control and Prevention, Atlanta, GA.

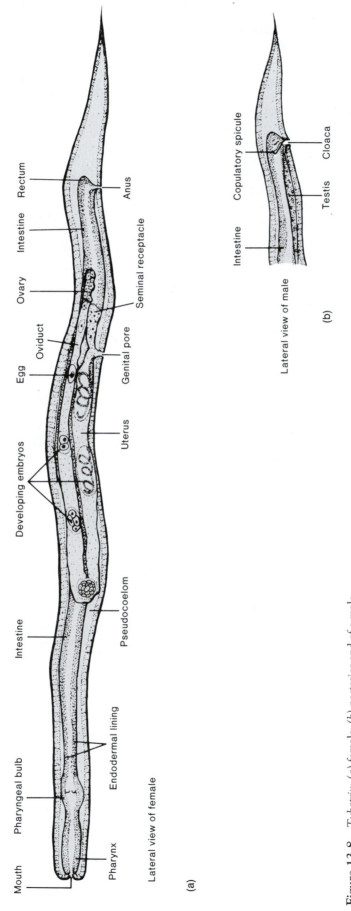

Figure 13.8 *Tubatrix:* (*a*) female; (*b*) posterior end of a male.

192

▼ Make a wet mount of the fluid from the top of a culture dish. Examine it under low and high power. Note particularly the thrashing movements characteristic of nematodes. What is the basis for these movements?

_____

_____ ▲

A handful of rich garden soil contains thousands of nematodes living off decaying organic matter. These nematodes are important in the decomposition of organic matter and serve as a source of food for soil arthropods, earthworms, and some fungi. Soil nematodes are essential to the energy flow and nutrient cycling in soil ecosystems. Soil nematodes can be collected using a Baermann apparatus.

▼ Make a wet mount of fluid drawn from a Baermann apparatus and examine under low and high power. ▲

PHYLUM NEMATOMORPHA

Nematomorphs are elongate worms called either horsehair worms or Gordian worms. Adults are dioecious and deposit eggs in water. Larvae that hatch from eggs are either eaten by, or penetrate, an arthropod host. The larva matures in the arthropod host and emerges as the adult worm when the insect host is near water.

▼ Observe demonstrations of nematomorph worms (see fig. 13.1). ▲

13.3 LEARNING OUTCOMES REVIEW—WORKSHEET 13.1

WORKSHEET 13.1   The Pseudocoelomate Body Plan: Aschelminths

13.1   Evolutionary Perspective—Learning Outcomes Review

1. Which of the phyla studied in this exercise belong(s) to the subkingdom Lophotrochozoa?
   a. Rotifera
   b. Nematoda
   c. Acanthocephala
   d. Nematomorpha
   e. Both a and c are correct.

2. Which of the following statements must be true of the pseudocoelom?
   a. It is a synapomorphy that unites all the aschelminths.
   b. It must have arisen more than once in animal evolution.
   c. It is unique to the Ecdysozoa.
   d. It is unique to the Lophotrochozoa.

13.2   Aschelminths That Do Not Molt—Learning Outcomes Review

3. All of the following are rotifer features except one. Select the exception.
   a. corona
   b. body consisting of head, trunk, and foot
   c. cuticle covering the body wall
   d. pseudocoelom
   e. parthenogenesis

4. In most rotifers, parthenogenesis occurs
   a. during spring and summer.
   b. in the winter.
   c. throughout the year.

5. After being drawn into the digestive tract of a rotifer by the _____, food is ground by a muscular

   _____.
   a. coelom; trunk
   b. ciliated corona; mastax
   c. mastax; corona

6. Parthenogenesis is
   a. the formation of a winter resting zygote.
   b. sexual reproduction involving both male and female rotifers.
   c. development of an unfertilized egg.

13.3   Aschelminths That Molt—Learning Outcomes Review

7. How can one tell a male from a female *Ascaris* based on external structure?
   a. The male has a hooked posterior end.
   b. The female lacks lateral lines.
   c. The male is smaller than the female.
   d. The male is somewhat flattened.
   e. Both a and c are correct.

8. Which of the following is true of both the female reproductive system and the male reproductive system of *Ascaris*?
   a. The genital opening is near the anterior third of the nematode.
   b. The reproductive system consists of a "Y-shaped" tube that is coiled and folded on itself in the body cavity.
   c. Gametes are produced in the ovaries or testes, which are the smallest-diameter tubules.
   d. The reproductive tract ends at the cloaca.

9. How would you distinguish the digestive tract from the tubule system of the reproductive tract of an *Ascaris*?
   a. The digestive tract is flattened and the reproductive tract is round in cross section.
   b. The digestive tract runs the entire length of the animal from mouth to anus. The reproductive tract does not run from mouth to anus.
   c. The reproductive tract is highly coiled and folded. The digestive tract is a more-or-less straight tube.
   d. All of the above are differences between the reproductive and digestive tracts of *Ascaris*.

   Questions 10–15. Imagine that you were given two cross sections from the midsection of *Ascaris*. One cross section is from a male and the other is from a female. Indicate whether the following would be observed in the male, the female, neither, or both.

10. a flattened digestive tract
    a. male only
    b. female only
    c. neither
    d. both

11. a muscle layer lining both the inner body wall and the digestive tract
    a. male only
    b. female only
    c. neither
    d. both

12. longitudinal and circular muscles lining the body wall just below the epidermis
    a. male only
    b. female only
    c. neither
    d. both

13. two large-diameter tubules packed with small circular structures
    a. male only
    b. female only
    c. neither
    d. both

14. numerous small-diameter tubules
    a. male only
    b. female only
    c. neither
    d. both

15. dorsal and ventral excretory canals
    a. male only
    b. female only
    c. neither
    d. both

16. Dogs and cats have ascarids with life cycles similar to the life cycle of the human ascarid. Dogs and cats are likely to pick up infections as a result of
    a. eating larvae encysted in animal flesh.
    b. eating embryonated egg capsules deposited with dog or cat feces in the environment.
    c. wading in infested water.
    d. being bitten by a mosquito carrying ascarid larvae.

17. Which of the following aschelminthes would be more closely related to members of the phylum Nematoda?
    a. Rotifera
    b. Nematomorpha
    c. Acanthocephala.

13.1   Evolutionary Perspective—Analytical Thinking

1. Distinguish between the triploblastic pseuocoelomate and the triploblastic coelomate organization.

13.2   Aschelminths That Do Not Molt—Analytical Thinking

2. Discuss the advantages of parthenogenesis for rotifers such as *Philodina*.

13.3   Aschelminths That Molt—Analytical Thinking

3. How do the differences between the triploblastic pseudocoelomate and triploblastic coelomate organization (from question 1) manifest themselves in comparing a cross section of *Ascaris* to a cross section of *Lumbricus* (Exercise 12)?

Exercise 14

Arthropoda

Prelaboratory Quiz

Study this week's laboratory exercise and then complete the following quiz to assess your preparation for the laboratory.

List the four arthropod subphyla containing living species and give an example of each.

Subphylum   **Example**

1. _____   _____
2. _____   _____
3. _____   _____
4. _____   _____

5. Which of the following word or phrase pairs is incorrectly associated?
   a. thorax; locomotion
   b. head; sensory and feeding functions
   c. abdomen; visceral functions
   d. chelicerae; sensory functions

6. Which of the following subphyla has members that are entirely extinct?
   a. Trilobitomorpha
   b. Chelicerata
   c. Crustacea
   d. Hexapoda

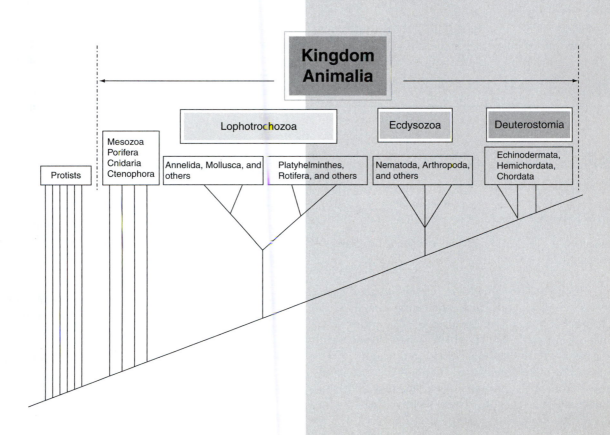

7. All of the following classes belong within the subphylum Chelicerata EXCEPT
   a. Arachnida.
   b. Malacostraca.
   c. Pycnogonida.
   d. Merostomata.

8. Biramous appendages are found in the following subphylum or subphyla.
   a. Crustacea only
   b. Myriapoda only
   c. Chelicerata and Crustacea
   d. Crustacea and Trilobitomorpha

9. Members of the subphylum Hexapoda include the
   a. insects.
   b. scorpions.
   c. millipedes.
   d. centipedes.

10. True/False   Millipedes have two pairs of appendages per apparent segment, whereas centipedes have one pair of appendages per segment.

11. True/False   In hemimetabolous development, the number of molts between hatching and death is variable within a species.

12. True/False   In holometabolous development, a pupal stage precedes the last molt into an adult insect.

13. True/False   Flies, such as the fruit fly, undergo hemimetabolous development.

14. True/False   The appendages of a crayfish are said to be serially homologous because they are evolved from a similar basic structure.

15. True/False   Book lungs, book gills, and tracheal systems are all modifications of the exoskeleton of arthropods to facilitate gas exchange.

16. True/False Compound eyes are found only in insects.

14.1 EVOLUTIONARY PERSPECTIVE

LEARNING OUTCOME

1. Describe the relationship of the Arthropoda to other animal phyla.

Arthropods—along with two closely related phyla, Onycophora and Tardigrada—form a monophyletic taxon sometimes called the Panarthropoda. They have a secreted cuticle, which is periodically shed during growth—thus they are ecdysozoans (see p. 199). Arthropods are represented in 600-million-year-old fossil deposits. Taxonomists currently recognize four subphyla containing extant arthropods. These subphyla include the

Chelicerata (the spiders, scorpions, ticks), Crustacea (crayfish, lobsters, crabs), Hexapoda (insects), and Myriapoda (centipedes and millipedes). A fifth subphylum, Trilobitomorpha, is entirely extinct but had a very long 255-million-year history. Ancestral crustaceans were probably the first arthropods. Trilobites and chelicerates were the next to appear, with the latter moving onto land. They were perhaps the first land animals. They myriapods and hexapods appear in the fossil record about 400 million years ago. Modern crustacean lineages arose at various times during this evolutionary history, and therefore this subphylum is probably paraphyletic.

14.1 LEARNING OUTCOMES REVIEW—WORKSHEET 14.2

14.2 PHYLUM ARTHROPODA

LEARNING OUTCOMES

1. Describe characteristics of members of the phylum Arthropoda.
2. Explain the importance of the exoskeleton in the success of the arthropods.

The phylum Arthropoda is the largest phylum in the animal kingdom. More than three-fourths of all known species are members. Considering how many remain to be described, the true figure is probably even higher.

A characterization of the phylum revolves around two features of their organization: **metamerism** and the **exoskeleton.** Like the annelids, the arthropods are metameric, though they show a much higher degree of tagmatization. The specialization of body regions is usually associated with three functions: sensory and feeding functions **(head),** locomotion **(thorax),** and visceral functions **(abdomen).** As we will see, there is considerable variation in degrees of tagmatization within the arthropod classes.

The exoskeleton is often cited as the major reason for arthropod success. It provides

1. structural support,
2. impermeable surfaces for the prevention of water loss,
3. a system of levers for muscle attachment and movement, and
4. protection from mechanical injury and environmental chemicals.

The exoskeleton necessitated a variety of adaptations that would allow the arthropod to live and grow within the confines of this rigid skeleton. These adaptations include

1. modifying the exoskeleton to provide for flexibility at joints;
2. modification of parts of the exoskeleton into highly developed sensory structures;

3. ecdysis, or molting, to allow for increased size; and
4. modification of the exoskeleton for gas exchange (tracheal system, book lungs, book gills).

A final characteristic of arthropods that has contributed to their success is reduced competition through **metamorphosis.** Larval forms are often very different in morphology and habits from the adults. This prevents larvae and adults from competing with one another for limited resources.

These adaptations have allowed the arthropods to fill virtually every conceivable habitat on the earth. Their impact on human history has been tremendous. Arthropods spread diseases and compete with humans for food. At the same time, they serve as food, produce useful products, and are essential for the pollination of about 65% of all plant species.

CLASSIFICATION

Subphylum Trilobitomorpha
Subphylum Chelicerata
    Class Merostomata
    Class Pycnogonida
    Class Arachnida
Subphylum Crustacea*
    Class Malacostraca
    Class Branchiopoda
    Class Maxillopoda
Subphylum Myriapoda
    Class Diplopoda
    Class Chilopoda
Subphylum Hexapoda*
    Class Insecta

*Classes listed in these subphyla are limited to those usually studied in introductory zoology laboratories.

14.2 LEARNING OUTCOMES REVIEW—WORKSHEET 14.2

14.3 SUBPHYLUM TRILOBITOMORPHA

LEARNING OUTCOME

1. Recognize trilobite fossils as representatives of the suphylum Trilobitomorpha.

Characteristics: All extinct, from Cambrian to Carboniferous periods; body divided into three longitudinal lobes; head, thorax, and abdomen; biramous appendages; one pair of antennae.

Because trilobites are all extinct, the only specimens available for study are fossilized.

▼ Examine any fossils available. Note the three lobes, head, (cephalon) compound eyes, thorax or trunk (the attached

Figure 14.1 A trilobite fossil.

biramous appendages are ventral and will not be visible), and the tail-like pygidium (fig. 14.1). ▲

14.3 LEARNING OUTCOMES REVIEW—WORKSHEET 14.2

14.4 SUBPHYLUM CHELICERATA

LEARNING OUTCOMES

1. Characterize members of the subphylum Chelicerata.
2. Describe the structure of members of the class Merostomata.
3. Recognize members of the class Pycnogonida.
4. Describe characteristics of members of the class Arachnida.
5. Identify structures studied in the garden spider.

The chelicerate body is divided into two regions. A feeding, sensory, and locomotor tagma is called the **prosoma.** Visceral functions are carried out in the **opisthosoma.** The first pair of appendages, called chelicerae, may be pincerlike, or chelate, and are used for feeding and defense. The second pair of appendages are **pedipalps** and are often sensory. They may also be used in feeding, locomotion, or reproduction. Pedipalps may also be chelate.

CLASS MEROSTOMATA— HORSESHOE CRABS

Characteristics: With book gills on the opisthosoma. Horseshoe crabs (*Limulus*) are marine, living along the Atlantic and Gulf coasts. Their body is very well adapted for pushing through sand and mud of coastal regions. In the process they scavenge for food. The hard exoskeleton provides excellent protection from all predators. Their body form apparently represents a nearly ideal combination of characters for the stable environment in which they live; the fossil record indicates that these arthropods have remained unchanged for over 250 million years.

▼ The head and the thorax are fused into one piece, the **prosoma** or cephalothorax. It is covered by a hard **carapace.** On the prosoma is a pair of **compound eyes.** Anteriorly and medially are two smaller **simple eyes.** The **opisthosoma** is located behind the prosoma. It has a series of spines around its margin and a long **telson** at its posterior aspect (fig. 14.2).

Note the appendages covering the ventral surface of the prosoma. The first pair of appendages is the **chelicerae,** used in food handling. Following them are the **pedipalps**

and four pairs of **walking legs.** All but the last pair of appendages are chelate (pincerlike). The fourth pair of walking legs has leaflike plates at the terminal end and is used for locomotion and digging. Behind the legs is a series of plates. The anterior- most plate is the **genital operculum,** covering the genital pores. The series of plates posterior to the genital operculum are the **book gills.** In the living state, book gills are highly vascularized and serve for gas exchange. ▲

CLASS PYCNOGONIDA—SEA SPIDERS

Characteristics: First pair of appendages modified to form chelicerae; four pairs of legs; one pair of pedipalps; lack antennae; body divided into prosoma and reduced abdomen; common in all oceans.

▼ Examine any specimens on demonstration. ▲

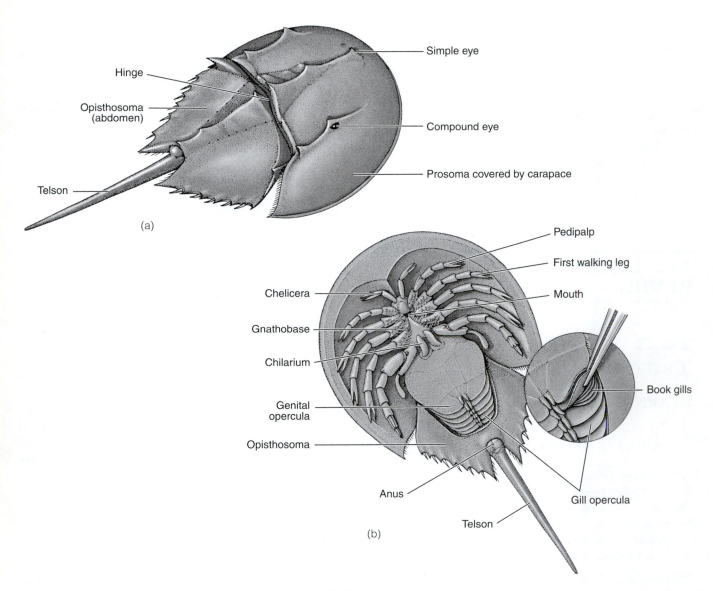

Figure 14.2 The horseshoe crab, *Limulus.* *(a)* Dorsal view. *(b)* Ventral view.

CLASS ARACHNIDA—SPIDERS, SCORPIONS, TICKS, MITES, HARVESTMEN, AND OTHERS

Characteristics: First pair of appendages modified to form chelicerae; four pairs of legs; one pair of pedipalps; lack antennae; body divided into prosoma and opisthosoma. The Arachnida are primarily terrestrial and show a number of modifications for that life-style. Three modifications are particularly important. First, the book gill is modified into a book lung, an internal series of vascular lamellae used for gas exchange with the air. Second, appendages are modified for more efficient locomotion on land. Third, water conservation is enhanced by more efficient excretory structures (coxal glands and Malpighian tubules). The impermeable exoskeleton went a long way in preadapting the arthropods for life in the terrestrial environment.

The Garden Spider

▼ Examine a preserved garden spider. Note that the body is divided into two regions, the **prosoma** and **opisthosoma** (fig. 14.3). The two body regions are joined by a slender **pedicel.** Using a hand lens or dissecting microscope, note eight **eyes** on the anterior-dorsal surface of the prosoma. Ventrally, note six pairs of appendages. Anterior-most are the **chelicerae** that possess fangs used for injecting a paralyzing toxin into the prey. Posterior to the chelicerae are **pedipalps** that are used in manipulating food and, in the male, in sperm transfer. Posterior to the pedipalps are four pairs of **walking legs.** On the opisthosoma note the slitlike openings to the **book lungs** (anterior-ventral aspect). Book lungs are invaginations of the exoskeleton and are used in gas exchange. Spiders also possess a tracheal system similar to that subsequently described for insects. Note also three pairs of **spinnerets** used in producing silk (posterior-ventral aspect). Silk is secreted as a fluid and hardens on contact with air. ▲

Other Arachnids

▼ Examine other specimens on demonstration, including

harvestmen,
mites and ticks,
scorpions, and
other spiders. ▲

14.4 LEARNING OUTCOMES REVIEW—WORKSHEET 14.2

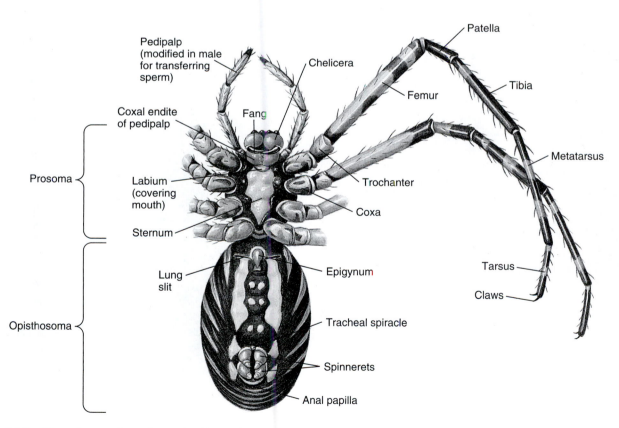

Figure 14.3 Ventral view of a garden spider (*Argiope*).

14.5 SUBPHYLUM CRUSTACEA

LEARNING OUTCOMES

1. Describe characteristics of members of the subphylum Crustacea.
2. Identify structures studied in a representative of the class Malacostraca, the crayfish.
3. Identify members of the class Maxillopoda.

With the exception of some members of one order (Isopoda), crustaceans are aquatic. Most are marine, although many are freshwater. They are readily recognized by their two pairs of antennae (all other arthropods have one pair or none) and their series of biramous appendages. The **biramous appendage** of a generalized crustacean consists of a basal series of segments called the **protopodite** with two rami attached distally. The medial ramus is the **endopodite,** and the lateral ramus is the **exopodite.** The ancestral condition was one pair of appendages on each segment, with all appendages showing similar structure. With the specialization of body regions (tagmatization), appendages became modified for functions associated with the region of the body to which they were attached. Thus head appendages are now involved with sensory and feeding functions, and thoracic appendages are involved with locomotion and defense. The specialization in function is reflected in specialization in structure. This will be obvious during the crayfish observations.

Because all appendages are presumably evolved from a similar basic structure, they are said to be **serially homologous.**

Class Malacostraca—Crayfish, Lobsters, and Crabs

Characteristics: Appendages modified for crawling along the substrate, or the abdomen and body appendages are used for swimming.

▼ The crayfish is a common inhabitant of freshwater lakes and streams all over the world. Obtain a preserved specimen. The exoskeleton is made up of lipids, proteins, and chitin (a polysaccharide). Hardening results from the deposition of calcium carbonate.

Note that the body is divided into two major regions, the **cephalothorax** and the **abdomen.** The cephalothorax is covered by an arching **carapace** (fig. 14.4). Anteriorly, note the stalked **compound eyes** attached just below a pointed extension of the carapace, the **rostrum.** Observe the surface of the compound eye under a dissecting microscope. Note the many facets that are lenses for individual photoreceptors, or **ommatidia.** The compound eye is especially adapted for detecting movement. Note also the short **first antennae** and longer **second antennae.** The sides of the carapace cover the **gill chamber.** Gently lift the carapace at the base of the legs and note the gills underneath. You will remove the portion of the carapace covering the gills on the left side of the crayfish.

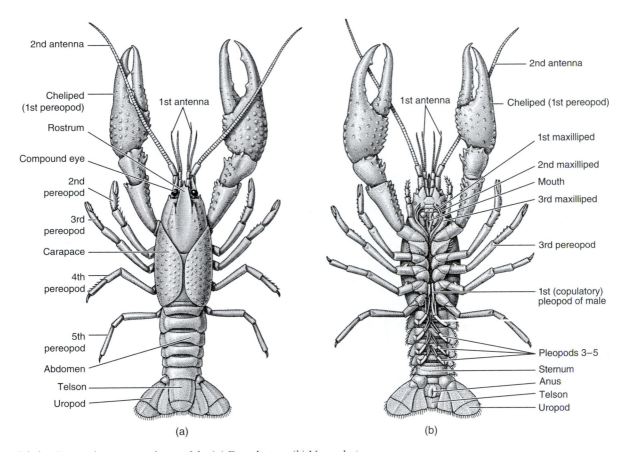

Figure 14.4 External structure of a crayfish. (*a*) Dorsal view. (*b*) Ventral view.

Carefully insert the tip of your scissors just below the carapace at the dorsolateral margin where the cephalothorax and abdomen meet. Cut anteriorly, being careful not to disturb the underlying gills. When you reach the groove that represents the point of fusion of the head and thorax, turn laterally and ventrally, continuing your cut until the lateral portion of the carapace can be lifted off. Note the gills attached to the base of the legs (fig. 14.5). Medial to the gills is the body wall. Water is circulated through the gill chamber by a process on the second maxilla. This process will be observed later.

The abdomen consists of six segments. Inside, powerful **flexor muscles** are responsible for the escape response of the crayfish. When alarmed, the crayfish thrusts the tip of its abdomen ventrally and anteriorly. These powerful, rapid strokes propel the crayfish posteriorly through the water. This can be demonstrated if living specimens are available. At the tip of the abdomen is a flat process, the **telson,** which bears the **anus** ventrally.

The appendages of the crayfish are biramous (fig. 14.6). Recall the terminology associated with the generalized

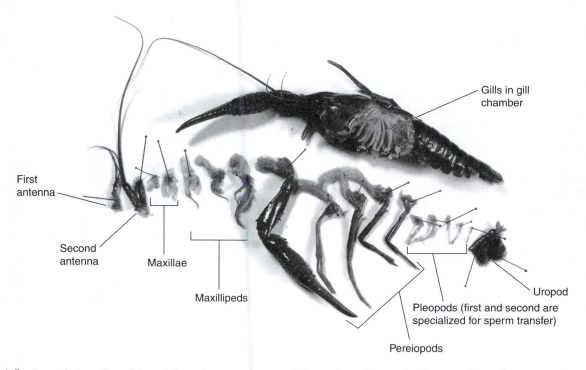

Figure 14.5 Lateral view of crayfish with lateral carapace removed. Appendages dissected and arranged in order.

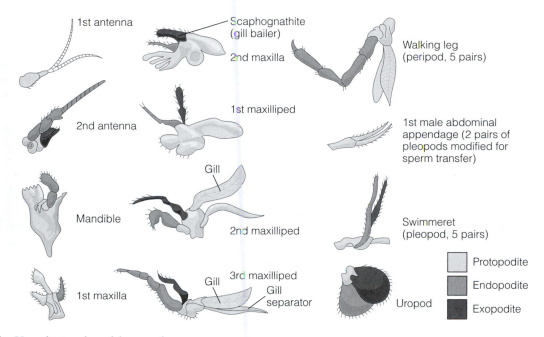

Figure 14.6 Homologies of crayfish appendages.

appendage. Beginning at the posterior tip of the abdomen, examine the **uropods** (see figs. 14.4 and 14.5). They arise from the last abdominal segment on either side of the telson. Notice the two leaflike segments (exopodite and endopodite) and the single basal segment (protopodite) to which they attach. The remaining abdominal segments have **pleopods** (swimmerets). These are typical biramous appendages. Note the three regions. In the male, the two pairs of pleopods closest to the cephalothorax are modified for sperm transfer. They are enlarged, hardened, and lie forward between the bases of the last walking legs. In the female, the anterior two pleopods are reduced in size, and all pleopods are used in drawing water across the eggs that develop attached to the ventral surface of the abdomen.

The cephalothorax bears walking legs, mouthparts, and antennae. These are best studied if removed. To remove an appendage, grasp it with forceps near its base. Loosen the appendage by probing at its attachment to the body wall with an insect pin or teasing needle. Then, with the forceps, manipulate the base of the appendage back and forth, pulling gently.

Working on the left side of the crayfish, and beginning at the posterior end of the cephalothorax, remove the next five appendages (see fig. 14.5). These are **pereiopods** (walking legs). In all of them the exopodite is lost. They are, therefore, secondarily uniramous. Note that the anterior three appendages are **chelate** (equipped with an opposable segment used in grasping). The anterior-most pair is greatly enlarged into defensive and offensive **chelae.** Note that, with some of the pereiopods, gills are attached at the basal segment and were removed with the leg. With others, the gills remain attached to the body wall. Observe the **genital pores** located at the basal segment of the fifth (male) or third (female) walking legs.

Five pairs of appendages are associated with the mouth. They have sensory, food-handling, and grinding functions. The posterior three pairs of mouth appendages are **maxillipeds.** Remove them posterior to anterior. Be sure to get the entire appendage. Note that the second and third bear a gill, but the first does not. Next are two pairs of **maxillae.** On the second maxilla, note that the exopodite bears a long, flat blade: the **gill bailer.** It functions in drawing water currents over the gills. The most anterior pair of mouth appendages are the **mandibles.** These are the primary organs of mastication. They are heavy, triangular structures bearing teeth along the inner margins. Probe around the base and carefully remove the mandible.

Look again at the first and second antennae. Both flagella of the first antenna derive from the exopodite. The long flagellum of the second antenna is the endopodite of that appendage, and a triangular plate just below the eye is the exopodite. At the base of the first antenna is the **statocyst,** an organ of equilibrium and balance. A sand grain, the **statolith,** within the cavity of the statocyst, stimulates these hairs in response to the pull of gravity. At the base of the second antenna is a ventral opening called the **nephridiopore.** It is the opening of the excretory system.

Internal Structure

Remove the carapace covering the dorsal cephalothorax by cutting 1 cm to the right of midsagittal. Extend the cut up to the rostum, then to the left, ending under the left compound eye. Keep the inner blade of your scissors close to the body wall at all times. Lift the carapace carefully, teasing it free from underlying tissue. As you gently lift, look underneath the carapace. About 1 cm posterior to the left compound eye you will see a large **mandibular muscle** that will need to be teased free of the carapace. Remove the portion of the body wall medial to the gills by cutting with scissors near the points of attachment of the legs (fig. 14.7).

Muscular System The mandibular muscles noted previously insert on and operate the mandibles. Just medial to the mandibular muscles, note the **gastric muscles** associated with stomach movements. Laterally, a band of muscles originates that runs posteriorly and dorsally, and eventually disappears under the dorsal aspect of the abdomen. These are **extensor muscles** of the abdomen. When they contract, they bring the abdomen back into line with the cephalothorax. They oppose the powerful **flexor muscles** of the abdomen. To see the remaining portion of the extensor and flexor muscles, remove the dorsal portion of the abdomen by cutting along each side of the abdomen to the base of each uropod. Lift the dorsal portion of the abdomen carefully, freeing it from underlying tissue. The extensor muscles attach to each abdominal segment and will tend to come up with the exoskeleton. Free them with a teasing needle as you proceed. After locating extensor muscles, note the large flexor muscles nearly filling the remainder of the abdomen. Alternate contraction of flexor muscles and extensor muscles cause the abdomen to be alternately flexed and extended, which propels a crayfish posteriorly in its escape response.

Circulatory System The **heart** is located in the posterior, dorsal third of the cephalothorax. It may be injected with red latex. The circulatory system of arthropods is open. Blood is pumped from the heart into vessels and then released into large tissue spaces called sinuses. After leaving these sinuses, blood flows to the gills, where it is oxygenated. Blood combines with the copper-containing pigment hemocyanin. Hemocyanin is dissolved in the plasma rather than being contained in cells. Carbon dioxide diffuses out of the blood at the gills. Surrounding the heart is a sac, the **pericardium.** It encloses the heart in a **pericardial sinus.** Blood from the gills is released into the pericardial sinus and then enters the heart through small openings, **ostia.** Remove the pericardium and any latex covering the heart. We will not attempt to trace the delicate vessels associated with the heart, although they can be seen in figure 14.7.

Reproductive System Paired **gonads** are just ventral to, and extend posterior to, the heart. Gonads are cream-colored and have a somewhat more uniform texture than the

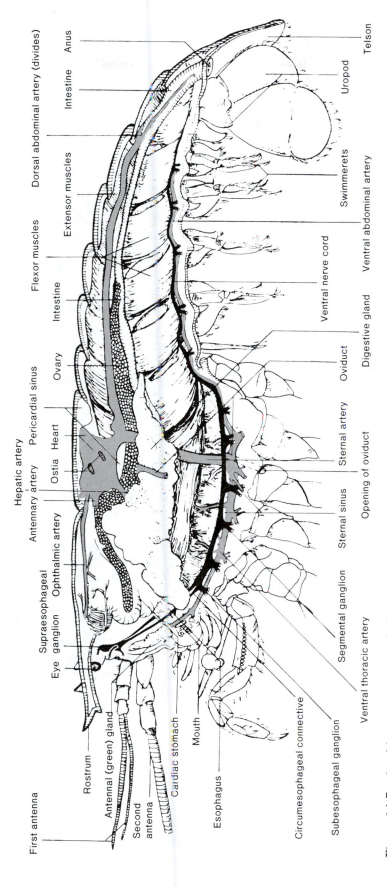

First antenna

Rostrum

Antennal (green) gland

Second antenna

Cardiac stomach

Mouth

Esophagus

Circumesophageal connective

Subesophageal ganglion

Ventral thoracic artery

Eye ganglion

Supraesophageal

Ophthalmic artery

Hepatic artery

Antennary artery

Ostia   Heart

Pericardial sinus

Ovary

Intestine

Flexor muscles

Extensor muscles

Dorsal abdominal artery (divides)

Intestine   Anus

Telson

Uropod

Swimmerets

Ventral abdominal artery

Ventral nerve cord

Digestive gland

Oviduct

Sternal artery

Opening of oviduct

Sternal sinus

Segmental ganglion

Figure 14.7 Crayfish, internal anatomy of a female.

207

digestive gland, which is just anterior to them. In the male, the sperm duct from each testis extends back to the genital opening at the base of the fifth pereiopod. In the female, the ovarian duct from each ovary leads to the genital opening at the base of the third pereiopod.

Digestive System The mouth leads to a short **esophagus** and then to the **stomach.** The stomach attaches just behind the rostrum anteriorly and by the gastric muscles posteriorly. By pushing gently on the lateral margin of the stomach, one can see the esophagus extending down to the mouth. Cut open the stomach middorsally. Inside, note the **gastric mill.** It consists of a set of teeth, two lateral and one dorsal, used in grinding food. These are modifications of the exoskeleton, which lines the anterior portion of the digestive tract. On the stomach wall you may also find calcareous **gastroliths.** These are stores of calcium carbonate. Prior to ecdysis, calcium carbonate is withdrawn into the blood from the old exoskeleton and stored at the stomach walls. After ecdysis, the calcium carbonate is reused in hardening the new exoskeleton. The large **digestive gland** noted earlier secretes digestive enzymes into the stomach. The **intestine** passes under the heart, then posteriorly just under the extensor muscles. Trace it to the telson. Locate the anus ventrally on the telson.

Excretory System Excretory organs of the crayfish are called **antennal glands,** or **green glands.** They are the dark-colored glands lying against the inner body wall at the base of the second antenna. Recall your observations of the nephridiopore on the ventral surface of the basal segment of the second antenna.

Nervous System In removing the pereiopods at their base, you have exposed the ventral nerve cord in the thoracic region. Looking from the left side, follow the nerve cord posteriorly. Follow it into the abdomen by freeing flexor muscles from the lateral and ventral body wall. Separate the nerve cord carefully from the muscle. Note the **segmental ganglia** (swellings) and the lateral branches that come off each ganglion. Note that you can now see the **ventral abdominal artery** running below the nerve cord. Return to the thoracic region and follow the ventral nerve cord anteriorly. Just posterior to the esophagus is a larger ganglion, the **subesophageal ganglion.** From it, two circumesophageal connectives go around either side of the esophagus and end at the **supraesophageal ganglion** (loosely called the brain). The supraesophageal ganglion can be seen inside the anterior head wall. The segmental ganglia are integrative areas for sensory and motor functions in their respective regions. The supraesophageal and subesophageal ganglia result from the fusion of multiple-head ganglia and are important in the integration of sensory and motor functions of the mouthparts, antennae, compound eyes, statocysts, and so forth. Recall your earlier observations of these structures.

Observe demonstrations of other malacostracans. ▲

Other Crustaceans

▼ Observe demonstrations of other crustaceans, including the following:

Class Branchiopoda—fairy shrimp, water fleas, and brine shrimp. Note flattened, leaflike appendages used in filter feeding, respiration, and often locomotion.

Class Maxillopoda, Subclass Thecostraca—barnacles. In barnacles, a planktonic larval stage is followed by settling to the substrate and metamorphosis to a sessile adult. Observe any living or preserved specimens. Can you see why they were thought to be molluscs by some early zoologists?

Class Maxillopoda, Subclass Copepoda—copepods. Copepods are among the most abundant animals in the world. They comprise an important component of marine food webs. ▲

14.5 LEARNING OUTCOMES REVIEW—WORKSHEET 14.2

14.6 SUBPHYLUM MYRIAPODA

LEARNING OUTCOMES

1. Describe characteristics of the subphylum Myriapoda.
2. Recognize members of the classes Diplopoda and Chilopoda.

Myriapods are distinguished from other arthropods by having one pair each of antennae, mandibles, and uniramous appendages.

Class Diplopoda—Millipedes

Millipedes are characterized by having one pair of antennae, two pairs of short legs per apparent segment, and a body that is round in cross section. Millipedes live on decaying plant and animal material and thus are very important in the decomposition of forest-floor litter. They are widely distributed terrestrial animals.

▼ Examine any demonstrations available. ▲

Class Chilopoda—Centipedes

Centipedes are active predators, feeding on a variety of other invertebrates. They have one pair of antennae, one pair of relatively long legs per segment, and their body is oval in cross section. The appendages of the first segment of the body are formed into poison claws that are used in feeding. Most are harmless to humans. Centipedes are often found beneath fallen trees on the forest floor. One group is common in human dwellings in North America.

▼ Examine any demonstrations available. ▲

14.6 LEARNING OUTCOMES REVIEW—WORKSHEET 14.2

14.7 SUBPHYLUM HEXAPODA

LEARNING OUTCOMES

1. Describe characteristics of members of the subphylum Hexapoda.
2. Identify external structures of the grasshopper or cockroach.
3. Identify representatives of the common insect orders.
4. Describe three forms of insect development.

Members of the subphylum Hexapoda are characterized by a body divided into head, thorax, and abdomen. They have five pairs of head appendages and three pairs of uniramous appendages on the thorax.

CLASS INSECTA

Members of the class Insecta have mouthparts that are exposed and project from the head. Their mandibles have two points of articulation. Members of a second hexapod class, Entognatha, have mouthparts that are hidden within the head and mandibles with a single point of articulation.

If the success of any animal group is measured by numbers of species and ecological adaptability, the insects are the most successful eukaryotes on this planet. With the exception of the marine environment, they have successfully adapted to nearly every habitat the earth has to offer. As with all arthropods, the exoskeleton is the primary reason for their success. Adaptations to the terrestrial environment include the incorporation of waxes into the exoskeleton, which waterproofs the exoskeleton and helps prevent desiccation. Many insects can thus live with little or no external source of water. In addition, insects are the only nonvertebrates to exploit air; the exoskeleton is modified to form wings without sacrificing any appendages.

In spite of their success, the insect body plan is remarkably constant. There are three tagmata: **head, thorax,** and **abdomen.** On each of the three segments of the thorax is a pair of appendages (hence the subphylum name, Hexapoda). Generally, two pairs of wings are carried on the second and third thoracic segments. The head bears one pair of compound eyes; zero, two, or three ocelli; one pair of antennae; and a series of mouthparts, three pairs of which derive from appendages associated with the primitive metameres of the head. In some orders, mouthparts are specialized for various feeding habits but are still built upon the same basic plan. The abdomen lacks walking legs and is associated with visceral functions (reproduction, digestion, and excretion).

The Grasshopper and Cockroach

Traditionally, in introductory courses the grasshopper is used to study insect structure. Your instructor may ask you to use a cockroach in addition to the grasshopper inasmuch as the cockroach is less specialized. The following discussion can be applied to either.

External Structure ▼ On your specimen, note the division of the body into three tagmata: head, thorax, and abdomen (fig. 14.8). The exoskeleton of the insect is very similar to that of the crustacean. Note the hardened plates (sclerites) with membranous areas between them. The hardening of the cuticle of the insect is different from that of the crustacean. Hardening is the result of polymerization of protein (it is said to be sclerotized), rather than the deposition of calcium salts.

On the head, note the single pair of **antennae,** the **compound eyes,** and the **mouthparts** (fig. 14.9). Using a dissecting microscope, examine the compound eyes closely. Note the many facets that make up each eye. Examine the area between the compound eyes for **ocelli.** As with the crayfish, the compound eyes are adapted to detect movement. The ocelli are used to detect changes in light intensity and do not form images.

Examine the mouthparts using a dissecting microscope. We will remove them as we did with the crayfish appendages. As with the crayfish, the mouthparts derive from segmental appendages. The lower lip is the **labium.** It represents the fusion of one pair of appendages. Note the **labial palps** on each side. The labium is primarily sensory. Remove the labium as follows. Fold the labium posteriorly with forceps and pierce the exoskeleton along its joint with the head. Gently pull it free using forceps. Just above the labium is a pair of **maxillae.** Note the palp on each maxilla. Maxillae are cutting and sensory structures. Remove them by probing at their point of articulation with the head and note the cutting surface on the inner margin. Above the maxillae are the primary grinding structures, the **mandibles.** The mandibles are highly sclerotized and difficult to remove. You will need to probe around the base of one mandible with a teasing needle to loosen it. Note the toothed margin. The upper lip is the **labrum.** It is sensory and not derived from paired appendages. Remove the labrum as you did with the labium. Inside the mouth cavity a tonguelike structure, the **hypopharynx,** can now be seen. It is also sensory.

Examine the thorax (see fig. 14.8). It consists of three segments devoted primarily to locomotion. The **prothorax** bears the first pair of legs. The second segment, the **mesothorax,** bears the second pair of legs and the first pair of wings. In both the grasshopper and the cockroach, the first pair of wings is narrow and leathery. The third segment, the

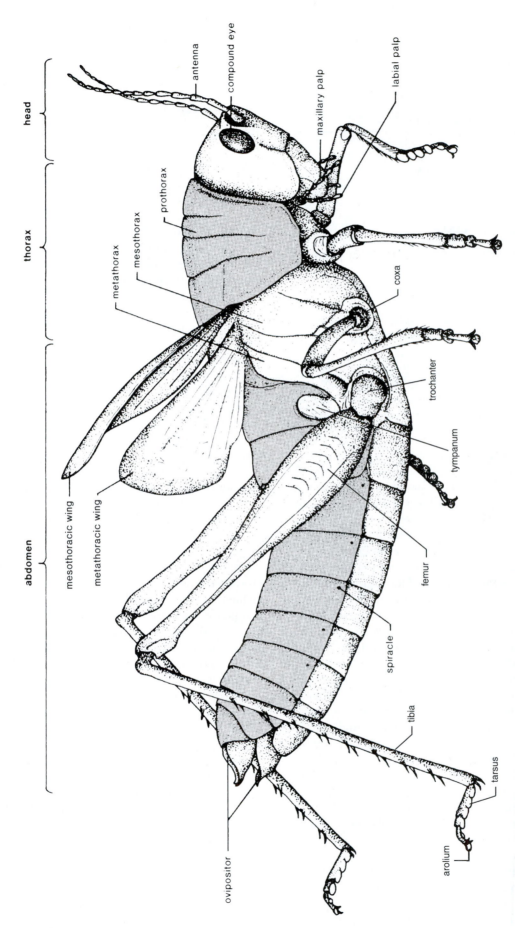

head

thorax

abdomen

antenna

compound eye

maxillary palp

labial palp

prothorax

mesothorax

metathorax

coxa

trochanter

tympanum

femur

spiracle

tibia

tarsus

arolium

ovipositor

mesothoracic wing

metathoracic wing

Figure 14.8   Grasshopper, lateral view of a female *Romalea*.

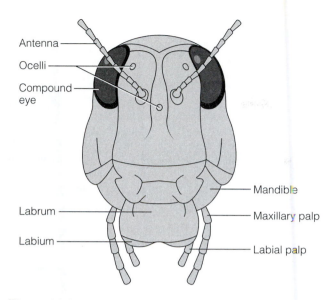

Figure 14.9 Grasshopper, anterior view of the head.

Antenna

Ocelli

Compound
eye

Labrum

Labium

Mandible

Maxillary palp

Labial palp

metathorax, bears the third pair of legs and the second pair of wings. These wings are broad and membranous. The legs of an insect are divided into five main segments joined together by membranous joints (see fig. 14.8). The **coxa** is a small segment that attaches at the ventrolateral margin of the body wall. The next three segments are the **trochanter, femur,** and **tibia.** The last segment is the **tarsus.** It is usually subdivided into smaller tarsal segments and bears terminal claws (two in the grasshopper and cockroach). Between the claws of the grasshopper is a fleshy pad, the **arolium,** which is used to help grip smooth surfaces. The legs of insects are variously modified with spines, setae, and ridges that may be used for specialized functions (e.g., gripping surfaces, swimming, and making sound). In grasshoppers the metathoracic leg is enlarged and muscular and is used for jumping. In male *Romalea* grasshoppers, the inner surface of the metathoracic leg bears a row of small spines that is rubbed against the lower edge of the front wing to produce sounds used in mating. Other insects produce sounds by rubbing wings together or other legs against wings. This rubbing form of sound production is called **stridulation.**

The abdomen of insects usually consists of 10 or 11 segments. All except the terminal segments are free of appendages. Note the openings on the lateral surface of each abdominal segment. These openings are called **spiracles.** They are the external openings of the insect's gas-exchange system. Spiracles lead to chitin-lined tubes called **tracheae,** which branch and rebranch throughout the insect's body. A tympanum is located laterally on each side of the first abdominal segment. It is the auditory organ of the grasshopper. The terminal segments often bear sensory projections called **cerci** (sing. cercus). Both the cockroach and grasshopper show these structures. An ovipositor is frequently present in females. The female grasshopper has an ovipositor consisting of pairs of lateral pointed plates at the tip of the abdomen.

The Insect Orders

Examine demonstrations showing representatives of the following major insect orders. Note the characteristics that distinguish each. The mouthparts of insects have been especially important in their success. Note where mouthparts have been adapted for different feeding methods (e.g., piercing, sucking, lapping). ▲

Thysanura Bristletails. Wingless; tapering abdomen; scales on the body; terminal tail filaments; long antennae.

Ephemeroptera Mayflies. Elongate; abdomen with two or three tail filaments; wings held above the body at rest; short antennae.

Odonata Dragonflies and damselflies. Elongate; membranous wings; abdomen long and slender; compound eyes occupying most of the head.

Mantodea Mantids. Prothorax long; prothoracic legs long and armed with strong spines for grasping prey.

Blattaria Cockroaches. Body oval and flattened; head concealed from above by a shieldlike extension of the prothorax.

Isoptera Termites. White and wingless (except in dispersal stages); antlike.

Dermaptera Earwigs. Elongate; chewing mouthparts; threadlike antennae; abdomen with unsegmented, forcepslike cerci.

Phasmida Walking sticks, leaf insects. Body elongate and sticklike; wings reduced or absent; some tropical forms are flattened and leaflike.

Orthoptera Grasshoppers, crickets. Forewing long, narrow, and leathery; hind wing broad and membranous; cerci present.

Phthiraptera Sucking and chewing lice. Small, wingless ectoparasites of birds and mammals; body dorsoventrally flattened.

Hemiptera Bugs, cicadas, leafhoppers, aphids. Piercing-sucking mouthparts; mandibles and first maxillae styletlike and lying in a grooved labium; wings membranous; hemimetabolous metamorphosis.

Thysanoptera Thrips. Small bodied; sucking mouthparts; wings narrow and fringed with long setae; plant pests.

Neuroptera Lacewings, snakeflies, ant lions. Wings membranous and net veined; hind wings held rooflike over body at rest; antennae long and threadlike.

Coleoptera Beetles. Forewings sclerotized, forming cover over abdomen (elytra).

Trichoptera Caddisflies. Mothlike with setae-covered antennae; chewing mouthparts; wings covered with setae and held rooflike over abdomen at rest; larvae aquatic and often dwell in cases that they construct.

Lepidoptera Moths and butterflies. Wings broad and covered with scales; mouthparts formed into a sucking tube.

Diptera Flies. Mesothoracic wings well developed, metathoracic wings reduced to knoblike halteres; mouthparts various modified, never chewing.

Siphonaptera Fleas. Laterally flattened; sucking mouthparts; jumping legs; parasites of birds and mammals.
Hymenoptera Ants, bees, and wasps. Anterior portion of abdomen constricted into a threadlike "waist"; wings (when present) membranous.

Insect Metamorphosis

Evolution of insects has resulted in the divergence of immature and adult body forms and habitats. If immature and adult insects do not compete for the same food resources and the same space, greater numbers of individuals can be supported in an area. Developmental patterns of insects reflect degrees of divergence between immatures and adults and are classified into three (four by some zoologists) categories.

Ametabolous metamorphosis occurs in the Thysanura and other orders. The primary differences between adults and immatures are body size and degree of sexual development. Both adults and immatures are wingless. The number of molts in the ametabolous development of a species is variable, and molting continues after sexual maturity is reached.

Hemimetabolous metamorphosis involves a species-specific number of molts between egg and adult stages. The immature stages **(instars)** gradually take on the adult form. This includes the development of external wing pads (except in those insects, such as the lice, that have secondarily lost wings), the attainment of adult body size and proportions, and the development of genitalia. No further molts normally occur once adulthood is reached. Except for their smaller size, sexual immaturity, and incompletely developed wings, immatures (called **nymphs**) usually resemble adults. When immature stages are aquatic, they often have gills (e.g., mayflies, order Ephemeroptera; dragonflies, order Odonata). These immatures are called **naiads.** Those insects listed on page 211 between, but not including, the Thysanura and the Neuroptera are hemimetabolous.

Holometabolous development occurs in the Neuroptera through the Hymenoptera in the list on pages 211 and 212. Eggs hatch into **larvae** that are very different from the adult in body form, behavior, and habitat. Larvae molt through a species-specific number of larval instars. The last larval molt forms the **pupal stage.** The pupal stage is often visualized as a period of apparent inactivity, but is actually a time of radical cellular change. The pupal stage may be enclosed in a protective case (cocoon) or by the last larval exoskeleton (chrysalis, puparium). The **adult,** which is usually winged, emerges after the pupal stage.

Hemimetabolous Metamorphosis ▼ You will observe paurometabolous development of the hemipteran *Oncopeltus fasciatus* (milkweed bug). Your instructor will provide you with some eggs in a cotton substrate. Preserve two to five eggs in a vial of 70% ethanol. Place the remaining eggs into a container; provide the container with raw, shelled sunflower seeds and water (as directed by your instructor). Cover tightly with a double layer of cheesecloth. Watch your culture over the next eight weeks. An approximate schedule of molts is outlined in table 14.1 in worksheet 14.1,

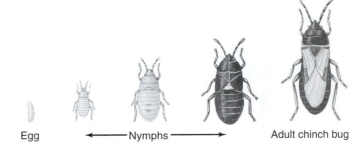

Egg ◄——— Nymphs ———► Adult chinch bug

Figure 14.10 Life cycle of the chinch bug *Blissus leucopterus*.

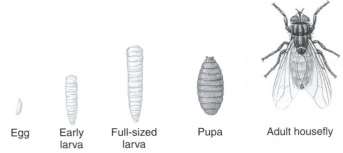

Egg   Early larva   Full-sized larva   Pupa   Adult housefly

Figure 14.11 Life cycle of the house fly *Musca domestica*.

and a similar sequence of molts is shown in figure 14.10. As the various instars appear, record the date in table 14.1 and remove and preserve one representative of each stage. At the end of the developmental cycle, you will have eggs, five nymphal instars, and the adult. Be prepared to hand in your vial when the exercise is completed.

Holometabolous Metamorphosis You will use the fruit fly *Drosophila melanogaster* to illustrate holometabolous metamorphosis. Figure 14.11 illustrates holometabolous metamorphosis of the housefly (*Musca domestica*). Your instructor will give you a container of medium and provide you with anesthetized fruit flies (see fig. 4.1). Place the flies into the container and cover. Eggs and first and second instars will be found in the medium. Third instars and pupae will be found on the sides of the container. The life cycle is relatively short (two weeks from egg to adult), and larval instars proceed rapidly. You will need to check your cultures frequently (as indicated in table 14.2 in worksheet 14.1). Because eggs are small and the first two instars crawl through the medium, these stages are difficult to find. As the life cycle proceeds, you may see larvae next to the side of your vial. You will certainly see tracks the larvae make while crawling through the medium. Use a dissecting microscope to help with these observations. Because egg and larval stages are difficult to find, you will not be asked to hand in a representative of each stage. ▲

14.7 LEARNING OUTCOMES REVIEW—WORKSHEET 14.2

14.8 EVOLUTIONARY RELATIONSHIPS

LEARNING OUTCOME

1. Describe what is known of the evolutionary relationships among the arthropod subphyla.

▼ The arthropods—along with two closely related phyla, the Onychophora and Tardigrada—form a monophyletic lineage often designated by the name Panarthropoda. Precise evolutionary relationships within this lineage are controversial. The phylogenetic tree shown in figure 14.12 shows one interpretation of these relationships. Examine the diagram and answer Analytical Thinking questions 5–7 in worksheet 14.2.▲

14.8 LEARNING OUTCOMES REVIEW—WORKSHEET 14.2

WORKSHEET 14.1   Insect Development

TABLE 14.1   Development of *Oncopeltus fasciatus*

Week Number/Date	Instar/Body Form	Description
1–2 _____	Egg	Light yellow early; bright orange prior to hatching.
2–3 _____	First Instar	Small, orange, no wing pads obvious.
3–4 _____	Second Instar	Larger with four wing pad spots on dorsal thorax.
4–5 _____	Third Instar	Larger with wing pads just reaching metathoracic legs.
5–6 _____	Fourth Instar	Longer with wing pads just beyond metathoracic legs. Two dorsal black spots at tip of abdomen.
6–8 _____	Fifth Instar	Wing pads reaching well beyond metathoracic legs. Three black spots on dorsal surface of posterior abdomen.
8–9 _____	Adult	Wings fully developed.

TABLE 14.2   Development of *Drosophila melanogaster*

Day/Date	Instar/Body Form	Description
1–2 _____	Egg	Small, white, with two fingerlike projections at one end.
2–4 _____	First Instar	Small maggot in media.
4–5 _____	Second Instar	Larger maggot in media.
5–7 _____	Third Instar	Larger maggot, found on sides of vial.
7–14 _____	Pupa	Attached to side of vial. Turns from light to dark as development proceeds.
14–16 _____	Adult	Wings and adult body form fully developed.

WORKSHEET 14.2   Arthropoda

14.1   Evolutionary Perspective—Learning Outcomes Review

1. Members of which one of the following phyla are most closely related to members of the Arthropoda?
   a. Rotifera
   b. Nematoda
   c. Annelida
   d. Mollusca

14.2   Phylum Arthropoda—Learning Outcomes Review

2. A region of an arthropod's body devoted to locomotion is often called the
   a. thorax
   b. abdomen
   c. opisthosoma

3. In the arthropods, adults are often different in form as compared to larvae. The change from the larval form to the adult form reduces competition for resources and is called
   a. tagmatization.
   b. metamerism.
   c. metamorphosis.

14.3   Subphylum Trilobitomorpha—Learning Outcomes Review

4. All of the following are trilobite characteristics except one. Select the exception.
   a. body divided into three longitudinal lobes.
   b. cephalon, trunk, pygidium
   c. two pairs of antennae
   d. biramous appendages

14.4   Subphylum Chelicerata—Learning Outcomes Review

5. All of the following structures are associated with the prosoma of a horseshoe crab except one. Select the exception.
   a. compound eyes
   b. simple eyes
   c. book gills
   d. chelicerae
   e. pedipalps

6. Which of the following appendages of a horseshoe crab is not chelate?
   a. pedipalps
   b. chelicerae
   c. first walking leg
   d. last walking leg

7. The prosoma and opisthosoma of a spider are joined by
   a. a pedicel.
   b. the pedipalps.
   c. spinnerettes.
   d. book lungs.

8. Spiders are members of the class _____, which also includes harvestmen, ticks, and scorpions.
   a. Pychnogonida
   b. Arachnida
   c. Araneae
   d. Merostomata

14.5 Subphylum Crustacea—Learning Outcomes Review

9. Which of the following characteristics is sufficient to identify an arthropod (extant or extinct) as a crustacean?
   a. tagmatization
   b. exoskeleton and jointed appendages
   c. two pairs of antennae
   d. biramous appendages

10. This portion of a biramous appendage attaches to a crustacean's body wall.
    a. exopodite
    b. endopodite
    c. protopodite

11. The gill chamber of a crayfish is located
    a. just inside the body wall of a crayfish.
    b. outside the body wall but beneath an arching carapace.
    c. in the abdominal region.

12. In a male crayfish, these appendages are modified for sperm transfer and can be used to distinguish males from females.
    a. The first two pairs of periopods.
    b. The maxillipeds.
    c. The last two pairs of pleopods.
    d. The first two pairs of pleopods.

13. The anus of a crayfish is located on the
    a. uropods.
    b. telson.
    c. basal segment of the fifth periopod.
    d. basal segment of the second antenna.

14. A structure called the scaphognathite is located on the _____ and is used in circulating water through the gill chamber.
    a. mandibles
    b. second maxilliped
    c. second maxilla
    d. first pair of gills

15. Gastroliths are calcium carbonate stores associated with the crayfish
    a. stomach.
    b. digestive gland.
    c. intestine.
    d. antennal glands.

14.6 Subphylum Myriapoda—Learning Outcomes Review

16. Members of the class _____ are oval in cross section, possess one pair of legs per segment, and are fast moving predators.
    a. Chilopoda
    b. Diplopoda
    c. Insecta

14.7 Subphylum Hexapoda—Learning Outcomes Review

17. The anterior to posterior order of mouthparts of a grasshopper is
    a. labium, labrum, mandibles, maxillae.
    b. maxillae, mandibles, labium, labrum.
    c. labrum, mandibles, maxillae, labium.
    d. mandibles, maxillae, labrum, labium.

18. The wings of a grasshopper are attached to the
    a. prothorax and mesothorax.
    b. mesothorax and metathorax.
    c. prothorax and metathorax.

19. Grasshoppers and crickets are members of the insect order
    a. Odonata.
    b. Blattaria.
    c. Orthoptera.
    d. Coleoptera.

20. Immatures of insects with hemimetabolous metamorphosis are called
    a. nymphs.
    b. larvae.
    c. naiads.
    d. Either a or c is correct depending on the insect order being considered.

14.8　Evolutionary Relationships—Learning Outcomes Review

21. Arthropods, along with the _____ form a monophyletic group referred to as Panarthropoda.
    a. nematodes
    b. rotifiers
    c. onychophorans
    d. tardigrades
    e. Both c and d are correct.

14.1　Evolutionary Perspective—Analytical Thinking

1. The Annelida and the Arthropoda were once considered to be closely related. What features of body organization in the two groups may have suggested that conclusion and how would one explain those similarities in light of modern cladistic and molecular analyses?

14.4 Subphylum Chelicerata—Analytical Thinking

2. Compare the structures and functions of the first two pairs of appendages found in members of the classes Merostomata and Arachnida. Use your observations of the horseshoe crab and the garden spider as a basis for your comparisons.

14.5 Subphylum Crustacea—Analytical Thinking

3. Compare the concepts of serial homology and homology. How is serial homology reflected in the structure of crayfish appendages that you studied in this exercise?

14.7 Subphylum Hexapoda—Analytical Thinking

4. Compare the mouthparts of the grasshopper to those of the crayfish. Note major similarities and differences. Do you think that the similarities reflect homologies in these structures?

14.8 Evolutionary Relationships—Analytical Thinking

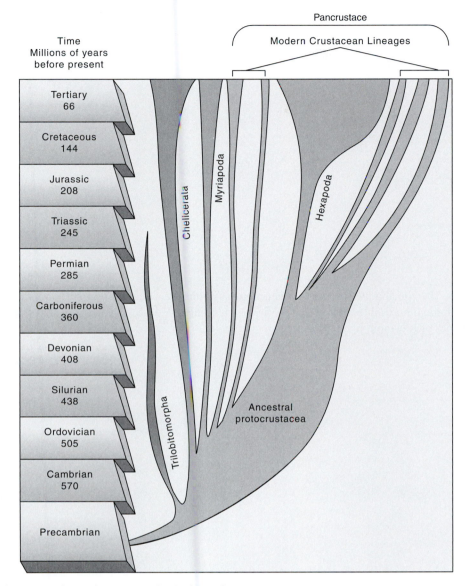

Figure 14.12 Evolutionary relationships among the Arthropoda.

5. Older interpretations of arthropod phylogeny suggested that the trilobites were ancestral to other arthropods, at least the Crustacea and the Chelicerata. What does the phylogeny represented in figure 14.12 suggest regarding the status of the trilobites in arthropod evolution?

6. The Myriopoda and the Hexapoda show numerous structural similarities including tracheal systems, uniramous appendages, and Malpighian tubules. Based on the phylogeny in figure 14.12, do you think that these characters represent homologies or is it more likely that they are convergent characters? Explain.

7. Assess the validity of the subphylum Crustacea based on the phylogeny in figure 14.12.

Exercise 15

Echinodermata

Prelaboratory Quiz

Study this week's laboratory exercise and then complete the following quiz to assess your preparation for the laboratory.

1. The echinoderms are most closely related to the
   a. Annelida.
   b. Chordata.
   c. Arthropoda.
   d. Mollusca.

2. The echinoderms possess
   a. biradial symmetry.
   b. bilateral symmetry.
   c. pentaradial symmetry.
   d. asymmetry.

3. All of the following are characteristics of the echinoderms EXCEPT
   a. water-vascular system.
   b. calcium carbonate endoskeleton.
   c. oral and aboral surfaces.
   d. cephalization.

4. Members of the class Ophiuroidea include the
   a. basket stars.
   b. sea stars.
   c. sea cucumbers.
   d. sea urchins.

5. True/False   Members of the class Holothuroidea include the sea cucumbers.

6. True/False   A chewing apparatus called Aristotle's lantern is found in members of the class Ophiuroidea.

7. True/False   Pyloric cecae are digestive glands found in the arms of members of the class Asteroidea.

8. True/False   In the water-vascular system of echinoderms, tube feet attach to the lateral canal by way of the ring canal.

9. True/False   The madreporite of sea stars is the external opening of the water-vascular system.

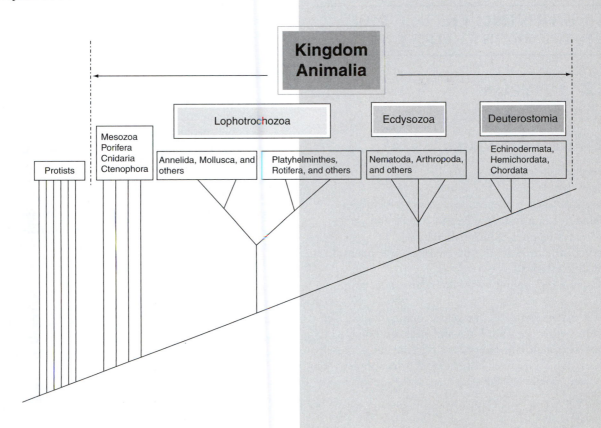

15.1 EVOLUTIONARY PERSPECTIVE

LEARNING OUTCOME

1. Describe the relationship of the echinoderms to other animal phyla.

Most zoologists believe that echinoderms share a common ancestry with the chordates. Much of the evidence for these evolutionary ties comes from similarities in the early embryological development of members of these phyla and molecular data. These shared developmental characteristics are the basis for grouping these phyla (as well as a few other smaller phyla) together as **deuterostomes** (see p. 223). Unfortunately, no fossils have been discovered that document a common ancestor for these phyla or explain how the deuterostomate lineage derived from ancestral diploblastic or triploblastic stocks.

Adult echinoderms have a form of radial symmetry called pentaradial symmetry (see the following discussion). Ancestors of all of the deuterostomate phyla were bilaterally symmetrical. Some zoologists believe that the earliest echinoderms were also bilaterally symmetrical. Evidence for this comes from the observation that echinoderm larval stages are bilaterally symmetrical. In addition, certain extinct echinoderms, called carpoids, were bilaterally symmetrical or asymmetrical. Other zoologists are reconsidering this conclusion. Precambrian-age fossils, which appear to be echinoderm fossils, have recently been discovered that show evidence of pentaradial symmetry.

15.1 LEARNING OUTCOMES REVIEW—WORKSHEET 15.1

15.2 PHYLUM ECHINODERMATA

LEARNING OUTCOMES

1. Describe characteristics of members of the phylum Echinodermata.
2. Characterize each of the classes studied, including habitat, feeding and reproductive habits, and distinctive features of body organization.
3. Identify structures studied on preserved specimens and slide materials.
4. Recognize representatives of the echinoderm classes.

The echinoderms share many embryological features with the chordates and thus are believed to be the major nonchordate phylum most closely related to the chordates. Based on the morphology of living forms, however, they are very different from any other phylum.

The echinoderms are marine and include sea stars, sea urchins, sea cucumbers, and sea lilies. All echinoderms show **pentaradial symmetry** arranged around an oral-aboral axis. Radial symmetry was described earlier as it applied to the Cnidaria. In the echinoderms, body parts are arranged in "fives" around the oral-aboral axis. Echinoderms are dioecious with external fertilization. In development, a bilateral larva metamorphoses into the radial adult. This is evidence that the echinoderm pentaradial symmetry is a secondary adaptation to a sedentary lifestyle and that echinoderms are descended from bilateral ancestors.

Two other functionally related characteristics of the echinoderms are a **water-vascular system** and a **calcium carbonate endoskeleton** made of numerous interlocking ossicles. The ossicles frequently protrude through the thin body wall, hence the name Echinodermata (spiny skin). The ossicles may form a rigid test (as in the sea urchin) or may be reduced to microscopic plates (as in the sea cucumber). The water-vascular system is a series of water-filled canals to which tube feet attach. Tube feet protrude through the calcareous skeleton and may be modified for locomotion, food gathering, and attachment, and they also function in respiration. The skeleton anchors the tube feet firmly in the body wall so they can be effective locomotor structures.

CLASSIFICATION (MAJOR LIVING FORMS)*

Class Crinoidea
Class Asteroidea
Class Ophiuroidea
Class Echinoidea
Class Holothuroidea

CLASS ASTEROIDEA

Characteristics: Star-shaped; rays not distinctly set off from central disk; ambulacral grooves with tube feet; suction disks on tube feet; pedicellariae present. Example: *Asterias*. The asteroids are the familiar sea stars. They are often found on hard substrates in marine environments, but some species are also found in sand or mud. They feed on snails, bivalves, crustaceans, polychaetes, corals, detritus, and a variety of other food items. Sea stars are often brightly colored in red, orange, blue, or gray. *Asterias* is a common orange sea star found along the Atlantic coast of North America and is studied in the following exercise.

External Structure

▼ Examine a preserved specimen of *Asterias* and note the **oral** and **aboral surfaces** (fig. 15.1). On the aboral surface, note the opening to the water-vascular system, the **madreporite.** The body is covered by a thin epidermis through

*This listing reflects a phylogenetic arrangement; however, in the exercises that follow, the echinoderms that are familiar to most students are studied first.

Figure 15.1 Sea star, oral view of *Asterias*.

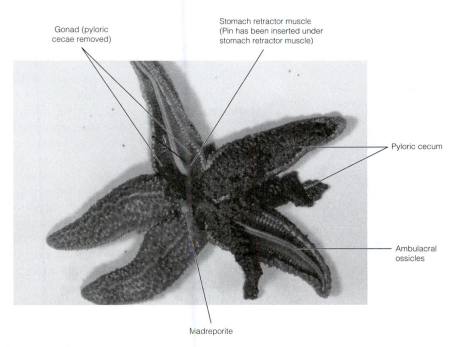

Figure 15.2 Internal structure of a sea star.

which modifications of the skeleton protrude. Using a dissecting microscope, note the calcareous **spines** and, around their bases, the pincerlike **pedicellariae.** The pedicellariae are primarily used in cleaning the body surface. Fleshy projections on the surface of the body are **dermal branchiae.**

These, along with the tube feet, allow for gas exchange. A small, pigmented **eyespot** is located at the tip of each ray but is difficult to see in preserved specimens.

Turn your specimen over to its oral surface. Note that each ray has a groove running down its length. Along either

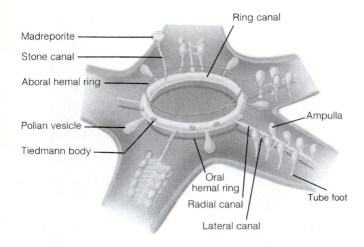

Figure 15.3 Sea star water-vascular system.

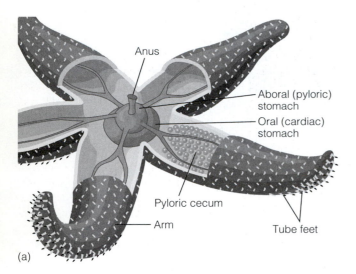

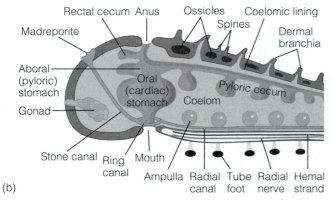

Figure 15.4 Digestive structures in a sea star. A mouth leads to a large cardiac (oral) stomach and a pyloric (aboral) stomach. Pyloric cecae extend into each arm. (*a*) Aboral view. (*b*) Lateral view through the central disk and one arm.

side of this **ambulacral groove** are rows of **tube feet.** Note the suction disks at their tips. Part the tube feet along the midline to see the groove more clearly. The central mouth is surrounded by larger, movable **oral spines.**

Internal Structure

Again, place the specimen aboral side up. With a pair of sharp scissors, pierce the body wall along the lateral surface of one ray. Carefully cut along the entire margin of that ray and then around the top of the central disk, leaving the madreporite attached to the body. Lift the body wall gently, freeing any tissue that adheres to the portion of the body wall being removed using the blade of your scalpel. Use figures 15.2 through 15.4 to aid in the following observations.

Note the large **pyloric cecum** (digestive glands) in the ray. In following the pyloric cecum toward the central disk, note that they unite with a membranous sac, the **stomach.** The stomach consists of two parts. Pyloric cecae attach to the **pyloric portion** of the stomach. The pyloric portion may have been removed with the body wall covering the central disk. The larger **cardiac portion** is easily seen. A rudimentary intestine and anus are usually left attached to the portion of the body wall removed.

When a sea star feeds, it wraps itself around the ventral margin of a bivalve. The tube feet attach to the outside of the shell, and the body-wall musculature forces the valves apart. When the valves are opened about 0.1 mm, the cardiac portion of the stomach is everted through the mouth and into the bivalve shell by increased coelomic pressure. Digestive enzymes are released from the pyloric cecae and partial digestion occurs within the bivalve shell. This weakens the bivalve's adductor muscles, and the shell eventually opens completely. Partially digested tissues are taken into the pyloric portion of the stomach, and into pyloric cecae, for further digestion and absorption.

Remove the pyloric cecae from the ray. Note the ridge of ossicles that represents the aboral side of the ambulacral groove. Lying along either side of the ambulacral ridge near the central disk is a thin strand of muscle. This muscle attaches to the ossicles of the ambulacral ridge and to the stomach wall. These **stomach retractor muscles** are used in returning the stomach to the body cavity after feeding (see fig. 15.2). At the base of the ray, note the **gonads.** Recall that sexes are separate. The sex of an individual cannot be determined by a general inspection of the gonads. Gonads open separately through aboral pores.

In the ray, note also the internal manifestations of the tube feet. Each tube foot consists of a tube extending to the outside, with a bulblike **ampulla** at its internal apex. Contraction of the ampulla forces water into the tube and causes the tube foot's extension. Each tube foot connects via a **lateral canal** to a **radial canal** running within the ambulacral ossicles. If injected, the radial canal can be seen in the ambulacral groove (oral surface). Each radial canal connects to a **ring canal** surrounding the stomach (see fig. 15.4b). The madreporite connects to the ring canal via a **stone canal.**

Also present in the ambulacral groove are a **radial nerve** and a **radial blood vessel.** These can be seen in a prepared cross section. Observe a slide under low power, noting these and other structures described earlier (fig. 15.5).

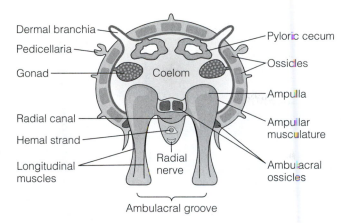

Figure 15.5 The body wall and internal anatomy of a sea star arm. A cross section through one arm of a sea star shows the structures of the water-vascular system and the tube feet extending through the ambulacral groove.

Figure 15.6 Class Echinoidea: Tests and live specimens of *Strongylocentrotus purpuratus*.

Observe a slide showing sea star larvae. The development that follows external fertilization results in a free-swimming bilateral larva called the **bipinnaria**. The bipinnaria larva is followed by a second larva, the **brachiolaria**. The mature brachiolaria larva attaches to the substrate and metamorphoses into an adult. ▲

CLASS OPHIUROIDEA—BRITTLE STARS, BASKET STARS

Characteristics: Arms sharply marked off from the central disk; body cavity limited to central disk; tube feet without suckers. Examples: *Ophiothris*, *Ophioderma*. Brittle stars and basket stars are located in all oceans, at all depths. Brittle stars have long, snakelike, unbranched arms; basket stars have arms that branch repeatedly. The arms contain a reduced coelomic space, ossicles, and large muscles. They are used to pull the animal along the substrate and sweep the substrate to collect prey or decaying animal matter. Some ophiuroids are filter feeders and wave their arms in the water to trap plankton on mucus-covered tube feet.

▼ Observe any demonstrations that are available. ▲

CLASS ECHINOIDEA—SEA URCHINS, SAND DOLLARS

Characteristics: Globular or disk-shaped, without rays; movable spines; skeleton (test) of closely fitting plates. Example: *Arbacia*. Echinoids are widely distributed in all oceans, at all depths. Sea urchins are specialized for living on hard substrates, often wedging themselves into crevices and holes in rock or coral. Sea urchins feed on algae, coral polyps, and dead animal remains. Food is manipulated by tube feet surrounding the mouth and passed to a chewing apparatus, called **Aristotle's lantern,** which projects from the mouth. Aristotle's lantern consists of about 35 ossicles and attached muscles and cuts food into small pieces for ingestion.

▼ Observe living or preserved sea urchins and a dried sea urchin skeleton or test (fig. 15.6). Look for evidence of pentaradial symmetry. Examine an Aristotle's lantern that has been removed from a test. ▲

Sand dollars and heart urchins usually live in sand or mud near the surface of the substrate. They are flattened and their symmetry is modified to a nearly bilateral or asymmetrical form. Sand dollars and heart urchins use tube feet to catch organic matter that settles on, or over, them. Sand dollars often live in very dense beds, which favor efficient reproduction and feeding.

▼ Examine any specimens available. ▲

CLASS HOLOTHUROIDEA—SEA CUCUMBERS

Characteristics: No rays; elongate along the oral-aboral axis; microscopic ossicles embedded in muscular body wall; circumoral tentacles. Example: *Stichopus*. Sea cucumbers are found at all depths in all oceans, where they crawl over hard substrates or burrow through sand or mud. They have no arms, they are elongate along the oral-aboral axis, and they lie on one side—which is usually flattened as a permanent ventral side. This results in sea cucumbers having a secondary bilateral symmetry. The oral end of a sea cucumber is surrounded by tentacles, which are really modifications of the water-vascular system. Tentacles are used in feeding by trapping food particles that settle on them or by sweeping across the substrate. The water-vascular system is also evidenced externally as five rows of tube feet that run between the oral and aboral ends of the animal.

▼ Examine preserved specimens for these aspects of external structure (fig. 15.7). Examine a demonstration slide showing cucumber ossicles. Normally these ossicles are embedded in the thick connective tissue of the body wall. A ring of larger

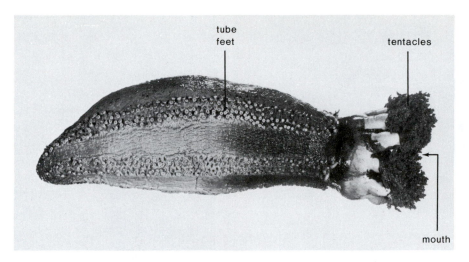

Figure 15.7 Sea cucumber, *Cucumaria*.

ossicles surrounds the oral end of the digestive tract and serves as a point of attachment for body-wall muscles. ▲

CLASS CRINOIDEA—SEA LILIES, FEATHER STARS

Characteristics: Rays present; aborally attached to substrate through a stalk of ossicles; spines, madreporite, and pedicellariae absent. Example: *Antedon*. Sea lilies are widely distributed and abundant. They are attached permanently to their substrate by a **stalk,** which bears a flattened disk, or rootlike extensions, at the attached end. The unattached end of a sea lily is called the **crown.** The aboral end of the crown is attached to the stalk and is supported by a set of ossicles, called the **calyx.** Five arms attach at the calyx, and the branching of the arms gives the sea lily a featherlike appearance. Feather stars are similar to sea lilies except they lack a stalk and are able to swim and crawl.

Crinoids feed by using their outstretched arms. When a planktonic organism contacts a tube foot, it is trapped and carried to the mouth by cilia within ambulacral grooves of the arm. This method of feeding is thought to reflect the original function of the water-vascular system.

▼ Observe any demonstrations that are available. ▲

15.2 LEARNING OUTCOMES REVIEW—WORKSHEET 15.1

15.3 EVOLUTIONARY RELATIONSHIPS

LEARNING OUTCOMES

1. Describe the relationships among extant echinoderm classes.
2. Identify synapomorphies used in distinguishing members of the echinoderm classes.

▼ A cladogram showing the evolutionary relationships among the echinoderm classes is shown in worksheet 15.1 (fig. 15.8). Examine the synapomorphies depicted on the cladogram. Based on your studies in this week's laboratory, you should be able to determine the position of each of the classes studied. Complete the cladogram and answer the questions that follow it. ▲

15.3 LEARNING OUTCOMES REVIEW—WORKSHEET 15.1

WORKSHEET 15.1   Echinodermata

15.1   Evolutionary Perspective—Learning Outcomes Review

1. Members of which one of the following phyla are most closely related to members of the Echinodermata?
   a. Rotifera
   b. Nematoda
   c. Arthropoda
   d. Chordata

2. Echinoderms are
   a. lophotrochozoans.
   b. ecdysozoans.
   c. deuterostomates.

15.2   Phylum Echinodermata—Learning Outcomes Review

3. All of the following are characteristics of members of the phylum Echinodermata except one. Select the exception.
   a. pentaradial symmetry
   b. water vascular system
   c. calcium carbonate exoskeleton
   d. bilaterally symmetrical larval stages

4. Sea stars are members of the class
   a. Crinoidea.
   b. Asteroidea.
   c. Ophiuroidea.
   d. Echinoidea.
   e. Holothuroidea.

5. Pincerlike structures found around the base of spines of sea stars are used for cleaning the body surface. These pincerlike structures are called
   a. dermal branchiae.
   b. ampullae.
   c. ambulacrae.
   d. pedicellariae.

6. During feeding on a bivalve, a sea star everts its
   a. pyloric cecae.
   b. cardiac stomach.
   c. pyloric stomach.
   d. madreporite.

7. During your observations of the sea star, you looked for a radial canal
   a. on the aboral surface of the ambulacral ossicles.
   b. within the ambulacral ridge on the oral surface of an arm.
   c. encircling the mouth.
   d. running between the ring canal and the madreporite.

8. Stomach retractor muscles run between
   a. the ambulacral ossicles and the pyloric stomach.
   b. the ambulacral ossicles and the cardiac stomach.
   c. the ring canal and the pyloric stomach.
   d. the ring canal and the cardiac stomach.

9. Sea urchins and sand dollars are members of the class
   a. Crinoidea.
   b. Asteroidea
   c. Ophiuroidea.
   d. Echinoidea.
   e. Holothuroidea.

10. A specialized set of ossicles used in feeding by a sea urchin is called
    a. the ampulla.
    b. the brachiolaria.
    c. the madreporite.
    d. Aristotle's lantern.

11. Sea cucumbers are members of the class
    a. Crinoidea.
    b. Asteroidea
    c. Ophiuroidea.
    d. Echinoidea.
    e. Holothuroidea.

12. The calyx of a sea lily
    a. supports the arms.
    b. attaches to a stalk of ossicles.
    c. bears modifications of the water-vascular system called tentacles.
    d. Both a and b are correct.

15.3   Evolutionary Perspective—Learning Outcomes

13. Which of the following pairs of taxa probably represent a monophyletic subset (clade)?
    a. Crinoidea and Holothuroidea
    b. Asteroidea and Holothuroidea
    c. Echinoidea and Holothuroidea
    d. Asteroidea and Echinoidea

14. Members of the class _____ are characterized by fusion of skeletal plates into a rigid test.
    a. Crinoidea.
    b. Asteroidea.
    c. Ophiuroidea.
    d. Echinoidea.
    e. Holothuroidea.

15.1   Evolutionary Perspective—Analytical Thinking

1. Investigate the term "Ambulacraria." How does this term inform our understanding of the relationships of the echinoderms to other animal phyla?

15.2 Phylum Echinodermata—Analytical Thinking

2. Discuss the evidence of pentaradial symmetry that you saw in members of the classes Echinoidea and Holothuroidea.

15.3 Evolutionary Relationships—Analytical Thinking

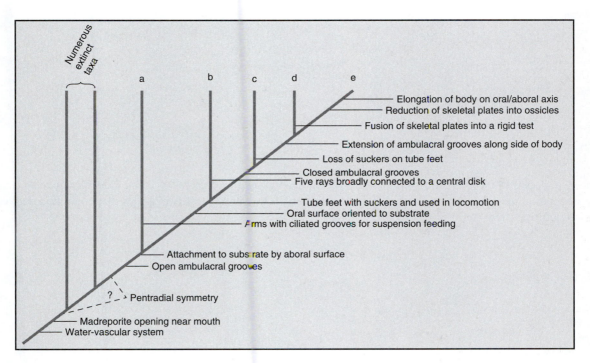

Figure 15.8 Evolutionary relationships among the Echinodermata.

Answer the following based on the figure 15.9 and you laboratory observations.

3. Fill in the class names.
   a. _____
   b. _____
   c. _____
   d. _____
   e. _____

4. Name one symplesiomorphic character for the phylum Echinodermata. _____

5. Name one synapomorphy that can be used to distinguish members of each of the following classes from other echinoderms.
   a. Crinoidea _____
   b. Asteroidea _____
   c. Ophiuroidea _____
   d. Echinoidea _____
   e. Holothuroidea _____

6. What does this cladogram imply regarding the ancestral status of pentaradial symmetry?

7. Some systematists think that the Asteroidea and Ophiuroidea should be depicted as being more closely related than they are represented to be in figure 15.8. What characteristic is shared by these two groups that could be a derived character uniting these two classes into a single clade?

Exercise 16

Chordata

Prelaboratory Quiz

Study this week's laboratory exercise and then complete the following quiz to assess your preparation for the laboratory.

1. All of the following are unique features of members of the phylum Chordata EXCEPT the
   a. dorsal tubular nerve cord.
   b. notochord.
   c. closed circulatory system.
   d. pharyngeal slits (pouches).
   e. endostyle or thyroid gland.

2. Members of the subphylum Urochordata include
   a. amphioxus.
   b. hagfish.
   c. tunicates.
   d. lampreys.

3. All of the following are members of the infraphylum Vertebrata EXCEPT
   a. lancelets.
   b. sharks.
   c. lampreys.
   d. frogs and toads.

4. Adult urochordates are
   a. fishlike predators.
   b. fishlike filter feeders
   c. sessile filter feeders.
   d. planktonic predators.

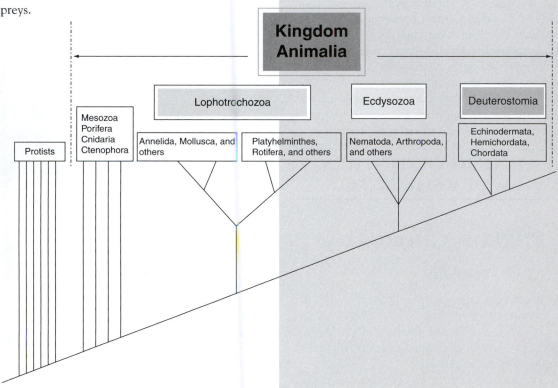

5. Which of the following is most dorsal in position in virtually all chordates?
    a. notochord
    b. nerve cord
    c. pharyngeal slits
    d. gut tract

6. True/False The amniotes include the amphibians, reptiles, and birds.

7. True/False Sharks, skates, and rays are members of the class Actinopterygii.

8. True/False Members of the class Petromyzontida possess a rasping tongue and a sucking mouth with teeth.

9. True/False The subphylum Cephalochordata is comprised of fishlike chordates that live partially buried in marine substrates and function as filter feeders.

10. True/False Both the birds and the mammals are endothermic and possess amniotic eggs.

16.1 EVOLUTIONARY PERSPECTIVE

LEARNING OUTCOME

1. Describe the relationships of members of the phylum Chordata to other animal phyla.

The relationship of chordates to echinoderms and the relationship of the echinoderm-chordate line to previous phyla were discussed in exercise 15. Zoologists have not yet found evidence documenting the origin of chordates from nonchordates. There are three subphyla within the phylum Chordata: Urochordata, Cephalochordata, and Craniata. Molecular evidence regarding the relationship of the tunicates, cephalochordates, and craniates is contradictory. Genomic evidence supports the close relationship of the tunicates and the craniates. Mitochondial and ribosomal evidence suggests cephalochordates are more closely related to the craniates.

16.1 LEARNING OUTCOMES REVIEW—WORKSHEET 16.1

16.2 PHYLUM CHORDATA

LEARNING OUTCOMES

1. Describe five unique characteristics of members of the phylum Chordata.
2. Describe members of the three chordate subphyla.
3. Identify the five unique chordate characteristics from slide preparations of Amphioxus.
4. Identify representatives of members of the craniate classes.

Because the remaining laboratory exercises will cover details of chordate structure and function, this exercise is intended to introduce the major characteristics of the phylum and classes.

It is relatively easy to characterize the phylum Chordata. Even though they show remarkable diversity of form and function, all chordates at some time in their life history have five easily recognized, unique features: the **notochord, pharyngeal slits (pouches),** a **dorsal tubular nerve cord, endostyle,** and a **postanal tail** (fig. 16.1).

The notochord is a supportive rod that extends most of the length of the animal along the dorsal body wall. It consists of a connective tissue sheath packed with vacuolated cells. It supports the body along the anterior-posterior axis, yet is flexible enough to allow freedom of body movement. In many vertebrate adults, the notochord is partly or entirely replaced by bone.

Pharyngeal slits are a series of openings between the digestive tract, in the pharyngeal region, and the outside of the body. Although the term "gill slits" is often used to refer to these openings, it is misleading. The earliest chordates used the openings for filter feeding and did not have gills associated with these pharyngeal slits. In some chordates that function is retained. Other chordates, however, have developed gills within the slits and use them for gas exchange. Like the notochord, the pharyngeal slits of higher vertebrates are mainly embryonic features. The eustachian tube of tetrapods, connecting the middle ear to the pharynx, is derived from a pharyngeal slit. The eustachian tube is used in equalizing pressure within the middle ear to prevent damage to the eardrum.

The dorsal tubular nerve cord and its associated structures are largely responsible for the success of the chordates. It runs along the dorsal axis of the body, just above the notochord, and is usually modified anteriorly into a brain.

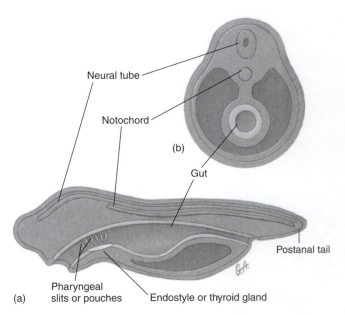

Figure 16.1 The generalized chordate body plan: (a) sagittal section; (b) cross section.

These nervous structures have allowed the development of the most complex animal systems for sensory perception, integration, and motor response.

The endostyle is present along the ventral aspect of the pharynx. It secretes mucus, which is used in filter feeding. In chordates that do not filter feed, the endostyle is transformed into the thyroid gland and is involved with producing and secreting iodine containing hormones involved with regulating metamorphosis and metabolism.

The fifth characteristic is a tail that extends posterior to the anal opening. The tail is generally supported by the notochord or by the vertebral column, which replaces the notochord in many adult chordates. In addition, chordates are coelomate, triploblastic, and bilaterally symmetrical.

CLASSIFICATION

The three chordate subphyla are listed below. The classification within the subphylum Craniata includes the name "Aves" because of the strong tradition associated with this name. Zoologists agree that the characteristics used to distinguish the birds first appeared within the reptiles, and that birds are a part of the reptilian lineage.

Subphylum Urochordata
Subphylum Cephalochordata
Subphylum Craniata
    Infaphylum Hyperotreti
        Class Myxini
    Infraphylum Vertebrata
        Class Petromyzontida

Class Chondrichthyes
Class Actinopterygii
Class Sarcopterygii
Class Amphibia
Class Reptilia
"Class Aves"
Class Mammalia

SUBPHYLUM UROCHORDATA— TUNICATES

Characteristics: Notochord, nerve cord, and postanal tail present only in free-swimming larvae; pharyngeal slits present in larvae and adults; pharyngeal basket; adults sessile, or occasionally planktonic, and enclosed in a cellulose-containing tunic. Example: *Ciona.*

Members of the subphylum Urochordata are the tunicates or sea squirts. Most are sessile as adults and are either solitary or colonial. Others are planktonic. Sessile urochordates attach their saclike bodies to rocks, pilings, the hulls of ships, and other solid substrates.

▼ Examine tunicates on demonstration (fig. 16.2). Note the body wall, which encloses a cavity called the atrium. At the unattached end are two siphons. The incurrent siphon brings seawater into an expanded pharyngeal basket, which filters food from the water. The endostyle aids in filter feeding. It is a ciliated groove that produces a mucous sheet for trapping food particles and moving them to the intestine. (The endostyle is the evolutionary forerunner of the vertebrate thyroid gland.) Water leaves the tunicate via the atrial (excurrent) siphon.

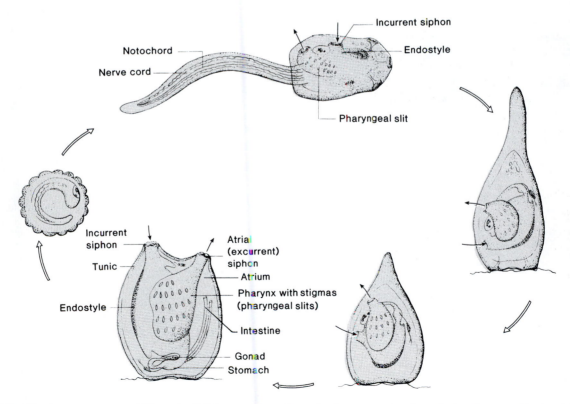

Figure 16.2 Tunicate structure and life cycle. Solid arrows show the path of water moving through the pharyngeal basket (pharynx).

Tunicates are monoecious. Gametes are shed through the atrial siphon for external fertilization, or eggs may be retained within the atrium for fertilization and early development. Cross-fertilization is the rule. Development results in the formation of a tadpole-like larva that possesses all five chordate characteristics (see fig. 16.2). After a brief free-swimming existence, the larva settles to a firm substrate and attaches by adhesive papillae located near the mouth. During metamorphosis, there is a 180° rotation of internal structures that results in the development of the adult body form (see fig. 16.2). Examine a slide showing a tunicate larva under the compound microscope. ▲

SUBPHYLUM CEPHALOCHORDATA

Characteristics: Body laterally compressed and transparent, fishlike; all five chordate characteristics persist throughout life; buccal apparatus used in feeding.

The Cephalochordata (lancelets) is a relatively small group of fishlike chordates. This group is used extensively in introductory courses because it shows features characteristic of the phylum Chordata. These features are probably similar to their appearance in ancestral chordates.

Lancelets are marine, spending most of their life with their posterior half buried in sandy substrates of shallow water (fig. 16.3a). They feed by filtering suspended organic particles from water in a manner similar to that described for tunicates. We will study the genus *Branchiostoma*, commonly known as amphioxus.

Whole Mount

▼ Obtain an amphioxus whole mount. Examine it under a dissecting microscope or low power of a compound microscope. Compare your observations to figure 16.3b. Note that anteriorly there is an **oral hood** with a terminal **rostrum** and fingerlike **oral cirri** that surround the mouth. These structures form a buccal apparatus used in feeding. Oral cirri are largely sensory, with some straining functions. Cilia lining the oral hood draw water currents into the gut tract, where food particles are caught in mucus secreted by the **endostyle** (to which the vertebrate thyroid gland has been homologized). Note the **pharyngeal slits** margined by supportive **pharyngeal bars.** Water passing into the digestive tract during filter feeding passes through pharyngeal slits to the **atrium,** the cavity surrounding the pharyngeal slits. Pharyngeal slits of cephalochordates, as with urochordates, are primarily involved with filter feeding. From the atrium, water moves to the outside through the ventral **atriopore.** Follow the digestive tract posteriorly. Just behind the pharyngeal slits, note that a diverticulum off the gut tract, called the **hepatic (gastric) cecum,** extends anteriorly. Enzymes secreted by this diverticulum partially digest food particles before absorption. Digestion is both extracellular and intracellular. The digestive tract ends at the **anus.** Locate the anus and note the postanal tail. The notochord and dorsal tubular nerve cord are difficult to see in whole-mount preparations. They are obscured by segmentally arranged bundles of muscles that make up most of the dorsal aspect of the body wall. These muscle bundles are called **myomeres,** which are easily observed.

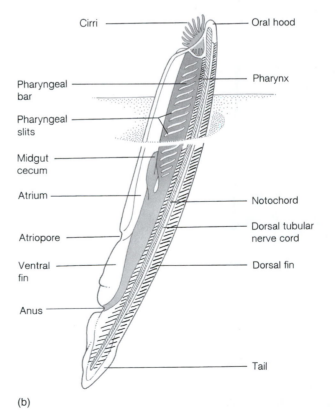

(a)

Cirri
Oral hood
Pharyngeal bar
Pharynx
Pharyngeal slits
Midgut cecum
Atrium
Notochord
Atriopore
Dorsal tubular nerve cord
Ventral fin
Dorsal fin
Anus
Tail

Figure 16.3 Subphylum Cephalochordata. (*a*) Amphioxus in its partially buried feeding position. (*b*) Amphioxus structure.

(b)

Cross Section

Obtain a slide showing one or more cross sections of amphioxus. The slide is likely to show sections through representative regions of the body. If so, locate a section through the pharyngeal region similar to that shown in figure 16.4a. Locate those structures described as being unique to the phylum: tubular nerve cord, notochord, pharyngeal slits, and endostyle. The endostyle is at the ventral margin of the pharynx. The coelom is restricted to the space dorsal and lateral to the pharynx and the space below the endostyle. Most of the chamber around the pharynx is the atrium. Lateral to the pharynx, you should also see **gonads.** (Cephalochordates are dioecious.) An ovary has large, nucleated cells, whereas a testis has much smaller germ cells that appear as darkly stained dots. Gametes are shed into the atrium and leave through the atriopore. External fertilization leads to a bilaterally symmetrical larva.

Examine the other sections on your slide and associate them with their respective body regions (fig. 16.4b). ▲

SUBPHYLUM CRANIATA

Characteristics: Skull surrounds the brain, olfactory organs, eyes, and inner ear. Unique embryonic tissue, the neural crest, contributes to a variety of adult structures, including sensory nerve cells and some skeletal and other connective tissue structures.

The craniates include the hagfish and eight classes of vertebrates. In subsequent laboratory exercises, you will examine aspects of vertebrate evolution, structure, and function. It is important for you to be able to recognize major vertebrate classes.

▼ Examine specimens of the following classes, noting as many of the characteristics as possible. ▲

Infraphylum Hyperotreti—Fishlike; skull consisting of cartilaginous bars; jawless; no paired appendages.
  Class Myxini—Hagfishes
    Mouth with four pairs of tentacles. Olfactory sacs open to the mouth cavity. Five to 15 pairs of pharyngeal slits.
Infraphylum Vertebrata—Vertebrae surround nerve cord and serve as primary axial support. Two or three semicircular canals.
  Class Petromyzontida—Lampreys
    Fishlike; jawless; no paired appendages; cartilaginous skeleton; sucking mouth with teeth and rasping tongue.
  Class Chondrichthyes—Sharks, skates, rays, ratfishes
    Tail with large upper lobe. Cartilaginous skeleton. Most lack opercula, swim bladder, and lungs.
  Class Actinopterygii—Ray-finned fishes
    Bony fishes having paired fins supported by dermal rays; basal portions of paired fins not especially muscular; tail fin with approximately equal upper and lower lobes; blind olfactory sacs; pneumatic sacs function as swim bladder.

Class Sarcopterygii—Lobe-finned fishes (lungfishes)
  Bony fishes having paired fins with muscular lobes; pneumatic sacs function as lungs; atria and ventricles at least partially divided.
Class Amphibia—Frogs, toads, salamanders
  Skin with mucoid secretions. Usually undergo metamorphosis. Moist skin functions as a respiratory organ.
Class Reptilia—Snakes, lizards, alligators, turtles
  Dry skin with epidermal scales. Amniotic egg.
"Class Aves"—Birds (Avian Reptiles)
  Feathers. Skeleton modified for flight. Metabolic heat used in temperature regulation (endothermic). Amniotic egg.
Class Mammalia—Mammals
  Have hair, mammary glands. Endothermic temperature regulation. Amniotic egg.

Members of the last three classes are relatively independent of water, partly because of the presence of amniotic eggs. Amniotic eggs possess a series of membranes that form bags for the nontoxic storage of nitrogenous wastes and prevent the embryo from desiccating. In addition, the eggs of amniotes other than most mammals possess a shell that protects the embryo from mechanical injury. This common feature, along with other adaptations to terrestrial environments, has resulted in reptiles, birds, and mammals being combined into a lineage called **Amniota.**

16.2 LEARNING OUTCOMES REVIEW—WORKSHEET 16.1

16.3 EVOLUTIONARY RELATIONSHIPS

LEARNING OUTCOMES

1. Discuss relationships among the craniate classes.
2. Describe key characters used in distinguishing craniate class representatives.

▼ A cladogram showing the evolutionary relationships among the chordates is shown in worksheet 16.1 (fig. 16.5). Examine the synapomorphies depicted on the cladogram. Based on your studies in this week's laboratory, you should be able to determine the position of each of the taxa studied. Complete the cladogram by matching the taxon or character listed in worksheet 16.1 questions 3 and 4, with the appropriate blank on the cladogram. Answer the questions that follow the cladogram. ▲

16.3 LEARNING OUTCOMES REVIEW—WORKSHEET 16.1

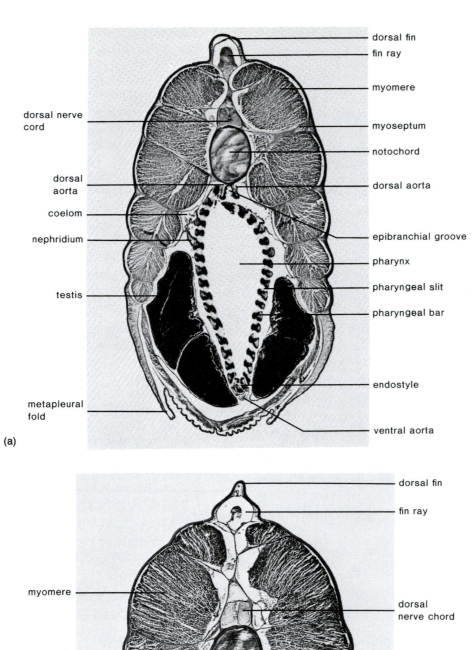

(a)

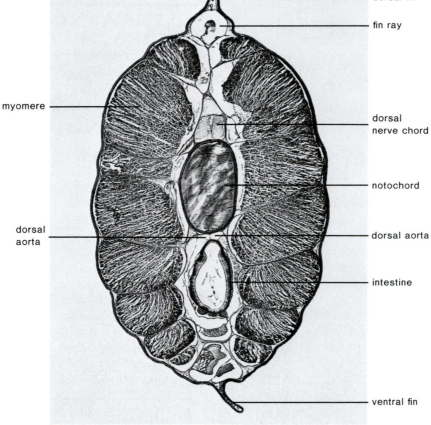

(b)

Figure 16.4 Amphioxus. (*a*) Cross section through the pharyngeal region. (*b*) Cross section through the intestinal region.

WORKSHEET 16.1 Chordata

16.1 Evolutionary Perspective—Learning Outcomes Review

1. Members of which one of the following phyla are most closely related to members of the Chordata?
   a. Rotifera
   b. Nematoda
   c. Arthropoda
   d. Echinodermata

2. Chordates are
   a. lophotrochozoans.
   b. ecdysozoans.
   c. deuterostomates.

16.2 Phylum Chordata—Learning Outcomes Review

3. All of the following are unique chordate characteristics except one. Select the exception.
   a. notochord
   b. pharyngeal slits
   c. nerve cord
   d. postanal tail
   e. endostyle or thyroid gland

4. Members of the subphylum _____ include the tunicates.
   a. Vertebrata
   b. Craniata
   c. Urochordata
   d. Cephalochordata

5. Where would you observe more of the unique chordate characteristics?
   a. In a urochordate adult.
   b. In a urochordate larva.

6. After passing through pharyngeal slits of amphioxus water enters the
   a. pharynx.
   b. intestine.
   c. atrium.
   d. hepatic cecum.

7. Which of the following structures are easily seen in a prepared whole mount of amphioxus?
   a. notochord
   b. pharyngeal slits
   c. postanal tail
   d. dorsal tubular nerve cord
   e. All of the above are easily seen.
   f. b and c (only) are easily seen.

8. Where would you look to observe the endostyle of amphioxus when viewing a cross section through its pharynx?
   a. At the dorsal aspect of the pharynx.
   b. At the ventral aspect of the pharynx.
   c. In the atrium lateral to the pharynx.
   d. Just dorsal to the notochord.

9. Segmentally arranged muscle bundles observed in amphioxus slide preparations are called
    a. metameres.
    b. longitudinal muscle bands.
    c. myomeres.
    d. pharyngeal muscle bands.

10. Name the vertebrate class whose members have dry skin with epidermal scales.
    a. Actinopterygii
    b. Chondrichthyes
    c. Amphibia
    d. Reptilia
    e. Mammalia

11. Which of the following vertebrate classes is a traditional class designation, but probably should be eliminated as a class name?
    a. Actinopterygii
    b. Chondrichthyes
    c. Aves
    d. Reptilia
    e. Mammalia

12. Members of this group of vertebrate fishes lack swim bladders and opercula.
    a. Actinopterygii
    b. Chondrichthyes
    c. Amphibia
    d. Sarcopterygii
    e. Myxini

13. Members of these groups comprise all of the Amniota.
    a. Actinopterygii and Sarcopterygii
    b. Mammalia and Amphibia
    c. Reptilia, Aves, and Mammalia
    d. Reptilia and Aves

16.3 Evolutionary Relationships—Learning Outcomes Review

14. All of the following have two or three semicircular canals and vertebrae surrounding the spinal cord except one. Select the exception.
    a. reptiles
    b. ray-finned fishes
    c. sharks
    d. hagfish
    e. lampreys

15. Which of the following are included in a clade characterized by having a muscular gizzard?
    a. amphibians and birds
    b. mammals and birds
    c. crocodilians and birds
    d. crocodilians and mammals

16.1 Evolutionary Perspective—Analytical Thinking

1. Molecular evidence regarding the relationships of the Urochordata, Cephalochordata, and Craniata is contradictory. Describe the contradiction. What might be responsible for contradictory data and what kind of evidence could resolve the contradiction?

16.2 Phylum Chordata—Analytical Thinking

2. Amphioxus is a classic animal that has been studied by many generations of zoology students during their introduction to the Chordata. This use of amphioxus is in spite of the fact that it is only one, very small representative of its phylum. Based on your observations in the laboratory, why do you think it has become so important to introductory zoology laboratories?

16.3 Evolutionary Relationships—Analytical Thinking

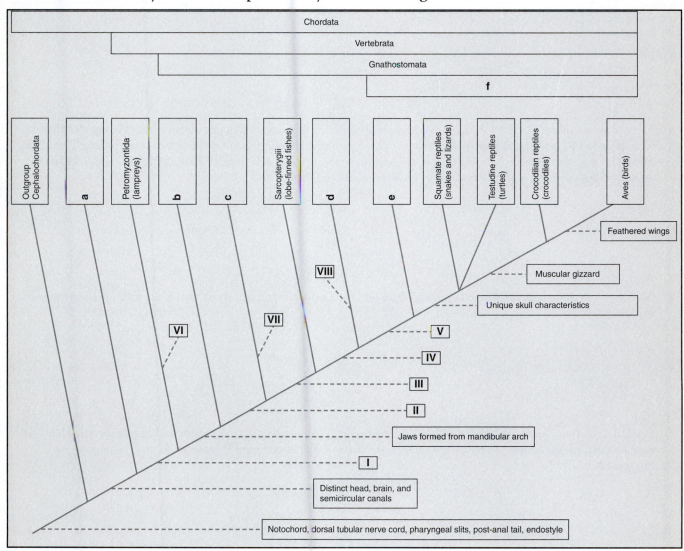

Figure 16.5 Evolutionary relationships among the craniates. See questions 3–8 on page 242.

3. Fill in the name of the chordate group associated with the following blanks in the cladogram in Figure 16.5.

   a. _____

   b. _____

   c. _____

   d. _____

   e. _____

   f. _____

4. Match the following characters to appropriate blanks on the cladogram indicated by the numbers below.

   I. _____   g. two or three semicircular canals

   II. _____   h. lungs or swim bladders

   III. _____   i. four limbs adapted for terrestrial locomotion

   IV. _____   j. amniotic membranes

   V. _____   k. muscular lobes at bases of appendages

   VI. _____   l. skin with mucoid secretions

   VII. _____   m. pneumatic sacks function as swim bladders, fins without muscular lobes

   VIII. _____   n. no paired appendages, sucking mouth with teeth, rasping tongue

5. Name and describe five symplesiomorphic characters for members of the phylum Chordata.

6. The taxonomy of the phylum Chordata is currently unsettled. The traditional eight-class system is being reevaluated because of recent cladistic analysis. Notice in the cladogram that the reptiles are divided into three orders: Testudines (turtles), Crocodylia (crocodiles and alligators), and Squamata (snakes and lizards). How is the cladogram shown in figure 16.5 inconsistent with the traditional class designations for birds and reptiles?

7. Birds are sometimes called "glorified reptiles." Based on the cladogram, do you think that is an accurate description? Explain your answer.

8. Similarly, the bony fishes have recently been reclassified into two classes, Actinopterygii and Sarcopterygii. They were formerly classified into a single class called Osteichthyes. Based on figure 16.5, what do you think prompted this reclassification?

UNIT

III

An Evolutionary Approach to Vertebrate Structure and Function

Unit III is a study of the structure and function of vertebrate organ systems. The "organ system–by–organ system" format of this unit is intended to help you understand vertebrate structure and function and some aspects of vertebrate evolution. Although time and space do not allow the examination of evolutionary changes in detail, you will gain an appreciation of the diversity of structure and function within the vertebrates. The rat is used as the primary animal for student dissection. Demonstration dissections of the shark and other vertebrates are used for comparative purposes. In addition, selected physiological exercises are designed to introduce you to the functions of some systems.

Exercise 17

Vertebrate Musculoskeletal Systems

Prelaboratory Quiz

Study this week's laboratory exercise and then complete the following quiz to assess your preparation for the laboratory.

1. At each end of a long bone is a region called the
   a. osteon.
   b. diaphysis.
   c. epiphysis.
   d. matrix.

2. Bone cells, or osteocytes, are found within
   a. canaliculi.
   b. lacunae.
   c. osteonic canals.
   d. the matrix.

3. Which of the following vertebrate classes has members whose vertebral column consists of cervical, trunk, sacral, and caudal regions?
   a. Osteichthyes
   b. Amphibia
   c. Reptilia
   d. Aves
   e. Mammalia

4. The axial skeleton consists of all of the following EXCEPT the
   a. skull.
   b. ribs.
   c. girdles.
   d. vertebral column.

5. Skeletal support for the gills of fishes is provided through
   a. visceral arches.
   b. dorsal ribs.
   c. the sternum.
   d. clavicles.

6. True/False   The vertebral column of reptiles, birds, and mammals consists of cervical, thoracic, lumbar, sacral, and caudal regions.

7. True/False   The dark bands in skeletal muscle are called I bands and are regions of overlap between actin and myosin.

8. True/False   Histologically, cardiac muscle differs from skeletal muscle in that the former has intercalated disks and is highly branched. The latter lacks intercalated disks and is unbranched.

9. True/False   The immovable point of attachment of a muscle on a skeleton is the muscle's insertion.

10. True/False   An action that increases the size of the angle between the anterior surfaces of the articulated bones is an extension.

In the following laboratories, we will be studying the major vertebrate organ systems. Of these, it is appropriate to study skeletal and muscular systems first, as they are the fundamental systems around which the body is built. We will deal with muscular and skeletal systems as a unit because they are functionally interdependent.

17.1 BONE STRUCTURE

LEARNING OUTCOME

1. Identify structures of a vertebrate long bone.

We will begin our study of the skeletal system by examining a typical long bone that has been sectioned longitudinally (fig. 17.1a).

▼ Examine the outer, uncut surface. The bone is divided into three regions. At each end is an **epiphysis,** which is separated from the **diaphysis** (shaft) by **epiphyseal lines.** In life, each epiphysis is covered by **articular cartilage.** The articular cartilage represents the region of contact with the adjacent bone. The epiphyseal lines indicate the position of **epiphyseal plates.** During growth, the epiphyseal plate is cartilaginous. New bone is laid down in these regions during elongation. If available, examine a demonstration showing the epiphyseal plates of a fetus. In this preparation, all body tissues have been cleared and the bones stained. In life, the outer and inner surfaces of the bone are covered with connective tissue (periosteum and endosteum, respectively). These tissues are involved with bone growth, maintenance, and repair.

Examine the inner cut surface of the bone. Within the diaphysis note the **marrow cavity.** In life, the marrow cavity is filled with yellow marrow that functions in fat storage. The bone of the diaphysis is very compact and hard. It is called **compact bone.** In the epiphysis there is a network of bone called **cancellous bone.** Individual bony filaments are

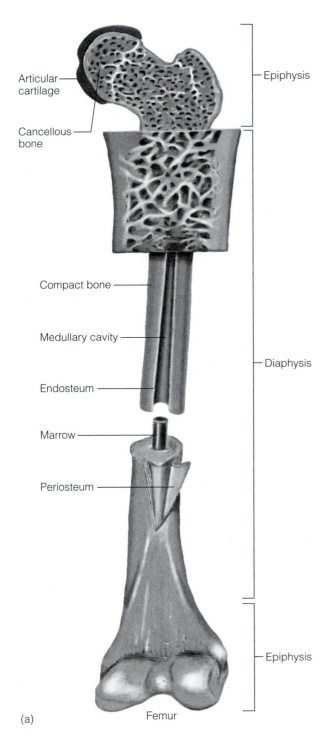

(a) Femur

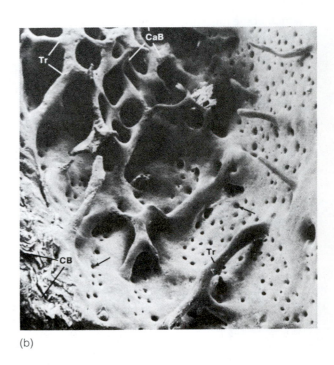

(b)

Figure 17.1   Bone structure. (*a*) The structure of a long bone. (*b*) Diaphyseal region of the rat femur. Cancellous bone (CaB), trabeculae (Tr), compact bone (CB). Arrows indicate blood vessel pores (×105). (*b*) From Richard Kessel & Randy Kardon/Visuals Unlimited.

trabeculae (fig. 17.1b). Red marrow fills the spaces between trabeculae and functions in blood cell production. ▲

Bone and Cartilage Histology

Bone and cartilage are special connective tissues that were not covered in the histology exercises. They will be covered at this time.

▼ Examine a slide of ground, compact bone under low power of a compound microscope (fig. 17.2). Note that the tissue is organized into a series of circular **osteon systems.** In the center of each is an **osteonic canal** that contains a blood vessel in the living bone. Turn to high power and note the layers of circular **lamellae** that surround the osteonic canal. **Lacunae** are the small cavities in bone and cartilage that contain bone and cartilage cells. In bone, they are the dark spaces between lamellae and contain **osteocytes** that function in maintenance and repair. The small lines radiating from the lacunae are **canaliculi.** Cell processes of osteocytes communicate with adjacent osteocytes through canaliculi. Lamellae are made of collagenous fibers impregnated with calcium phosphate. This nonliving matrix gives the bone its strength. ▲

Cartilage takes on a variety of forms depending on the type of fibers present and the nature of the matrix. It is typically found on the ends of bones (articular cartilage), between bones (intervertebral discs), connecting bone to bone (costal cartilage), and giving support to structures such as the nose and pinna of the ear. It is made of cells and fibers embedded in a nonliving matrix (fig. 17.3).

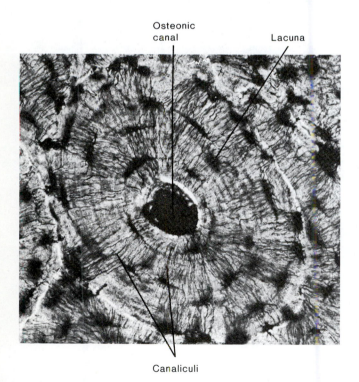

Figure 17.2   Compact bone, the osteon.

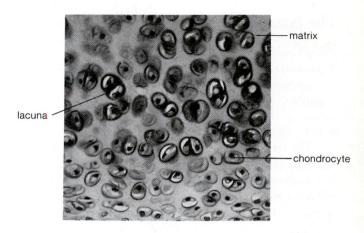

Figure 17.3   Hyaline cartilage.

▼ Observe a slide of hyaline cartilage. Examine it under low and then high power. Note the semitransparent **matrix.** Fibers embedded in the matrix cannot be seen. **Chondrocytes** (cartilage cells) are contained within clear lacunae. ▲

17.1 LEARNING OUTCOMES REVIEW—WORKSHEET 17.2

17.2 THE SKELETON

LEARNING OUTCOMES

1. Describe variations in the vertebrate axial skeleton and the adaptive significance of these variations.
2. Describe the basic structure of the vertebrate skull.
3. Describe variations in the appendicular skeleton within the vertebrates.

The vertebrate skeleton can be divided into axial and appendicular components. The **axial skeleton** includes those parts of the skeleton located along the longitudinal axis of the body (skull, vertebrae, ribs, and sternum). The **appendicular skeleton** consists of the more laterally located skeletal elements (pelvic and pectoral girdles and limbs). In the following study, you will examine the skeletons of representative vertebrates. The goal is not to learn every bone in what might be represented as a "typical" vertebrate, but to look at evolutionary changes in major components of the vertebrate skeleton. **During this exercise, you will be asked to record observations in worksheet 17.1. You should stop and think about how aspects of what you are seeing are adaptive for the animals in question. Unless your instructor tells you otherwise, feel free to "brainstorm" with neighboring students.**

The Axial Skeleton—Vertebral Column

Fishes

▼ Examine the skeleton of a fish (fig. 17.4). If you are looking at a shark (*Squalus*), remember that the skeleton is cartilaginous. This is atypical for most adult vertebrates and represents the retention of an embryonic condition rather than being a primitive situation.

The vertebral column of fishes consists of **trunk** and **caudal (tail)** vertebrae. These two regions can be distinguished by the absence of ribs in the tail region. Note that the vertebral column consists of a series of similar bones called **vertebrae.** Each vertebra contains a dorsal (neural) arch and, in the caudal region, a ventral (hemal) arch. The dorsal arches protect the spinal cord, and the ventral arches protect blood vessels and nerves of the tail. **Record your observations by answering questions 1 and 2 in worksheet 17.1 now.** ▲

Amphibians

▼ Examine the skeleton of *Necturus* (the mud puppy). Its axial skeleton is divided into four regions. Amphibians possess a neck, and the skull attaches to the axial skeleton via a single **cervical (neck) vertebra.** This permits a vertical nodding movement of the head. Posterior to the neck is the trunk region with its attached ribs. In the evolution of the amphibians, the last trunk vertebra developed into a **sacral vertebra.** This vertebra serves as the point of attachment for the hind limbs via a modified rib. Caudal vertebrae follow the sacral vertebra. ▲

Amniotes: Reptiles, Birds, and Mammals

All amniotes have a similar organization of the vertebral column (fig. 17.5). Use specimens from each group, in the following observations.

▼ Note the proliferation of cervical vertebrae. The first two are specialized into an **atlas** (articulates with the skull and the axis) and an **axis** (articulates with the atlas and the next cervical vertebra). Examine an atlas and an axis that have been disarticulated. Fit them together and then articulate them with the condyles of a skull. How does the combination of skull, atlas, and axis move in nodding the head? In turning the head? Examine disarticulated cervical vertebrae from a goose or turkey. How are they different from those of a mammal? **Record your observations by answering questions 3–6 in worksheet 17.1 now.**

The trunk region is subdivided into thoracic and lumbar regions in amniotes. The **thoracic vertebrae** have ribs attached that protect the heart and lungs. Locate the thoracic vertebrae and ribs on the skeletons available. The large dorsal (neural) spines of anterior thoracic vertebrae are points of origin of muscles that insert on the back of the skull.

Lumbar vertebrae are present in the abdominal region of amniotes. Note that there are no ribs associated with this region. The large transverse processes are formed from the fusion of ribs and vertebrae.

There is an increased number of sacral vertebrae in amniotes (two to four) as compared to the single sacral vertebra

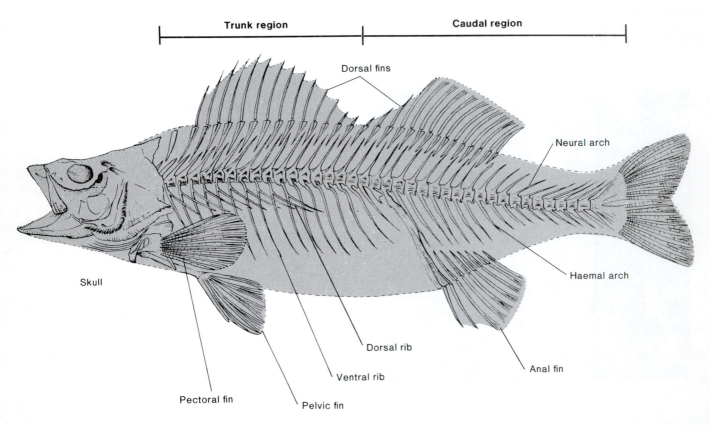

Figure 17.4 Lateral view of a perch skeleton.

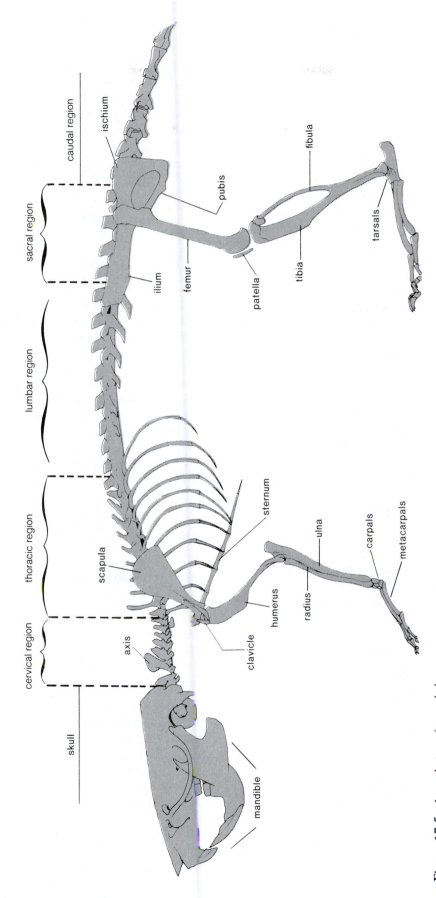

Figure 17.5 Lateral view of a rat skeleton.

in amphibians. Examine a sacrum and try to determine how many vertebrae are fused into this structure. In a bird skeleton, note that parts of the thoracic, lumbar, and sacral vertebrae are fused into a single unit, called the **synsacrum.** This fusion keeps the body rigid during flight and helps form an aerodynamic surface.

The caudal vertebrae of vertebrates are variable in number, size, and shape. Compare the caudal vertebrae of a rat, a bird, and a human. The fused caudal vertebrae of birds are called the **pygostyle,** which is where the tail feathers insert. Tail feathers and the pygostyle are used in maneuvering during flight. In humans caudal vertebrae are reduced to form the **coccyx.**

During the evolution of the amphibians, ribs attached ventrally to a plate of bone called the **sternum.** This arrangement is most easily seen in the amniotes. Examine the attachment of ribs to the axial skeleton of a mammal. Examine the sternum of a bird. The large keel on the sternum provides room for the attachment of flight muscles that also attach at the bird's wings. **Record your observations by answering questions 4–13 in worksheet 17.1 now.** ▲

The Axial Skeleton—Skull

The skull has a very complex evolutionary history and consists of many bones. In primitive vertebrates, the skull consisted of a cartilaginous trough, which housed the brain. (The skull of a shark is cartilaginous; however, this is the result of the retention of an embryonic, not a primitive, condition.) In later vertebrates, bone is found covering the top and sides of the skull, and the cartilaginous trough is replaced by bone. This resulted in additional support and protection for the brain. The vertebrate skull also includes gill supports in fishes and bones derived from gill supports in terrestrial vertebrates. Gill supports are called **visceral arches** and are thought to have been used originally for filter feeding and later supported gas exchange surfaces (gills). In the evolution of jawed fishes, bones of the first visceral arch were incorporated into bones of the upper and lower jaw. In terrestrial vertebrates, the visceral arches are incorporated into the hyoid bone for tongue support, the bones of the middle ear, and the cartilage supporting the larynx (voice box).

▼ Examine a shark skull and visceral arches. Note the similarity in the structure of the jaw and the visceral arches. Examine a human skeleton to see a hyoid bone and models and embedded mounts of the middle ear ossicles. ▲

The Appendicular Skeleton

The appendicular skeleton consists of pelvic and pectoral girdles as well as bones of the paired appendages. The organization of these bones is remarkably consistent throughout the vertebrates.

The Pectoral Girdle

The evolution of the pectoral girdle is complex. We will not describe it in detail. In fishes, the pectoral girdle consists of a number of bones and is usually not attached to the axial skeleton (see shark skeleton). In terrestrial vertebrates, the general trend is to reduce the number of bones and to attach these bones to the axial skeleton. In mammals, the result is a **scapula** that articulates with the arm at the **glenoid cavity.** The scapula serves as a point of attachment for muscles of the arm. The scapula is anchored through the **clavicle** and **sternum** to the axial skeleton. In the bird, the clavicles are fused into a single **furcula,** or wishbone, which serves for the attachment of flight muscles. In the cat, the clavicle does not anchor the scapula to the sternum. Instead, the clavicle is held embedded in muscles and tendons of the fore appendages.

▼ Examine skeletal material of a fish, an amphibian, a bird, a human or a rat, and a cat. ▲

The Pelvic Girdle

The evolution of the pelvic girdle is less complex. All terrestrial vertebrates have three pairs of bones that unite at the point of articulation with the hind limb, the **acetabulum.** Forming the anterior ventral border of each half of the pelvic girdle is the **pubis.** Along the posterior border and extending laterally to the acetabulum is the **ischium.** Extending dorsally from the acetabulum and attaching to the sacrum is the **ilium.** These three bones are fused in adults, and their individuality is obscured. The medial border of the pubis and ischium is joined with the same bone from the other side of the body through fibrocartilage. The immovable joint between the pubic bones is referred to as the **pubic symphysis.** The origin of the pelvic girdle is speculative.

▼ Examine skeletal materials provided. Note again the synsacrum of the bird, which shows fusion of the bones of the vertebral column and pelvic girdle. ▲

The Paired Appendages

The appendages of terrestrial vertebrates are all similarly organized. Homologies with bones of fishes have been suggested, but they are speculative. The organization of fore- and hind limbs is similar. In the discussion that follows, the forelimb will be described, and the analogous structure of the hind limb will be given parenthetically.

▼ The bone that articulates with the **glenoid cavity (acetabulum)** is the **humerus (femur).** Distally, the humerus (femur) articulates with two bones, the **ulna (tibia)** and **radius (fibula).** Note on the human skeleton the joint between the radius and humerus. This joint allows a rotation of the lower arm so that the radius crosses over the ulna distally. This allows one to turn palms up or down. Distal to these bones are the bones of the wrist, the **carpals** (the ankle, the **tarsals).** We will not learn the names of individual wrist (ankle) bones. Following these are the bones of the hand, the **metacarpals** (foot, **metatarsals)** and the bones of the fingers, the **phalanges** (toes, **phalanges).** Examine a variety of terrestrial skeletons for an appreciation of the consistent organization of the arm and leg bones of vertebrates. These similarities are a striking example of how homology is used in demonstrating evolutionary relationships. Examine a bird skeleton. Note the fusion and loss of appendage bones that occurred in the evolution of flight. In addition to the previously mentioned adaptations, the bones of birds are porous. **Record your observations by answering questions 14–18 in worksheet 17.1 now.** ▲

17.2 LEARNING OUTCOMES REVIEW—WORKSHEET 17.2

17.3 THE MUSCULAR SYSTEM—HISTOLOGY

LEARNING OUTCOMES

1. Identify histological features of skeletal muscle in microscopic preparations.
2. Explain the interaction of contractile proteins during contraction of skeletal muscle.

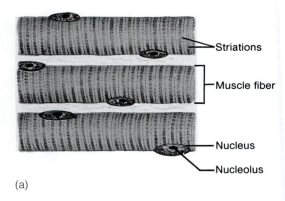

(a)

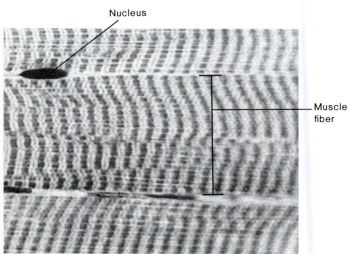

Nucleus

Muscle fiber

(b)

Figure 17.6 Skeletal muscle as viewed in the light microscope: (*a*) diagram; (*b*) photomicrograph. A small portion of these cells is shown in both (*a*) and (*b*).

3. Identify smooth muscle in microscopic preparations.
4. Identify histological features of cardiac muscle in microscopic preparations.

Because the structure of muscle is closely tied to its function, the study of muscle histology has been delayed until now. Muscles are classified by structure, location, and contractile characteristics.

Skeletal Muscle

Skeletal muscles, as the name implies, are associated with the skeleton. They can be consciously controlled, and they appear striated under the microscope.

▼ Observe a prepared slide of skeletal muscle under the low and high powers of a compound microscope (figs. 17.6 and 17.7). Skeletal muscle fibers are long and multinucleate (syncytial). Within each muscle cell are thousands of regularly arranged **myofilaments (actin** and **myosin)** that give the tissue a striated appearance. Note the banding.

Associate the banding seen in the compound microscope with the diagrammatic representation shown in figure 17.8. ▲

A muscle cell or muscle fiber consists of bundles of myofilaments called **myofibrils.** Surrounding myofibrils is a membrane system, actually modified endoplasmic reticulum, called **sarcoplasmic reticulum.** Associated with

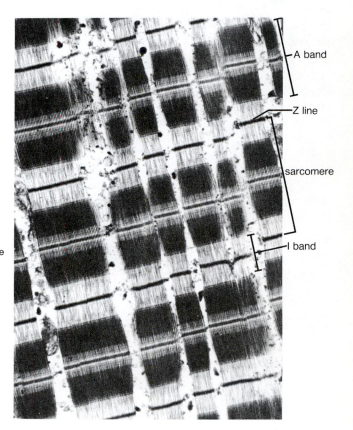

Figure 17.7 Skeletal muscle as viewed in a transmission electron micrograph. A small portion of a single, vertically oriented cell is shown here.

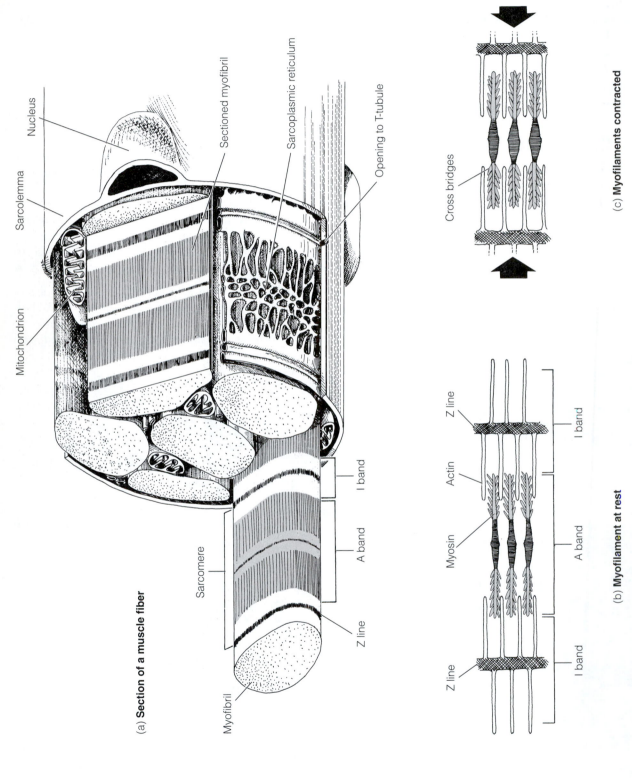

(a) **Section of a muscle fiber**

Nucleus

Sarcolemma

Mitochondrion

Sectioned myofibril

Sarcoplasmic reticulum

Opening to T-tubule

Sarcomere

Myofibril

Z line

A band

I band

Cross bridges

(c) **Myofilaments contracted**

Z line

Actin

Myosin

Z line

I band

A band

I band

(b) **Myofilament at rest**

Figure 17.8 Skeletal muscle ultrastructure. (*a*) Section of a muscle fiber (cell). A muscle fiber contains bundles of myofilaments, called myofibrils. Note the banding pattern of skeletal muscle shown in the myofibril. (*b*) Myofilaments (actin and myosin) at rest. (*c*) Contracted myofilaments.

the sarcoplasmic reticulum are invaginations of the sarco-lemma (muscle plasma membrane) called **transverse tubules (t-tubules).** The regular arrangement of the myofilaments, actin and myosin, results in the banding pattern observed in the microscope. The dark bands running across each muscle fiber are called **A bands.** They represent regions of overlap between actin and myosin. The light bands are called **I bands** and are regions of actin filaments. Down the middle of each I band is a **Z line** (not easily seen in the light microscope). The Z line is made of protein that runs across the muscle fiber and anchors longitudinally oriented actin filaments. The distance from one Z line to another is a single contractile unit called a **sarcomere.** As a muscle action potential (impulse) passes along the sarcolemma, it turns into the transverse tubules (t-tubules) at a Z line. As the action potential passes along the t-tubule, calcium ions are released from the sarcoplasmic reticulum. Calcium ions diffuse to the actin and myosin filaments, initiating contraction. During contraction, cross bridges on the myosin molecules interact with the actin to cause a sliding of actin relative to myosin, thus shortening the sarcomere. When multiplied over the length of the muscle fiber, the sliding produces a pronounced shortening of the muscle. Essential for the process is the enzymatic breakdown of ATP to ADP and Pi. The energy released facilitates the interaction of cross bridges and actin.

Smooth Muscle

Smooth muscles are associated with internal organs, cannot be consciously controlled, and are nonstriated.

▼ Examine a slide of smooth muscle (fig. 17.9). Myofilaments are not as regularly arranged in smooth muscle as in striated muscle, and so no banding is seen. Each muscle cell is spindle shaped and has a single nucleus. Try to find a region in your slide where muscle cells have been separated so individual cells can be seen. This is most likely to happen near one edge of the section. Observe under high power. ▲

Cardiac Muscle

Cardiac muscle is associated with the heart. Cardiac muscle shares characteristics of both smooth and striated muscle, being involuntary, striated, and uninucleate. Its cell boundaries between the ends of the cells are referred to as **intercalated disks.** These are plasma membranes specialized for conducting electrical impulses between cells. Cardiac muscle fibers are highly branched (fig. 17.10).

▼ Examine a slide of cardiac muscle. Under high power and reduced light, note the striations, the branched fibers, and the intercalated disks. ▲

17.3 LEARNING OUTCOMES REVIEW—WORKSHEET 17.2

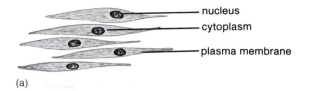

(a)

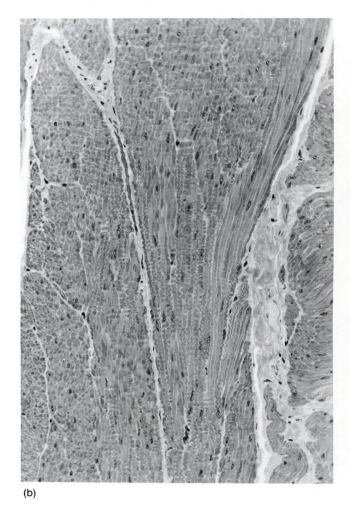

(b)

Figure 17.9 Smooth muscle: (*a*) diagram of smooth muscle with cells teased apart; (*b*) photomicrograph of smooth muscle in cross section (left) and longitudinal section (right).

17.4 MUSCULOSKELETAL INTERACTIONS

LEARNING OUTCOME

1. Predict actions of skeletal muscles based on knowledge of their points of origin and insertion on the skeleton.

The interdependence of skeletal muscle and the skeletal system is obvious. Neither could function without the other. A muscle functions by being attached at an immovable point at one end (the muscle's **origin**) and a movable point at the other end (the muscle's **insertion**).

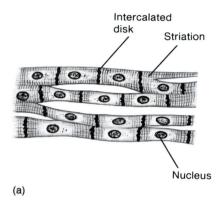

(a)

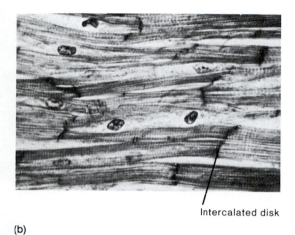

Intercalated disk

(b)

Figure 17.10 Cardiac muscle: (a) diagram;
(b) photomicrograph.

A muscle attaches on the skeleton at one or more of the tuberosities, trochanters, ridges, and processes. The action of a muscle is determined by the muscle's origin, insertion, and the nature of the joint over which it acts. The following terms are used to describe muscle actions:

Flexion   Decreases the size of the angle between the anterior surfaces of the articulated bones. Flexing movements are bending or folding movements.

Extension   Increases the size of the angle between the anterior surfaces of the articulated bones. Restores bones to the original position after flexing.

Abduction   Moves a bone away from the median plane of the body.

Adduction   Moves a bone toward the median plane of the body.

Rotation   The pivoting of a bone on its own axis.

▼ Examine a human skeleton on which origins and insertions of five muscles have been labeled. Describe the location of the origins, insertion, and actions of each of the muscles by filling in table 17.1 in worksheet 17.1. ▲

17.4 LEARNING OUTCOMES REVIEW—WORKSHEET 17.2

WORKSHEET 17.1 The Skeleton

1. Fishes have eyes on the sides of their heads and, thus, have excellent peripheral vision. Why is this advantageous for animals that lack a neck?

2. What must a fish do to "turn its head?"

3. How does the combination of skull, atlas, and axis move in nodding the head?

4. How does the combination of skull, atlas, and axis move in turning the head?

5. How are the cervical vertebrae of a turkey or goose different from those of a mammal?

6. Select axial skeleton illustrations that would characterize each of the vertebrate classes listed below. (Tr = trunk, Ca = caudal, S = sacral, Cv = cervical, L = lumbar, Th = thoracic)
   a. Chondrichthyes _____
   b. Actinopterygii _____
   c. Amphibia _____
   d. Reptilia _____
   e. Aves _____
   f. Mammalia _____

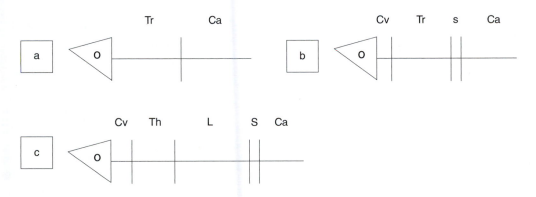

7. What is the function of each of the following regions of the vertebral column in terrestrial vertebrates?
   a. cervical
   b. thoracic
   c. lumbar
   d. sacral

8. Ribs of mammals attach to the axial skeleton at the _____ and _____. Do all ribs attach in the same manner? Explain.

9. Birds are sometimes said to have a fifth appendage (in addition to legs and wings). What is the fifth appendage, and why is it needed in addition to the other four?

10. Based on previous observations of live birds, what do you think are three functions of that fifth appendage?
    a. _____
    b. _____
    c. _____

11. Name two modifications of the axial skeleton of birds for flight.
    a. _____
    b. _____

12. Describe the functions of these modifications.

13. Why is the "hump" on the back of a bison so much larger than on most other grazing animals?

14. What functions does the clavicle play in the appendicular skeleton of most vertebrates?

15. How is the clavicle of a cat different from that of a human or a rat?

16. What are the implications of this arrangement for a cat that is walking? (Hint: What do you notice about the scapulae when a cat walks?)

17. What are the implications of this arrangement for a cat that jumps from a tree to the ground and lands on its front legs? (Hint: What happens to humans when we fall from a tree and land on our outstretched arms?)

18. What problems does the structure of the mammalian pelvic girdle present during the birth process?

Musculoskeletal Interactions

TABLE 17.1 The Origins, Insertions, and Actions of Human Muscles			
Muscle	Origin	Insertion	Action (Including Which Bone Moved)
1. Sternocleidomastoid			
2. Deltoid			
3. Biceps brachii			
4. Rectus femoris			
5. Adductor magnus			

WORKSHEET 17.2 Vertebrate Musculoskeletal Systems

17.1 Bone Structure—Learning Outcomes Review

1. Each end of a long bone is referred to as
   a. the diaphysis.
   b. an epiphysis.
   c. the osteon.
   d. the trunk.

2. Cancellous bone consists of a network of filaments called trabeculae. This type of bone tissue is found in
   a. the diaphysis.
   b. an epiphysis.
   c. the osteon.
   d. the trunk.

3. Circular layers of compact bone that surround an osteonic canal are called
   a. lacunae.
   b. canaliculi.
   c. lamellae.
   d. osteocytes.

17.2 The Skeleton—Learning Outcomes Review

4. If you discovered a new fossil vertebrate and its axial skeleton included a skull, a cervical vertebra, trunk vertebrae, a sacral vertebra, and caudal vertebrae, you should conclude that the fossil was a(n)
   a. bony fish.
   b. amphibian.
   c. reptile.
   d. bird.
   e. mammal.
   f. c–d could all be correct.

5. How could you distinguish the thoracic region of an axial skeleton from the lumbar region of an axial skeleton?
   a. The thoracic region would have ribs, the lumbar region would not have ribs.
   b. Lumbar vertebrae all have large transverse processes. Thoracic vertebrae have small transverse processes.
   c. Dorsal (neural) spines are usually larger on thoracic vertebrae—especially in the anterior thoracic region.
   d. All of the above are differences between thoracic and lumbar regions.

6. If you were trying to distinguish thoracic and lumbar regions of a skeleton you would be looking at a(n)
   a. bony fish.
   b. amphibian.
   c. reptile.
   d. bird.
   e. mammal.
   f. c–d could all be correct.

7. All of the following are adaptations of a bird's axial skeleton for flight except one. Select the exception.
   a. pygostyle
   b. synsacrum
   c. furcula

8. Side-to-side rotation of the skull of a mammal is accomplished through movements between these two bones (or sets of bones).
   a. skull and atlas
   b. atlas and axis
   c. axis and remaining cervical vertebrae
   d. cervical and thoracic vertebrae

9. Visceral arches are features of vertebrate skulls. Which of the following is false regarding these bones or cartilages?
   a. They form gill supports in fish.
   b. The first visceral arch was incorporated into the jaw of early vertebrates.
   c. They contribute to the hyoid bone of mammals.
   d. They are incorporated into the middle ear in the form of ear ossicles.
   e. All of the above are true.

17.3 The Muscular System—Histology—Learning Outcomes Review

10. Bundles of myofilaments in skeletal muscle are called
    a. actin.
    b. myosin.
    c. myofibrils.
    d. sarcomeres.

11. In a prepared slide of skeletal muscle you will see all of the following except one. Select the exception.
    a. multinucleate cells
    b. darkly stained A bands
    c. lighter I bands
    d. intercalated disks

12. One contractile unit of skeletal muscle is the distance between two adjacent Z lines. This contractile unit is called a
    a. sarcomere.
    b. myofiber.
    c. myofilament.
    d. I band.
    e. A band.

13. All of the following are characteristic of smooth muscle cells except one. Select the exception.
    a. spindle-shaped cells
    b. one nucleus per cell
    c. an absence of banding within the cells
    d. intercalated disks form cell boundaries

17.4 Musculoskeletal Interactions—Learning Outcomes Review

14. During muscle contraction the insertion of a muscle moves toward the origin of a muscle.
    a. True
    b. False

15. A movement that increases the angle between the anterior surfaces of articulated bones is called abduction.
    a. True
    b. False

17.1 Bone Structure—Analytical Thinking

1. Evaluate the statement that "bone is hard but not dead."

17.2   The Skeleton—Analytical Thinking

2. In 2004 a group of researchers discovered a fossil in the Canadian arctic. It was a fossil that showed fins, gills, and scales. It also had a widened, dorsoventrally flattened skull and forelimbs that had bones clearly homologous to the humerus, radius, ulna, and bones of the wrist and hand. In addition, this fossil had a pectoral girdle and a freely movable neck. These fossils were found in association with an ancient river bed. Assess the importance of this fossil, named *Tiktaalik*.

17.3   The Muscular System—Histology—Analytical Thinking

3. When a skeletal muscle contracts, would the following distances increase, decrease, or remain unchanged? Explain your answers.
   a. Z line-to-Z line distance

   b. A band width

   c. I band width

   d. sarcomere length

17.4   Musculoskeletal Interactions—Analytical Thinking

4. Skeletal muscles are usually arranged in pairs that oppose one another. Explain why this arrangement is necessary based on your knowledge of how muscles work.

Exercise 18

Vertebrate Nervous Regulation

Prelaboratory Quiz

Study this week's laboratory exercise and then complete the following quiz to assess your preparation for the laboratory.

1. A nerve impulse is carried toward the central nervous system through a(n)
   a. motor neuron.
   b. sensory neuron.
   c. effector organ.
   d. sensory receptor.

2. A(n) _____ initiates an impulse in a sensory neuron.
   a. effector organ
   b. motor neuron
   c. sensory receptor
   d. neurolemmocyte

3. The cell body of a multipolar (motor) neuron is associated with
   a. a sensory receptor.
   b. the central nervous system or a ganglion.
   c. an effector organ.
   d. Both a and c are correct.

4. In this laboratory you will be examining a prepared microscope slide of a neuromuscular junction. In this slide you will be seeing
   a. a synapse between a nerve and a muscle.
   b. a synapse between two nerves.
   c. the dendritic region of a motor neuron.
   d. the dendritic region of a sensory neuron.

5. In this exercise, you will be studying the brain structure of a(n)
   a. fish and a bird.
   b. bird and a mammal.
   c. amphibian and a mammal.
   d. fish and a mammal.

6. True/False The embryonic region of the vertebrate brain, called the metencephalon, develops into the cerebral hemispheres of the adult vertebrate.

7. True/False The region of the vertebrate brain that is continuous with the spinal cord is the medulla oblongata.

8. True/False The pituitary gland connects to the olfactory tract of the vertebrate brain.

9. True/False The embryonic diencephalon develops into the epithalamus, hypothalamus, and optic lobes of the adult vertebrate brain.

10. True/False The peripheral nervous system has two subdivisions: the somatic nervous system and the autonomic nervous system.

The regulation and coordination of physiological processes in the vertebrate body are the roles of two closely linked systems: the **nervous system** and the **endocrine system.** These coordinating systems function by allowing the body to respond to changes in the environment, resulting in the maintenance of constant internal conditions.

The nervous system functions through very rapid **nerve impulses** that travel along nerve cells (neurons). A response always involves a **sensory receptor** receiving a stimulus and initiating an impulse in a **sensory neuron** (fig. 18.1). The sensory neuron carries the impulse to the **central nervous system** (brain or spinal cord). After processing this sensory information, the central nervous system then initiates an impulse that travels via **motor neurons** to an **effector organ** (the responding structure).

The endocrine system has a similar basic organization. Replacing the nerve impulse, however, is a chemical called a **hormone,** and replacing the nerve cell is the bloodstream. The nervous system often acts as an intermediary between the environment and the endocrine system. That is, the nervous system receives information regarding environmental changes and conveys that information to the endocrine system. An endocrine gland then responds by secreting a hormone that initiates changes in the animal, which compensate for the environmental change.

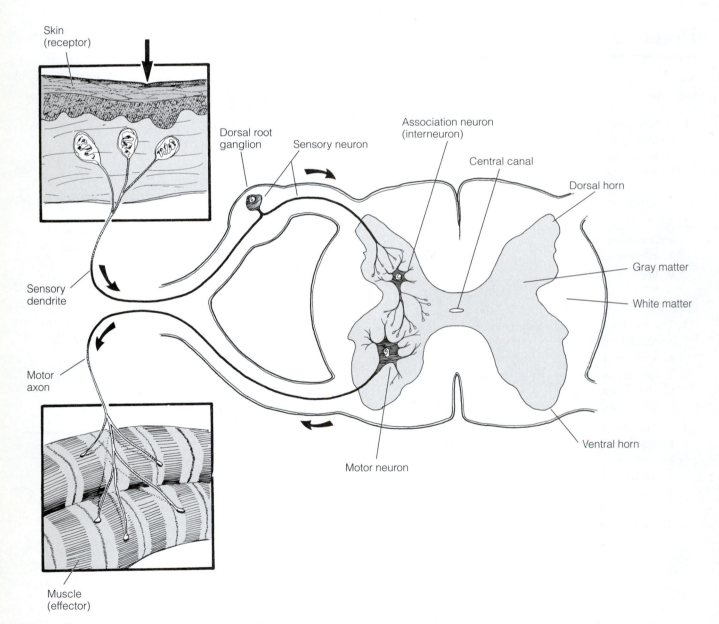

Figure 18.1   Cross section of the spinal cord showing components of a response. A response involves a sensory receptor receiving a stimulus and initiating an impulse in a sensory neuron, which conducts the impulse to the central nervous system. A motor neuron then conducts an impulse to the responding structure.

18.1 THE NEURON

LEARNING OUTCOMES

1. Identify parts of a neuron from slide preparations.
2. Describe the function of a synapse based on the observations of the neuromuscular junction.

The basic functional unit of the nervous system is the nerve cell, or **neuron.** Two of its variations are shown diagrammatically in figure 18.2. The most common of these will be used to illustrate neuron structure. The **multipolar neuron** is a motor neuron whose nerve cell body is usually associated with the central nervous system or a ganglion (see fig. 18.2a). It ends at an effector structure. The enlarged **nerve cell body** houses the nucleus and other cytoplasmic organelles involved with cell maintenance. A **dendritic region** has small processes radiating out from the nerve cell body. These fibers are where the impulse is generated.

▼ Examine a slide of an ox spinal cord smear. Under low power locate star-shaped nerve cell bodies of multipolar neurons. Switch to high power and locate a relatively clear nucleus containing a darkly stained nucleolus. The processes radiating out from the nerve cell body are mostly **dendrites** (fig. 18.3). Although it is difficult to distinguish, one of the processes radiating out from the nerve cell body is the axon. ▲

The **axon** propagates the impulse from the nerve cell body to the effector structure. The axon is covered with a fatty insulating material called myelin. **Myelin** is produced by and contained in **neurolemmocytes,** which wrap around the axon. Figure 18.4 shows a transmission electron micrograph of a myelinated axon. Myelin is a lipid (fat) found in the plasma membrane of the neurolemmocyte. The figure also shows diagrammatically how degrees of myelination can be achieved by the neurolemmocyte wrapping around the axon. Axons are said to be white, or myelinated, if they contain relatively large quantities of myelin. Axons are said to be gray, or nonmyelinated, if they contain less myelin. The significance of neurolemmocytes is that they increase nerve impulse velocity. This is called **saltatory conduction** and can be visualized as the impulse "jumping" from node (region between neurolemmocytes) to node.

▼ Examine a cross section of a medullated nerve under low power. Note that a nerve is a bundle of many neurons. To see individual neuron cross sections, focus on the interior of the nerve and switch to high power. Each neuron can be seen as a light dot surrounded by layers of darkly stained membranes. These membranes are the myelin-containing neurolemmocyte membranes. ▲

The axon of a multipolar neuron ends at the effector structure, where fine branches called terminal fibers end at a **synapse.** The synapse is the region where an impulse is transferred to an effector structure or another nerve cell. A motor neuron ending on a muscle cell has a synapse referred to as a **neuromuscular junction.** Here, the impulse reaching the synapse causes the release of a **neurotransmitter** from **synaptic vesicles.** The neurotransmitter diffuses across the **synaptic cleft** to initiate a muscle action potential and contraction of the muscle.

▼ Examine a slide showing the neuromuscular junction. Figures 18.5 and 18.6 show light and electron micrographs of this synapse. ▲

18.1 LEARNING OUTCOMES REVIEW—WORKSHEET 18.1

18.2 THE CENTRAL NERVOUS SYSTEM

LEARNING OUTCOMES

1. Identify major structures of the shark brain.
2. Identify major structures of the sheep brain.
3. Describe major evolutionary changes in the vertebrate brain between the fish and mammalian extremes.

The central nervous system, consisting of the **brain** and the **spinal cord,** is hollow and contains **cerebrospinal fluid.** Recall that a dorsal tubular nervous system is one of the unique characteristics of the Chordata. In the central nervous system, sensory information from a variety of sources is processed. This processing is called **integration** and occurs within the gray matter. The central nervous system is organized into regions of gray and white matter. **Gray matter** (usually inside) consists of nerve cell bodies and nonmyelinated axons. **White matter** is made up of myelinated fiber tracts carrying impulses from one part of the central nervous system to another.

The Spinal Cord

▼ Use lower power of a compound or dissecting microscope to examine a spinal cord sectioned through the dorsal root ganglion (see fig. 18.1). Distinguish between the white matter and gray matter. The gray matter is organized into an H-shaped central region. White matter lies outside the gray. Note the small central canal in the center of the gray matter. The **central canal** extends up to the brain, where it is continuous with the ventricles (canal system) of the brain. Note the roots of the spinal nerves associated with the spinal cord. In the mammal, the **dorsal root** has sensory neurons entering the spinal cord, and the **ventral root** has motor neurons leaving the spinal cord. The enlarged **dorsal root ganglion** is the location of nerve cell bodies of sensory neurons. ▲

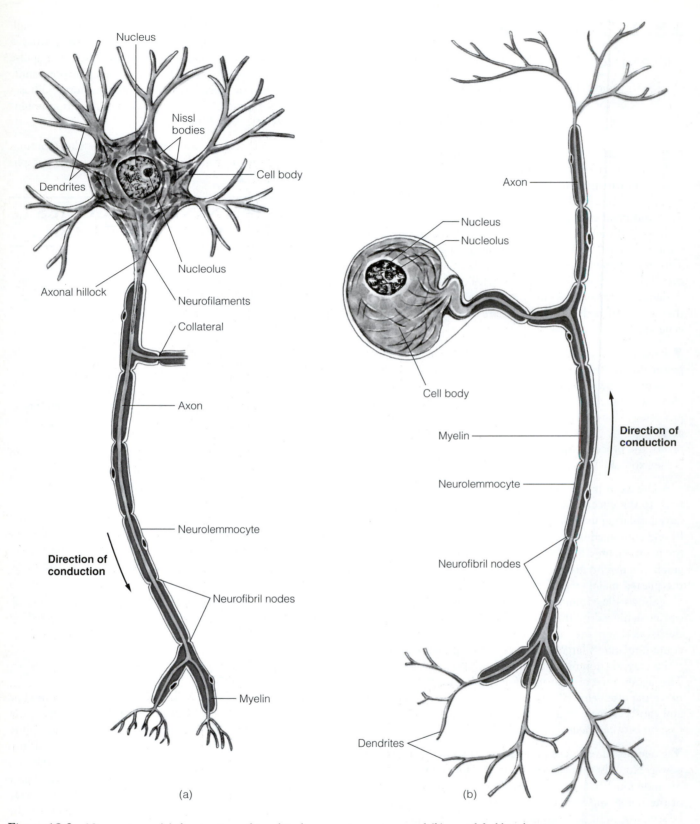

Figure 18.2 Neuron types: (*a*) the structure of a multipolar, or motor, neuron, and (*b*) a modified bipolar, or sensory, neuron.

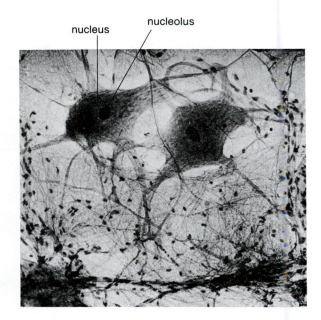

Figure 18.3 Photomicrograph of motor neurons from the vertebrate spinal cord.

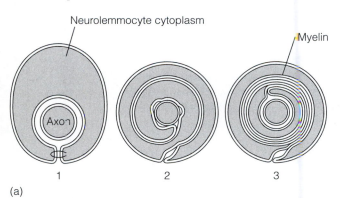

(a)

Figure 18.4 Cross section of an axon and neurolemmocyte. (a) Diagrammatic representation showing how degrees of myelination can be achieved. Successive wraps of a neurolemmocyte provide increasing myelination. (b) Transmission electron micrograph. The arrow points to the myelin sheath (×73,000).

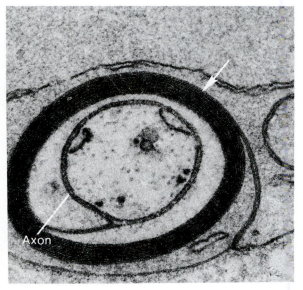

(b)

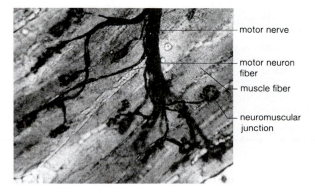

Figure 18.5 Photomicrograph of the neuromuscular junction (×175).

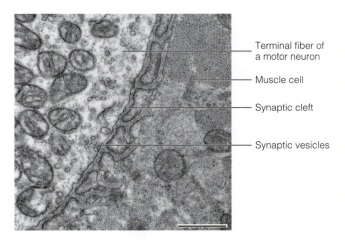

Figure 18.6 Transmission electron micrograph of a neuromuscular junction (×46,000). Synaptic vesicles contain a neurotransmitter (acetylcholine) that is released into the synaptic cleft in response to an action potential arriving at the synapse. A muscle action potential is initiated that results in muscle contraction.

The Brain

The primary area for integration within the central nervous system is the brain. Embryologically, the brain develops as three anterior swellings of the spinal cord. The **prosencephalon (forebrain)** develops as an anterior swelling associated with olfaction. The **mesencephalon (midbrain),** which develops just posterior to the prosencephalon, functions in visual integration. The posterior-most swelling is the **rhombencephalon (hindbrain).** It develops in association with the sense of equilibrium and balance and, later, hearing. Each of these regions then further differentiates as indicated in table 18.1.

The Shark Brain

The brain of *Squalus* is typical of many lower vertebrates (figs. 18.7 and 18.8). In it we can see the general form for the five primary regions of the brain. A demonstration dissection has been provided for you (figure 18.9).

▼ In the anterior-most region, note the large **nasal capsules,** which are the locations for sensory receptors associated with the sense of smell. Attached to the nasal capsules are the **olfactory bulbs,** which narrow posteriorly to form the **olfactory tract.** This tract represents the pathway for

TABLE 18.1   The Fate of the Three Embryonic Brain Regions of Vertebrates

Earlier Embryonic Structures	Later Embryonic Structures	Adult Structures
Prosencephalon (Forebrain)	Telencephalon	Cerebral hemispheres
	Diencephalon	Epithalamus, thalamus, hypothalamus
Mesencephalon (Midbrain)	Mesencephalon	Optic lobes (tectum, corpora quadrigemina)
Rhombencephalon (Hindbrain)	Metencephalon Myelencephalon	Cerebellum Medulla oblongata

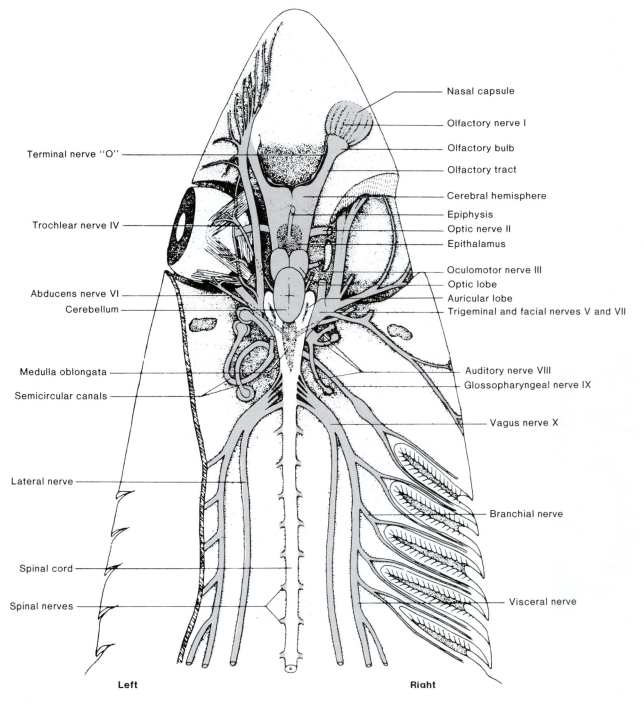

Left   Right

Figure 18.7   Shark head showing brain and cranial nerves (dorsal view).

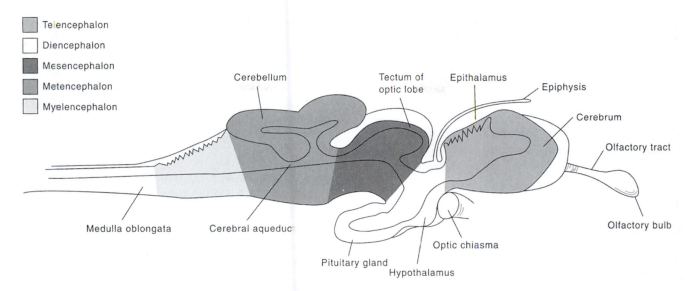

Figure 18.8   Sagittal section of a shark brain. Embryological regions are indicated by different shading patterns.

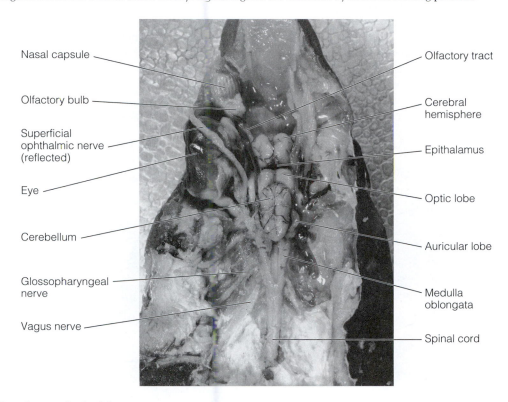

Figure 18.9   Dorsal view of a shark brain.

olfactory information coming back to the **cerebral hemispheres** (swellings located just posterior and slightly medial to the olfactory bulbs and olfactory tracts). In fishes the cerebral hemispheres process olfactory information.

Just posterior to the cerebral hemispheres is the **thalamus.** Its very thin roof is the **epithalamus. The epiphysis** is attached to the epithalamus. The epiphysis is a thin stalk of tissue that hormonally regulates functions influenced by photoperiod. This is very delicate and is probably missing from your specimen. The bulk of the thalamus is an area of synapse for sensory and motor information traveling between the cerebrum and the mesencephalon. The floor of

the diencephalon is the **hypothalamus.** If the brain you are looking at is still embedded in the chondrocranium, the hypothalamus will be impossible to see. The hypothalamus is the primary visceral control center. Attached to it ventrally is the **pituitary gland.** The hypothalamus is the area that coordinates nervous and endocrine systems.

Posterior to the thalamus is the midbrain. In lower vertebrates this region is also referred to as the **optic lobes.** Note the two large dorsal swellings. The roof of the midbrain is referred to as the **tectum.** In fishes it is the primary area for the integration of sensory information and coordination of motor responses.

Posterior to the midbrain is the **cerebellum.** The **body of the cerebellum** is the large, median, oval mass lying over the top of the brain stem. Extending laterally are the earlike **auricular lobes.** The body of the cerebellum serves as a center for muscle coordination, and the auricular lobes function in equilibrium and balance.

Posterior and ventral to the cerebellum is the **medulla oblongata.** It tapers into the spinal cord and functions as a conducting pathway between the spinal cord and higher brain centers. It is also the point of origin for a number of cranial nerves. ▲

The Mammalian (Sheep) Brain

▼ If your specimen has the **meninges** removed, observe a demonstration specimen that still has at least some of the meninges intact. The outer tough layer of connective tissue is the **dura mater.** Below the dura is a loose, weblike **arachnoid layer,** and tightly adhering to the surface of the brain is the **pia mater.** Between these layers **cerebrospinal fluid** circulates and forms a cushion for the brain. All of these layers extend down around the spinal cord as well.

The most obvious structure in the sheep brain is the **cerebrum** (fig. 18.10). Note that it is proportionately much larger in the sheep than the shark. This reflects its role in coordination and integration, which it has taken over from the tectum of fishes. Note the highly convoluted **cerebral cortex.** These convolutions greatly increase the surface area. The cerebral cortex consists of masses of gray matter pushed to the outside of the brain due to the increased size and importance of the cerebrum. The two **cerebral hemispheres** are separated by the **longitudinal cerebral fissure.** Posterior to the cerebral hemispheres is the **cerebellum.** The cerebellum of mammals remains a center for motor coordination and equilibrium and balance. Carefully spread the cerebellum from the cerebral hemispheres. Note the four swellings referred to as **corpora quadrigemina.** This is the roof of the midbrain. In the shark this was the tectum and represented the major site for nervous coordination and integration. In mammals the function is limited to optic and auditory reflexes. Extending posteriorly from the cerebellum is the **medulla.** In mammals the medulla functions as a respiratory center, awakening center, cardioregulatory center, and as a pathway for sensory and motor impulses going between the spinal cord and higher brain centers.

In a lateral view (see fig. 18.10), find the **olfactory bulb** on the anterioventral border of the cerebrum. Extending

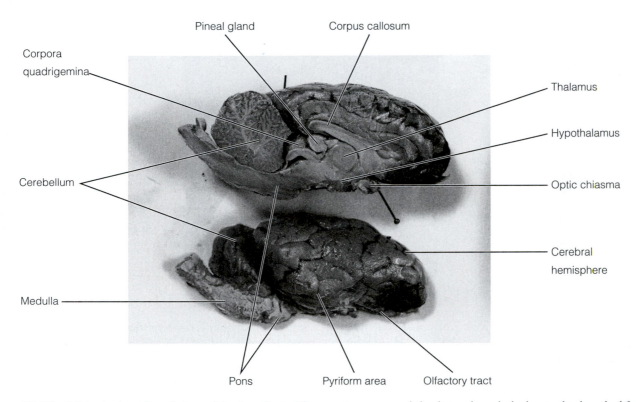

Figure 18.10 Midsagittal and lateral views of the sheep brain. These sections were made by slicing through the longitudinal cerebral fissure.

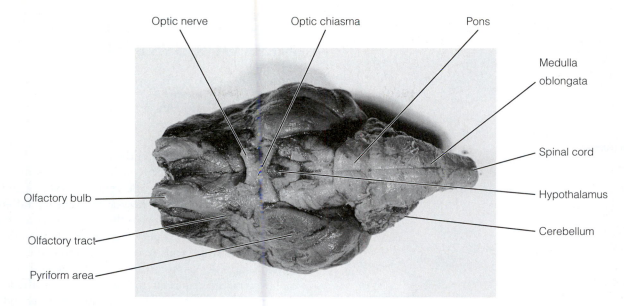

Optic nerve   Optic chiasma   Pons

Medulla oblongata

Spinal cord

Hypothalamus

Cerebellum

Olfactory bulb

Olfactory tract

Pyriform area

Figure 18.11   Ventral view of a sheep brain.

posteriorly is the **olfactory tract.** It ends in the **pyriform area** of the cerebral cortex. Olfactory information is processed here.

Turn the brain to a ventral view (fig. 18.11). Locate again the olfactory bulb, olfactory tract, and pyriform area. Medial and slightly anterior to the pyriform area is the **optic chiasma,** where **optic nerves** bring sensory information from the eyes to the brain through the diencephalon. Posterior to the optic chiasma is the main visceral control center, the **hypothalamus.** The **pituitary gland** connects to the brain at this point. It has most likely been broken off as the brain was removed from the skull. If available, examine a brain in which the pituitary gland has been left in place. Posterior to the hypothalamus is a broad enlargement that is derived from the metencephalon. This is the **pons.** The pons is a relay center for information going between the cerebrum and cerebellum. Posteriorly, the pons narrows into the medulla. Note the attachment points of one or more cranial nerves. Individual nerves will not be identified.

Finally, examine a midsagittal section of the sheep brain (see fig. 18.10). Find the structures previously mentioned that can be seen from this view. In addition, find

the **thalamus,** which serves as a relay site between upper and lower brain centers. Note also the canal system that is continuous with the central canal of the spinal cord. These canals contain cerebrospinal fluid. ▲

Brain Evolution

We have just examined two extremes in the evolution of the vertebrate brain. You may have noticed that two regions of the brain (the telencephalon and mesencephalon) underwent profound structural changes during vertebrate evolution.

▼ To appreciate these changes, examine figure 18.12 and any models or preserved brains available for study. Compare the structure and size of each brain region in the vertebrate classes represented. ▲

18.2 LEARNING OUTCOMES REVIEW—WORKSHEET 18.1

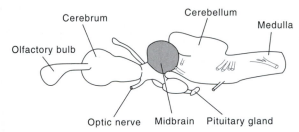

Fish (Shark)

Cerebrum · Cerebellum · Medulla · Olfactory bulb · Optic nerve · Midbrain · Pituitary gland

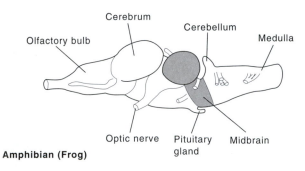

Amphibian (Frog)

Cerebrum · Cerebellum · Medulla · Olfactory bulb · Optic nerve · Pituitary gland · Midbrain

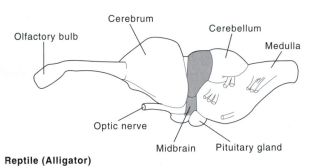

Reptile (Alligator)

Cerebrum · Cerebellum · Medulla · Olfactory bulb · Optic nerve · Midbrain · Pituitary gland

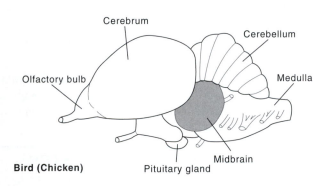

Bird (Chicken)

Cerebrum · Cerebellum · Medulla · Olfactory bulb · Midbrain · Pituitary gland

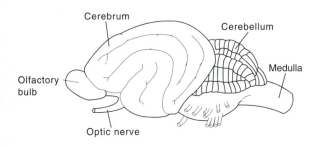

Mammal (Cat)

Cerebrum · Cerebellum · Medulla · Olfactory bulb · Optic nerve

18.3 THE PERIPHERAL NERVOUS SYSTEM

LEARNING OUTCOME

1. Describe the structure of the human peripheral nervous system.

The final component of the nervous system is the peripheral nervous system. It composed of all nerves and ganglia outside the central nervous system. It is comprised of two sets of nerves: the **sensory (afferent) nerves** transmit information to the central nervous system and the **motor (efferent) nerves** transmit information away from the central nervous system. The **autonomic nervous system** is a subset of motor nerves and ganglia that innervate visceral structures. It mediates parasympathetic and sympathetic or "fight or flight" responses. Sensory and motor nerves consist of bundles of neurons leaving the central nervous system and going to the structure being innervated. (Recall your observations of the nerve cross section earlier in this laboratory.) Some of these nerves are associated with the brain (**cranial nerves**), and other nerves are associated with the spinal cord (**spinal nerves**). Spinal nerves nerves exit the central nervous system between vertebrae in the cervical, thoracic, lumbar, and sacral regions.

▼ Examine a model of a human torso for **spinal nerves.** Notice in the cervical, upper thoracic, lumbar, and sacral regions that the nerves fuse with neighboring nerves, forming a **plexus**. A **cervical plexus** represents cervical spinal nerves fusing, then going to the neck and upper thorax. A **brachial plexus** (lower cervical and upper thoracic spinal nerves) goes to the pelvis and legs. A **lumbosacral plexus** (lumbar and sacral spinal nerves) goes to the pelvis and legs. Spinal nerves supplying the lower two-thirds of the thorax come independently to the structures they innervate. We will not distinguish anatomically between components of the somatic and autonomic nervous systems. ▲

18.3 LEARNING OUTCOMES REVIEW—WORKSHEET 18.1

Figure 18.12   Brains of representatives of the vertebrate classes. The midbrain (mesencephalon) is shaded in gray. Note the relative sizes of the brain regions in each class representative. The mesencephalon of the mammalian brain is covered by the expanded and highly convoluted cerebrum.

WORKSHEET 18.1   Vertebrate Nervous Regulation

18.1 The Neuron—Learning Outcomes Review

1. A motor neuron
   a. carries an impulse toward the central nervous system.
   b. begins at a sensory receptor.
   c. carries an impulse away from the central nervous system.
   d. ends at a sensory receptor.

2. Which of the following parts of a neuron were you able to observe in slide preparations in the laboratory?
   a. nerve cell body
   b. dendrites
   c. axons
   d. all of the above

3. Which of the following structures were associated with neurolemmocytes?
   a. nerve cell body
   b. dendrites
   c. axons
   d. all of the above

4. Synaptic vesicles release their contents into
   a. axon terminal fibers.
   b. dendrites.
   c. nerve cell bodies.
   d. the synaptic cleft.

18.2 The Central Nervous System—Learning Outcomes Review

5. White matter of the central nervous system is comprised mainly of
   a. nonmyelinated axons.
   b. synapses.
   c. myelinated axons carrying information between regions of the central nervous system.
   d. ganglia.

6. In the spinal cord cross section, gray matter was observed
   a. inside of white matter.
   b. outside of white matter.
   c. mixed evenly through the spinal cord section.

7. The prosencephalon of the embryonic vertebrate brain develops into all of the following adult structures except one. Select the exception.
   a. cerebral hemispheres
   b. thalamus
   c. epithalamus
   d. cerebellum

8. In the shark brain, the anterior portion of the cerebral hemispheres is associated with the
   a. optic lobes.
   b. epithalamus.
   c. olfactory tracts.
   d. cerebellum.

9. All of the following are true of the cerebellum of the shark brain except one. Select the exception.
   a. It originates embryologically from the metencephalon.
   b. It serves as a center for muscle coordination.
   c. It is located just anterior to the optic lobes.
   d. It includes auricular lobes that function in equilibrium and balance.

10. All of the following structures of the sheep brain are visible from a lateral view except one. Select the exception.
    a. olfactory tract and bulb
    b. cerebellum
    c. corpora quadrigemina
    d. pyriform area

11. These are swellings on the roof of the mesencephalon of the sheep brain. To see them you needed to carefully spread the cerebellum from the cerebral hemispheres. They function in optic and auditory reflexes in the sheep.
    a. corpora quadrigemina
    b. olfactory bulbs
    c. optic bulbs
    d. pyriform bulbs
    e. Both b and c are correct.

12. According to the laboratory manual, the major center for the integration of all sensory information and the coordination of motor responses in fishes is (are) the
    a. cerebral hemispheres.
    b. roof of the midbrain (tectum).
    c. medulla oblongata.
    d. cerebellum.

13. The function of the cerebral hemispheres in fishes is
    a. olfaction.
    b. vision.
    c. integration of all sensory information and coordination of motor responses.
    d. equilibrium and balance.

14. According to the laboratory manual, the major center for the integration of all sensory information and the coordination of motor responses in mammals is (are) the
    a. cerebral hemispheres.
    b. roof of the midbrain (tectum).
    c. medulla oblongata.
    d. cerebellum.

15. The function of the midbrain in mammals is
    a. visual and auditory reflexes.
    b. cardiac control.
    c. integration of all sensory information and coordination of motor responses.
    d. equilibrium and balance.

16. Which of the following statements is false regarding anatomical and functional comparisons of shark and sheep brains?
    a. Cerebral hemispheres of the sheep brain are greatly enlarged in comparison to the shark, and their size reflects their role in integration and coordination.
    b. Cerebral hemispheres of the shark brain are much larger relative to other brain regions because of their role in integration and coordination.
    c. The midbrain of the shark is relatively large in comparison to other shark brain regions, and its size reflects its important functions.
    d. A major trend in the evolution of the vertebrate brain is the shifting of integration and coordination functions from the midbrain (as seen in the shark) to the cerebral hemispheres (as seen in the sheep).

18.3   The Peripheral Nervous System—Learning Outcomes Review

17. Motor neurons that innervate visceral structures like the heart are part of the
    a. central nervous system.
    b. autonomic nervous system.
    c. thoracic nervous system.
    d. peripheral nervous system.
    e. Both b and d are correct.

18. The cervical plexus is the
   a. fusion of cranial nerves leaving the brain and innervating structures in the neck region.
   b. fusion of cervical spinal nerves that innervate structures in the neck and upper thorax.
   c. fusion of thoracic spinal nerves that innervate structures in the neck region.
   d. fusion of cranial nerves passing through the neck and innervating structures in the upper thorax.

18.1   The Neuron—Analytical Thinking

1. The knee-jerk reflex involves two neurons. A sensory neuron takes the stimulus from the knee to the spinal cord and a motor neuron takes the response from the spinal cord to the muscles of the leg. Trace this response pathway noting the kind of neurons involved, the portions of this pathway involving myelnated vs. nonmyelinated neuron processes, structures or regions associated with the spinal cord, and synaptic events involved within the spinal cord.

18.2   The Central Nervous System—Analytical Thinking

2. Regions of the brains of some vertebrates are specialized for sensory functions that have become very important in a particular group. Birds, for example depend on highly developed visual senses. Many fishes, such as lampreys and salmon depend on olfaction. Explain why vision is so important in birds and olfaction is so important in lampreys and salmon. What parts of the brain of birds and the brain of these fish do you think would be especially well developed in association with these sensory functions?

3. Explain why it would be difficult to associate a particular cervical spinal nerve with a particular receptor in the skin based on tracing the nerve away from the spinal cord.

Exercise 19

Vertebrate Circulation

Prelaboratory Quiz

Study this week's laboratory exercise and then complete the following quiz to assess your preparation for the laboratory.

1. The most abundant kind of white blood cell that functions in phagocytosis of foreign substances is a(n)
   a. neutrophil.
   b. eosinophil.
   c. lymphocyte.
   d. monocyte.
   e. basophil.

2. The white blood cell that functions in immune responses is a(n)
   a. neutrophil.
   b. eosinophil.
   c. lymphocyte.
   d. monocyte.
   e. basophil.

3. All of the following are present in the heart of a fish EXCEPT a
   a. sinus venosus.
   b. conus arteriosus.
   c. two-chambered ventricle.
   d. single-chambered atrium.

4. Exchanges of gases, nutrients, and wastes occur across
   a. venules.
   b. capillaries.
   c. arterioles.
   d. All of the above are correct.

5. Strands of connective tissue that anchor atrioventricular valves to the heart walls of mammals are called
   a. coronary strands.
   b. valve cords.
   c. ventricular strands.
   d. chordae tendineae.

6. True/False   Blood leaving the left ventricle of the mammalian heart enters the pulmonary artery.

7. True/False   The heart of a bird has completely divided atria and ventricles, and the sinus venosus is a large chamber that receives blood from the vena cavae and serves as a pacemaker.

8. True/False   In the electrocardiogram, the QRS complex corresponds to ventricular depolarization.

9. True/False   The dissection of the rat circulatory system in this exercise will begin by observing veins in the abdominal region of the rat.

10. True/False   Blood is pumped from the heart of a shark to the gills and immediately returns to the heart before circulating through the rest of the body.

The structure and function of vertebrate circulatory systems is complex. In this section we will be able to touch on major aspects of both structure and function in only a few representative organisms. The circulatory system is the means of transport for oxygen, carbon dioxide, food, metabolic wastes, and hormones in the vertebrate body. In addition, the circulatory system helps to maintain body temperature and proper pH and to fight disease.

The blood and lymph are functionally the most important parts of the circulatory system because they carry substances to and from all parts of the body. The heart circulates blood through a system of closed vessels. A **closed circulatory system** is one in which blood is confined to vessels throughout the system. On leaving the heart, blood enters **arteries.** Arteries break up into **arterioles** and finally **capillaries.** It is in the very tiny capillaries that exchange with the tissues takes place. Capillaries coalesce into **venules,** and venules into **veins.** Blood pressure in venules and veins is very low. Valves are present here to ensure one-way movement of blood back to the heart. The **lymphatic system** is an arrangement of vessels that provides a means for getting excess tissue fluids and proteins back to the bloodstream, as well as serving in body defenses. At the tissues, tissue fluids and proteins enter the lymphatic system through porous walls of blind-ending lymphatic capillaries. The lymphatic vessels have valves that ensure a one-way movement of fluids back toward the heart. In higher vertebrates, lymph nodes are located along lymphatic vessels. These serve as sites for the storage of white blood cells (lymphocytes) involved with the body's immune system. The spleen is similar to a lymph node, but it also functions with the cardiovascular system in storing red blood cells, destroying worn-out red blood cells, and producing red and white blood cells.

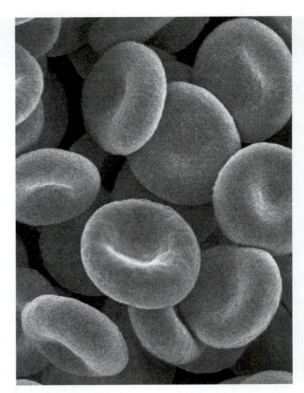

Figure 19.1 Human erythrocytes, scanning electron micrograph (×6,250).

19.1 FORMED ELEMENTS

LEARNING OUTCOMES

1. Describe the function of seven types of formed elements in vertebrate blood.
2. Identify red blood cells and white blood cells using a compound microscope.

Blood consists of **plasma** and **formed elements.** Plasma (55% by volume) consists of water, dissolved salts, suspended proteins, sugars, gases, and other suspended and dissolved materials. The remaining 45% of blood volume consists of formed elements—blood cells and platelets. Formed elements are classified as follows:

1. **Red blood cells (erythrocytes,** fig. 19.1). Contain hemoglobin that functions in carrying oxygen to and from the tissues.
2. **White blood cells (leukocytes,** fig. 19.2).
   Granular leukocytes—Contain cytoplasmic granules and irregular nuclei.

   a. **Neutrophils**—The most abundant leukocyte. Phagocytizes microbes or other injurious particles. About 60% of WBCs.
   b. **Eosinophils**—Phagocytize foreign proteins and immune complexes. Tend to proliferate in allergic reactions and parasitic infections. About 3% of WBCs.
   c. **Basophils**—Play a role in preventing blood clotting by releasing heparin. Promote inflammatory responses by releasing histamine. About 1% of WBCs.

   Nongranular leukocytes—Lack cytoplasmic granules. Regular nuclei present.

   a. **Lymphocytes**—Function in the body's immune response. Second most abundant leukocyte. About 30% of WBCs.
   b. **Monocytes**—Phagocytize microbes and other injurious particles. About 6% of WBCs.
3. **Platelets.** Cellular fragments containing several compounds that facilitate clotting.

Examining Blood Smears

▼ In this exercise you will examine a slide of human blood. Obtain a prepared slide of human blood. Examine it under low and high power of your compound microscope. Erythrocytes will be stained red, leukocyte nuclei dark blue.

Pseudopodia

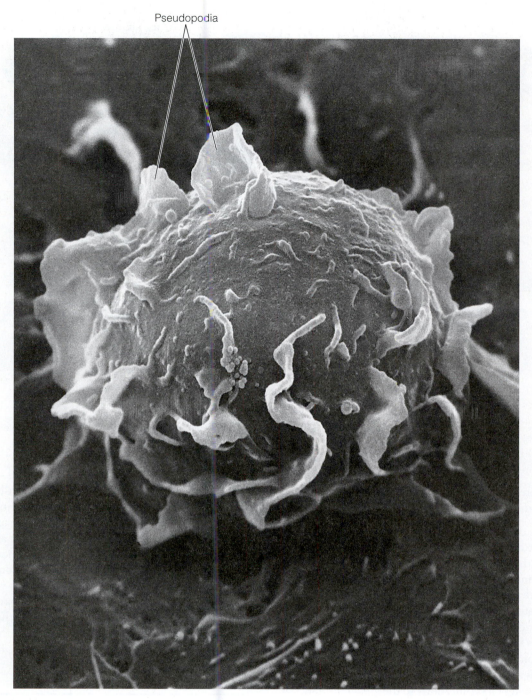

Figure 19.2   Monocyte (macrophage) from the human lung (×16,000).

Compare what you see to figure 19.3. Find as many kinds of white blood cells as you can. If your instructor directs, examine the slide under oil immersion. Try to appreciate the relative numbers of different cell types. You may also observe small, dark, irregular spindles or disks. These are platelets. ▲

19.1 LEARNING OUTCOMES REVIEW—WORKSHEET 19.1

19.2 SYSTEMIC CIRCULATION—THE SHARK

LEARNING OUTCOMES

1. Identify structures of the shark heart.
2. Trace the path of blood through the shark heart.
3. Trace the path of blood from the ventral aorta of a shark through the gills to the dorsal aorta.
4. Identify major veins of the shark.

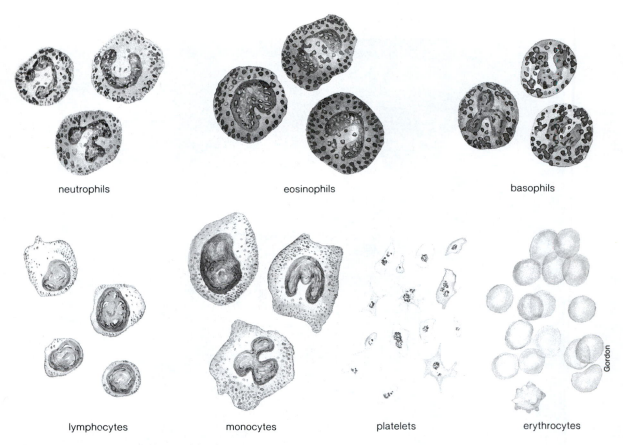

neutrophils   eosinophils   basophils

lymphocytes   monocytes   platelets   erythrocytes

Figure 19.3   Formed elements of human blood.

In this section you will use demonstration dissections and models to study circulation in fishes.

▼ Examine a dissected *Squalus* specimen (fig. 19.4). Locate the heart just anterior to the large gray liver. Blood enters the heart at the thin-walled sac at the posterior end; this is the **sinus venosus.** In mammals, the sinus venosus has been reduced to the **sinoatrial node.** In all vertebrates, the sinus venosus initiates the heartbeat. Blood then enters the single **atrium** (anterior and dorsal to the sinus venosus). The **ventricle** receives blood from the atrium. (The ventricle is the primary pumping structure of the heart.) The blood then enters the **conus arteriosus** and goes to the **ventral aorta.** The ventral aorta takes blood anteriorly to the gills. The branches to the gills are called **afferent branchial arteries.** Trace one or two of these vessels forward to the gills. In the gills, the vessels break into capillaries for gas exchange. Compare the dissection to figure 19.5 and any models available.

At this point, fold open the lower jaw as shown in figure 19.6. In doing so, notice the vessels on the roof of the mouth that come from the gills (fig. 19.7). These are **efferent branchial arteries.** They join medially to form the **dorsal aorta.** The dorsal aorta then runs posteriorly, delivering blood to the body. Trace the aorta a short distance. Blood returns to the heart through a series of veins that may not be injected in your specimen. Major vessels that return blood to the sinus venosus are the **anterior cardinals** (from the head), **posterior cardinals** (from the body),

and the **lateral abdominals.** The latter can be seen along the lateral body wall. The posterior cardinals empty as large sinuses into the sinus venosus and lie along the dorsal body wall. The anterior cardinals enter anteriorly and laterally into the sinus venosus. These cannot be seen easily because they are embedded in muscles of the head. All three of these veins unite to enter the sinus venosus at the **common cardinal.** The shark also has a **hepatic portal system** that delivers blood from capillaries of the intestine to the liver. **Hepatic veins** deliver blood from the liver to the sinus venosus. Find the hepatic portal vein, but do not try to find the hepatic veins. ▲

19.2 LEARNING OUTCOMES REVIEW—WORKSHEET 19.1

19.3 THE MAMMALIAN HEART

LEARNING OUTCOMES

1. Indentify external structures of the mammalian heart.
2. Identify internal structures of the mammalian heart.
3. Trace the path of blood through the mammalian heart.

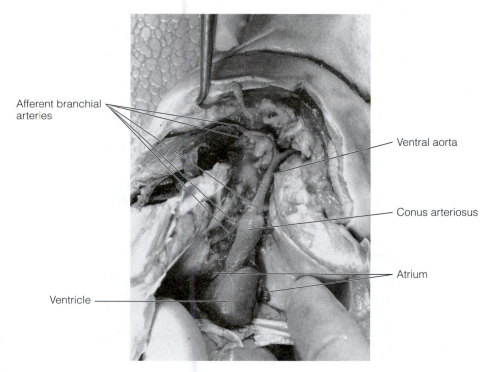

Afferent branchial arteries

Ventral aorta

Conus arteriosus

Atrium

Ventricle

Figure 19.4 Heart, ventral aorta, and afferent branchial arteries of a shark.

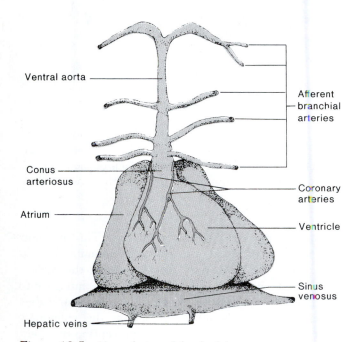

Ventral aorta

Conus arteriosus

Atrium

Hepatic veins

Afferent branchial arteries

Coronary arteries

Ventricle

Sinus venosus

Figure 19.5 Ventral view of the shark heart.

After studying mammalian circulation, you will study the hearts of other vertebrates to gain an appreciation of the evolutionary changes that led to the mammalian heart. The heart is responsible for pumping blood through arteries and arterioles to the capillaries, where exchanges take place. In studying the mammalian heart, you will use fresh or preserved cow, pig, or sheep hearts.

External Structure

▼ Examine a heart and note the division into two types of chambers (fig. 19.8). The large muscular portion consists of two **ventricles.** The two small earlike lobes on top of the ventricles are the **atria.** The widest part of the heart is the base, and the tapered portion is the apex. Before locating specific structures, one must be able to distinguish left from right on the heart. In doing this, examine the thickness of the ventricular walls. Because the left ventricle must pump blood over the entire body, its wall is much thicker than the wall of the right ventricle, which pumps blood a short distance to the lungs. The left ventricular wall is therefore firmer than the right ventricular wall. Feel the ventricular walls to determine right and left. Now orient the heart ventral surface up so the heart's right is on your left.

Before dissecting the heart, identify the pulmonary artery and the aorta. The **pulmonary artery** comes off the right ventricle and arches to the left. The **aorta** comes off the left ventricle and passes dorsal to the base of the pulmonary artery. The aorta has the thicker walls of the two vessels. It arches to the left and then posteriorly. Distend the walls of the aorta and note the elasticity. When the ventricles contract, a pulse of blood enters the aorta, distending its walls. The elastic rebound of the aorta helps propel blood through the vessel.

Internal Structure

Beginning at the base of the pulmonary artery, make an incision through the ventral wall of the right ventricle and

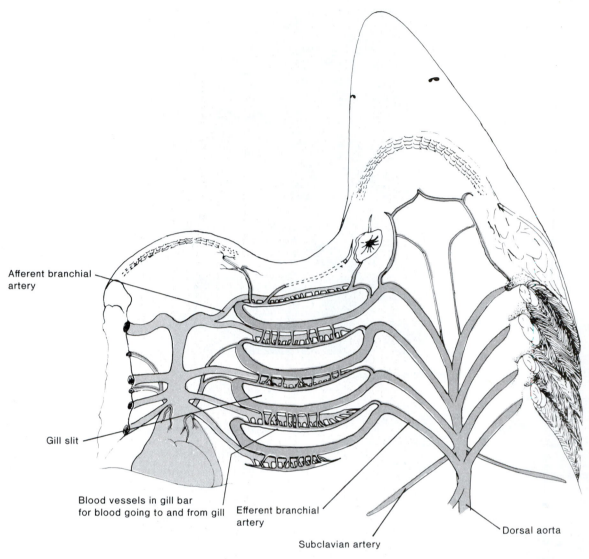

Afferent branchial
artery

Gill slit

Blood vessels in gill bar
for blood going to and from gill

Efferent branchial
artery

Subclavian artery

Dorsal aorta

Figure 19.6   Dorsal view of the heart, ventral aorta, and afferent branchial arteries. Ventral view of the efferent branchial arteries and dorsal aorta.

extend the incision toward the apex of the heart. Near the apex, continue the incision back toward the base on the dorsal aspect of the heart. You should then be able to lift the outer wall of the ventricle to expose internal structures (fig. 19.9). Do the same for the left ventricle, beginning near the base of the aorta and ending on the dorsal aspect at the base of the heart. Rinse out any clotted blood inside the ventricles.

Blood enters the heart through **vena cavae** that join the right atrium. These vessels are easiest to locate by probing into the right atrium through the right ventricle with a blunt probe. Blood passes from the right atrium through the **tricuspid (right atrioventricular) valve** into the right ventricle. Note the three flaps of connective tissue that make up the valve. Strands of connective tissue, called **chordae tendinae,** anchor the valve flaps to prevent the valve from everting into the right

atrium during ventricular contraction. This prevents the backflow of blood into the right atrium. Blood is forced into the **pulmonary artery** through the **pulmonary semilunar valve.** Extend your cut far enough into the pulmonary artery to observe the semilunar valve. When the ventricle relaxes, blood in the pulmonary artery falls back against this valve, preventing the reentry of blood into the right ventricle. Blood in the pulmonary artery moves toward the lungs. Note that a vessel is defined as an artery on the basis of the direction of blood flow (away from the heart).

Blood returns to the heart from the lungs through **pulmonary veins** that enter the heart at the left atrium. Probe into the left atrium through the left ventricle to find these vessels. Blood then passes into the left ventricle through the **bicuspid (left atrioventricular) valve.** It is similar to the tricuspid valve except it is made of two flaps

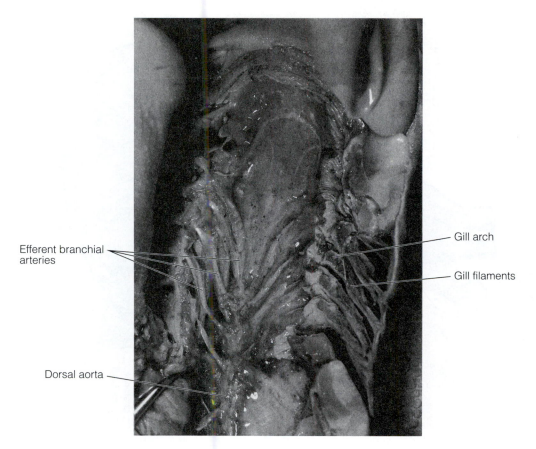

Figure 19.7 Efferent branchial arteries of the shark.

Labels on figure: Efferent branchial arteries; Gill arch; Gill filaments; Dorsal aorta

of connective tissue rather than three. From the left ventricle, blood moves to the **aorta** through the **aortic semilunar valve.** Find the aortic semilunar valve by cutting back toward the heart from the distal cut-end of the aorta. Just above the semilunar valve, note two small openings. These are the openings to the **coronary circulation.** During the relaxation period of the ventricles, blood falls back against the aortic semilunar valve and enters the vessels supplying the heart muscle with blood. Blood in the coronary circulation returns to the right atrium through a **coronary sinus,** which can be located by probing through the right atrium along the dorsal surface of the heart near where the atria and ventricles meet. Finally, place your left index finger into the right atrium and your right index finger into the left atrium and feel along the medial walls of the atria. You should detect a thin layer of connective tissue separating the two atria. In the fetus an opening called the **foramen ovale** shunts some blood from the right atrium to the left atrium, thus bypassing the lungs. Normally, this thin layer of connective tissue closes the foramen ovale after birth. ▲

19.3 LEARNING OUTCOMES REVIEW—WORKSHEET 19.1

19.4 SYSTEMIC CIRCULATION—THE RAT

LEARNING OUTCOMES

1. Identify major arteries of the rat circulatory system.
2. Identify major veins of the rat circulatory system.
3. Trace a drop of blood from the left ventricle of the rat heart to each of the following structures and then back to the right atrium of the rat heart: right foreleg, left foreleg, stomach, right hindleg.

We will begin our dissection of the rat circulatory system at the heart. Be very careful in this dissection, because arteries and veins are easily torn. Doubly or triply injected specimens will be used. Arteries are injected with red latex, veins with blue latex, and the hepatic portal system with yellow latex (if triply injected). This scheme is reversed in the pulmonary circulation.

▼ Reread the general instructions for dissection in exercise 11 (p. 158) and remember that you will be using your rat in most of the exercises that follow and while studying for laboratory tests and quizzes. Do not cut your rat unnecessarily or allow it to desiccate.

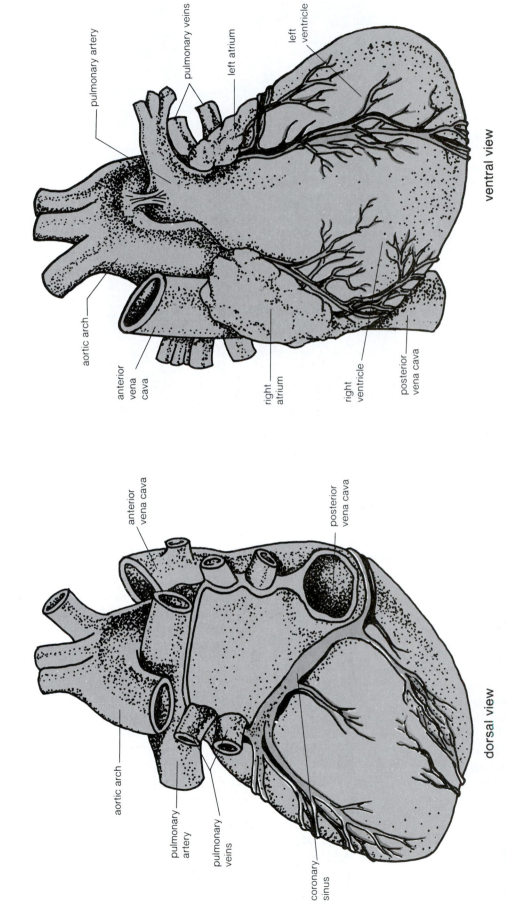

pulmonary artery

pulmonary veins

left atrium

left ventricle

aortic arch

anterior vena cava

right atrium

right ventricle

posterior vena cava

ventral view

anterior vena cava

posterior vena cava

aortic arch

pulmonary artery

pulmonary veins

coronary sinus

dorsal view

Figure 19.8 The mammal heart, external structure.

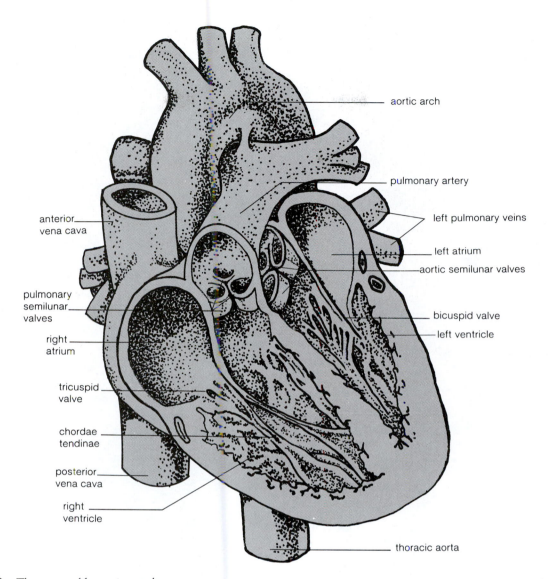

Figure 19.9 The mammal heart, internal structure.

With your rat positioned ventral side up on your dissecting pan, open the body cavity by making a midventral incision in the abdominal wall (fig. 19.10). Cut through the skin and body wall with a scalpel. After the initial opening is made, extend the incision anteriorly using scissors. Keep the lower blade of the scissors close to the body wall so as not to damage the viscera below. At the posterior margin of the rib cage, cut to either side of the sternum. (In the process, you will cut through the diaphragm, a thin muscular sheet that separates the abdominal and thoracic cavities.) Continue anteriorly until your cut joins the incision in the neck that was made during the injection process. Extend your incision in the abdomen posteriorly until you reach the pelvic region, but do not damage genitalia. Find the diaphragm and free it by cutting close to its attachment at the ventral and lateral body wall. The body wall can now be spread and pinned to the bottom of the pan. Lateral incisions through the body wall help keep the body wall spread.

In the following exercise, you will need to follow blood vessels to various internal organs. Use figure 19.10 to familiarize yourself with the location of major organs. Lift and separate organs gently in the process. Do not cut or tear any tissues.

Covering the heart and vessels in the upper chest and neck will be a large lymphatic gland, the **thymus.** Remove the thymus with forceps. Note that the heart is covered by a sac of connective tissue, the **pericardial sac.** It is composed of two layers. The outer sac is the **parietal pericardium,** and the membrane that tightly covers the heart and vessels is the **visceral pericardium.** Remove the parietal pericardium. Remove any fat covering the heart and then identify the four chambers and major vessels entering and leaving it (in the rat there are left and right anterior vena cavae as well as the single posterior vena cava).

The Arterial System

Follow the aorta away from the heart, removing connective tissue so you can see the vessels (fig. 19.11). The **innominate artery** is the first branch of the aorta. Follow

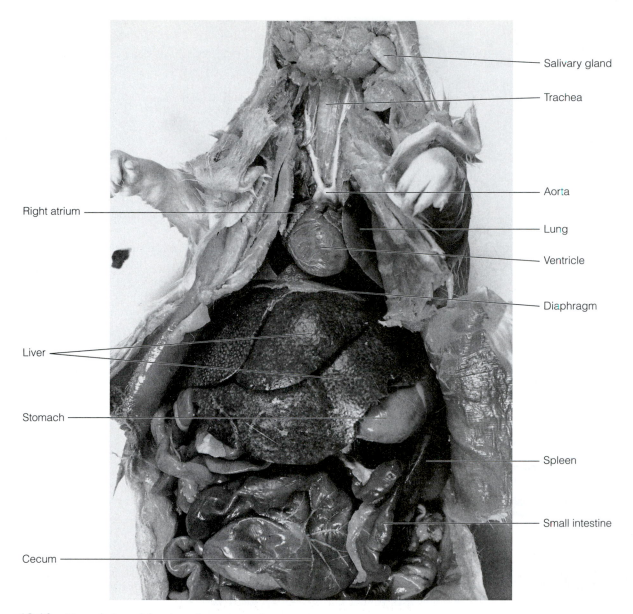

Right atrium

Liver

Stomach

Cecum

Salivary gland

Trachea

Aorta

Lung

Ventricle

Diaphragm

Spleen

Small intestine

Figure 19.10   Ventral view of the visceral structures of a rat.

the innominate anteriorly. It soon branches into two major arteries (fig. 19.12). The **right subclavian artery** branches off laterally, carrying blood to the foreleg. The medial branch is the **right common carotid artery.** It carries blood to the head. The common carotid branches in the upper region of the neck into an internal carotid (carrying blood to the brain) and an external carotid (carrying blood to the skin and muscles of the head). You need not find these branches.

Return to the aorta at the point of divergence of the innominate. Note that the aorta now curves dorsally and posteriorly. This curve is the **arch of the aorta.** The next two branches of the aorta are the **left common carotid** and the **left subclavian.** In contrast to those on the right side, these arteries come off the aorta individually. Locate and trace these branches a short distance. Trace the aorta posteriorly.

In the thoracic region, it gives off branches to the body wall. These are collectively referred to as intercostal arteries.

The next major branch of the aorta is posterior to the diaphragm (fig. 19.13). It is referred to as the **coeliac artery.** It gives off branches to the spleen, liver, and stomach. Posterior to the coeliac is the **superior mesenteric artery.** It delivers blood to the small intestine. The large arteries delivering blood to the kidneys are the **renal arteries.** Branching off the renal arteries, or occasionally the aorta, are smaller **adrenal arteries,** which take blood to the adrenal glands.

Following the aorta posteriorly, note small branches that go dorsally and laterally to the body wall. These are **iliolumbar arteries.** The **ovarian** or the **spermatic artery** branches off the aorta just posterior to the renal artery. Follow this artery to the ovary or the testis.

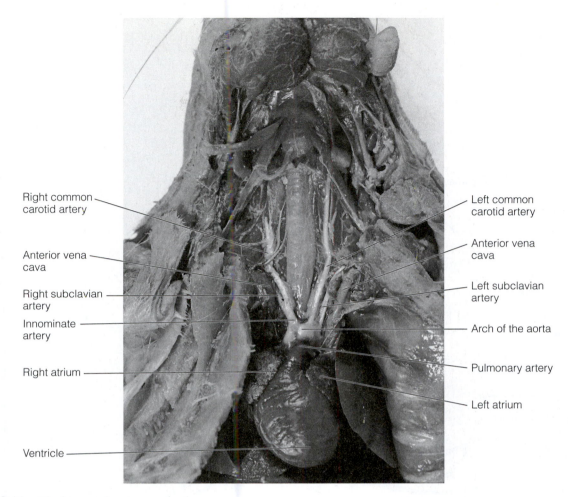

Right common carotid artery

Anterior vena cava

Right subclavian artery

Innominate artery

Right atrium

Ventricle

Left common carotid artery

Anterior vena cava

Left subclavian artery

Arch of the aorta

Pulmonary artery

Left atrium

Figure 19.11   The heart and major vessels of a rat.

At this time, if you have not already done so, continue your incision through the skin and body wall to expose the vessels lying on top of the muscles of one leg. Note that in the pelvic region the aorta branches into two main arteries carrying blood to the legs. These are the **common iliacs.** A third branch, the **caudal artery,** goes dorsally and posteriorly into the tail. Numerous smaller branches of these arteries carry blood to all parts of the tail and legs. They will not be located.

The Venous System

Return to the heart (see fig. 19.11). Again find the **left** and **right anterior vena cavae.** Follow them anteriorly. They branch into **subclavian veins,** bringing blood back from the forelegs, and **internal** and **external jugular veins,** bringing blood back from the head (fig. 19.14). The internal jugulars return blood from the brain. They are small and may not be injected. External jugulars bring blood back from the muscles and skin of the head. Entering into the left anterior vena cava very near the heart is the **azygous vein.** It returns blood to the heart from the chest wall. Follow this vessel to

the chest wall. Return to the posterior vena cava and follow it posteriorly. It runs through the diaphragm and the liver. Within the substance of the liver the **hepatic vein** joins the posterior vena cava. We will not attempt to dissect this vessel. Trace the **renal veins** (see fig. 19.13). Blood from the kidneys enters the vena cava through renal veins. Note that veins from the adrenal gland **(adrenal veins)** empty into the renal vein before reaching the vena cava. The **spermatic** or the **ovarian veins** enter the vena cava directly on the right side of the body (the next posterior vessel), but through the renal vein on the left side.

Continue following the vena cava posteriorly. The next branch following the right spermatic or ovarian vein is the **iliolumbar vein,** delivering blood from the body wall. The vena cava then branches as vessels from the hind legs **(common iliac veins)** unite. The common iliac veins receive numerous branches from all parts of the leg. We will not trace these branches.

Finally, recall that blood entered the gut tract through the coeliac and superior mesenteric arteries. We have not seen how this blood gets back to the vena cava. This is the job of the **hepatic portal system.** A portal system is a

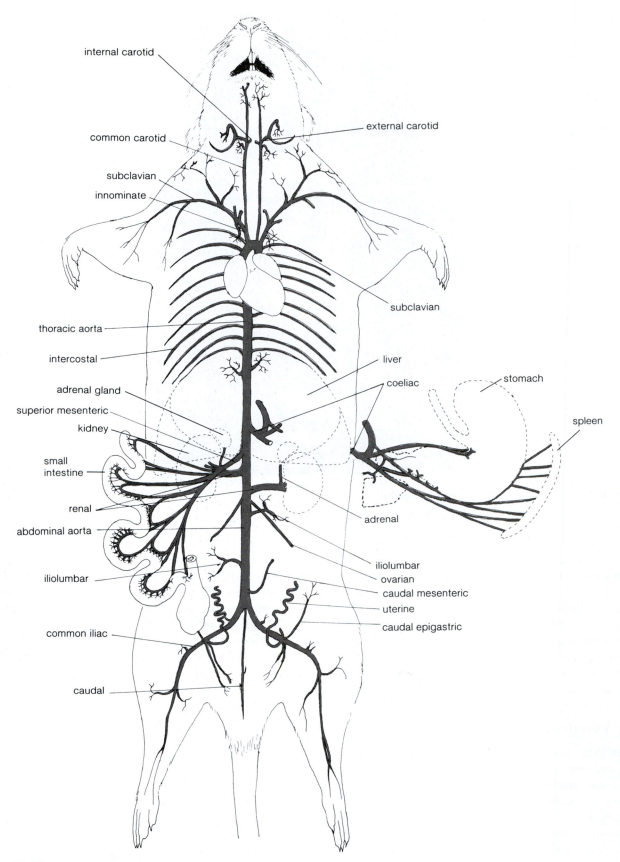

Figure 19.12   Major arteries of the rat.

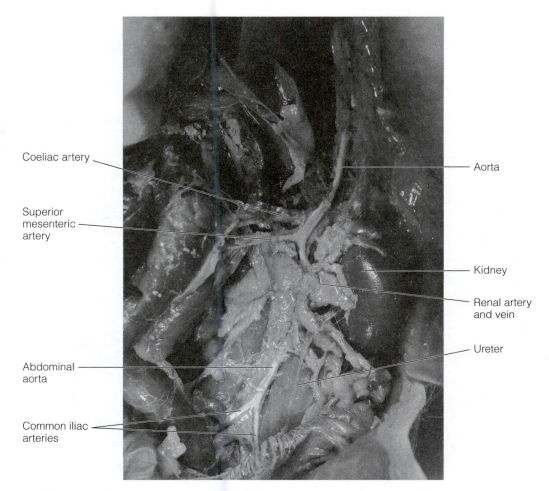

Figure 19.13 Abdominal vessels of a rat.

vascular system that begins in a capillary bed and ends in a capillary bed. The capillary beds in the hepatic portal system are in the gut tract at one end and the liver at the other. If your specimen is triply injected, the hepatic portal system will be yellow. If doubly injected, your specimen will have dried blood in the hepatic portal system. Trace the vessels from the intestine to the liver. Blood from the intestine passes to the liver, where it is cleaned of impurities picked up from the intestine. Blood is then passed to the hepatic vein and then to the posterior vena cava. ▲

19.4 LEARNING OUTCOMES REVIEW—WORKSHEET 19.1

19.5 COMPARATIVE HEART STRUCTURE

LEARNING OUTCOME

1. Describe evolutionary changes in the vertebrate heart.

You have just observed two extremes in vertebrate circulatory systems. The changes that take place in the heart between fishes and mammals reflect evolutionary relationships among the vertebrate classes. The following exercise is a study of fish, lungfish, reptile, and bird hearts and should help you understand these relationships.

▼ As you work through the following observations, consult any models, preserved hearts, and diagrams available. Also turn frequently to figure 16.5 in worksheet 16.1, which depicts evolutionary relationships among the vertebrates. The evolutionary changes that follow are summarized in table 19.1. ▲

Recall the four chambers you studied in the shark heart: sinus venosus, atrium, ventricle, and conus arteriosus. The following changes in the heart accompanied the adaptation of vertebrates to terrestrial environments by permitting gas exchange with air without excessive water loss.

Internal gas-exchange surfaces (lungs) provide large surfaces for gas exchange and at the same time help retard water loss because of the lungs' position within the body cavity. The development of lungs, however, was accompanied by a

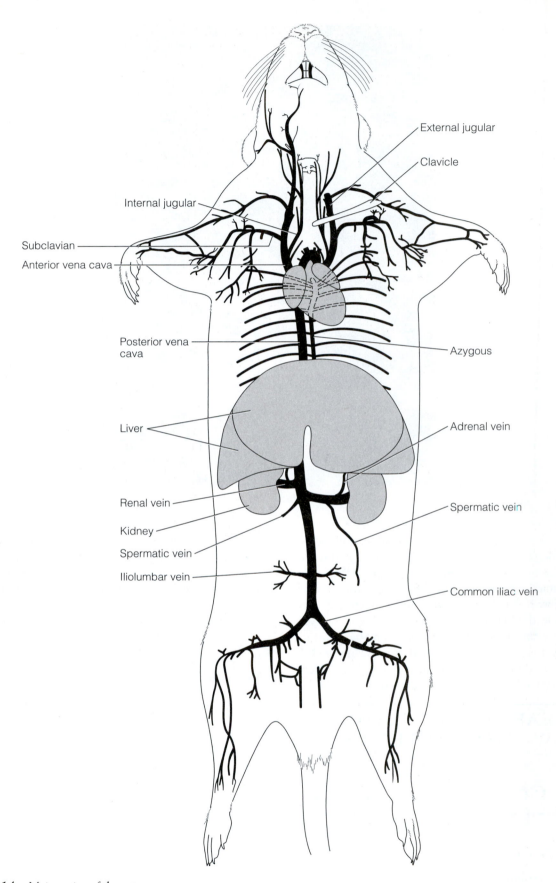

Figure 19.14 Major veins of the rat.

External jugular

Clavicle

Internal jugular

Subclavian

Anterior vena cava

Posterior vena cava

Azygous

Liver

Adrenal vein

Renal vein

Kidney

Spermatic vein

Spermatic vein

Iliolumbar vein

Common iliac vein

Table 19.1   Evolutionary Changes in the Vertebrate Heart

	Sinus Venosus	Atrium	Ventricle	Conus Arteriosus
Shark	Single	Single	Single	Single, leads to ventral aorta
Lungfish	Single	Partially divided, R. and L. to accommodate pulmonary circulation	Partially divided to help separate oxygenated and nonoxygenated blood	Single, with valve to help separate oxygenated and nonoxygenated blood
Reptile				
Crocodile	Smaller	Divided	Completely divided	Incorporated into bases of a single pulmonary artery and two systemic arteries
All others	Smaller	Divided	Partially divided	
Bird	Smaller	Divided	Complete separation of oxygenated and nonoxygenated blood	Same as reptiles, except left systemic is gone
Mammal	Gone as a chamber; represented by sinoatrial node	Divided	Same	Same except systemic arches to the left rather than right as in birds

change in circulatory pathways that brings blood into close contact with air. How this may have been accomplished is reflected in the structure of the lungfish circulatory system (fig. 19.15a, b). The lungfish is used here because it is a living vertebrate that shows a circulatory pattern that may be similar to that present in early amphibians. The lungfish respires primarily through lungs. Lungs allow it to survive both periodic droughts and the stagnated water in which it lives. In addition, the lungfish has gills to aid in gas exchange. It has a single sinus venosus but has a partially divided atrium. Blood from the sinus venosus passes into the right atrium, to the right side of a partially divided ventricle, and then through the conus arteriosus to the ventral aorta and gills. In addition, blood from the ventricle can be shunted to the lungs through a modified afferent arteriole. After oxygenation at the lungs occurs, blood returns to the left atrium. From the left atrium, blood goes to the left side of the partially divided ventricle and through the ventral aorta to the body. A valve in the conus arteriosus helps keep oxygenated and nonoxygenated blood separate.

From this basic pattern, the lineage leading to modern reptiles and birds diverged from that leading to modern mammals (fig. 19.15c). In the heart of the modern reptile, the sinus venosus is reduced in size (compared to that of the lungfish), the atria are completely divided, the ventricles are partially divided (except in the crocodile where they are completely divided), and the conus arteriosus is incorporated into the bases of the major arteries leaving the heart. These arteries result from the splitting of the ventral aorta. A single pulmonary artery now carries blood from the right ventricle to the lungs. Right and left systemic arteries (aortae) carry blood from the heart to the body. The separation of oxygenated and nonoxygenated blood is more efficient because a valve directs

oxygenated blood to the systemic arteries and nonoxygenated blood to the pulmonary arteries. This valve also diverts blood away from the lungs during periods of apnea (no breathing).

Birds, which evolved from ancient reptiles, have four completely separate heart chambers. However, the left systemic, present in reptiles, is absent, and the sinus venosus is further reduced in size (fig. 19.15d).

In the evolutionary pathway between ancient reptiles and mammals, the conus arteriosus and ventral aorta divided into two vessels rather than three (fig. 19.15e). Complete separation of oxygenated from nonoxygenated blood resulted from the development of a pulmonary artery, aorta (systemic artery), and completely divided atria and ventricles. In mammals, the sinus venosus is represented only by a small patch of tissue in the wall of the right atrium, the sinoatrial node. In all vertebrates, the sinus venosus (sinoatrial node) initiates the heartbeat.

19.5 LEARNING OUTCOMES REVIEW—WORKSHEET 19.1

19.6 CIRCULATORY PHYSIOLOGY—THE ELECTROCARDIOGRAM

LEARNING OUTCOMES

1. Explain the electrical events associated with depolarization and repolarization of the heart.
2. Interpret the basic waveforms of an electrocardiogram.

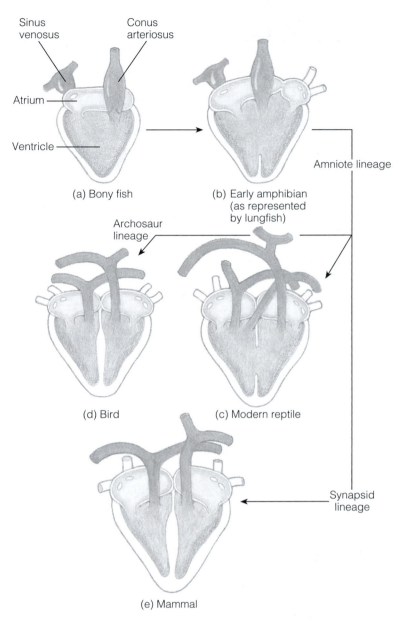

Figure 19.15   A possible sequence in the evolution of the vertebrate heart. (*a*) Fish heart. (*b*) In lungfish, partially divided atria and ventricles separate pulmonary and systemic circuits. The hearts of modern reptiles (*c*) were derived from the form shown in (*b*). (*d*) The archosaur and (*e*) synapsid lineages resulted in completely separated, four-chambered hearts.

The heart is a self-excitatory pumping structure. The rate at which the heart beats is controlled by the pacemaker, or **sinoatrial (SA) node.** The SA node initiates the heartbeat by sending out an impulse that very rapidly spreads over the heart through specialized conducting fibers. Because the SA node is located in the right atrium just below the point at which the vena cavae enter, the atria are first to receive this impulse. The impulse passes from the atria to the ventricles through the **atrioventricular (AV) node.** The impulse is delayed about 0.1 second at the AV node and then passes down the septum between the ventricles to the apex of the heart. It then proceeds up the sides to the base of the heart. This process is referred to as the **depolarization** of the heart. Immediately after depolarization, the muscle recovers **(repolarization)** and then contracts. The impulses moving over the heart set up electromagnetic fields that circulate through the body. They can be picked up by placing electrodes at various points on the body. What results is the **electrocardiogram (ECG).** A typical ECG is shown in figure 19.16. The various deflections in the ECG are labeled as shown in figure 19.16 and correspond to the following events:

P wave   atrial depolarization
QRS complex   ventricular depolarization
T wave   ventricular repolarization

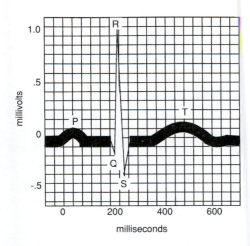

milliseconds

Figure 19.16 The electrocardiogram tracing.

Atrial repolarization is masked by ventricular depolarization. Your instructor will demonstrate an ECG. Note that these waves are produced from the impulses passing over the surface of heart muscle cells just prior to contraction. They are not a measurement of the contractions of heart muscle.

19.6 LEARNING OUTCOMES REVIEW—WORKSHEET 19.1

WORKSHEET 19.1 Vertebrate Circulation

19.1 Formed Elements—Learning Outcomes Review

1. Which of the following formed elements is most numerous on a blood smear preparation? These formed elements are phagocytic.
   a. neutrophils
   b. eosinophils
   c. basophils
   d. lymphocytes
   e. monocytes

2. These formed elements are whole cells, but lack nuclei when mature.
   a. neutrophils
   b. red blood cells
   c. basophils
   d. lymphocytes
   e. platelets

3. These formed elements have regular nuclei, lack cytoplasmic granules, and function in the body's immune response.
   a. neutrophils
   b. eosinophils
   c. basophils
   d. lymphocytes
   e. monocytes

4. In a blood smear, one can distinguish red blood cells from white blood cells by the presence of
   a. darkly stained nuclei in white blood cells.
   b. the larger size of red blood cells.
   c. the greater abundance of white blood cells.
   d. darkly stained nuclei in red blood cells.

19.2 Systemic Circulation—The Shark—Learning Outcomes Review

5. After leaving the sinus venosus of the shark heart, blood enters the
   a. ventral aorta.
   b. conus arteriosus.
   c. atrium.
   d. ventricle.

6. Blood leaving the ventricle of a shark heart is
   a. lacking oxygen.
   b. rich in oxygen.
   c. on the way to the gills.
   d. entering the conus arteriosus.
   e. a, c, and d are all correct.
   f. b and d only are correct.

7. After leaving efferent brachial arteries, blood of the shark next enters
   a. the ventral aorta.
   b. the dorsal aorta.
   c. the posterior cardinal veins.
   d. the lateral abdominal veins.

8. The hepatic portal system of the shark carries blood to the shark's
   a. brain.
   b. body wall.
   c. liver.
   d. gills.

19.3   The Mammalian Heart—Learning Outcomes Review

9. Blood in the pulmonary veins of the mammal heart is
   a. freshly oxygenated and entering the right atrium.
   b. deoxygenated and entering the right atrium.
   c. leaving the heart and flowing to the lungs.
   d. entering the left atrium.

10. The pulmonary artery originates from the
    a. right ventricle.
    b. left ventricle.
    c. right atrium.
    d. left atrium.

11. Chordae tendinae attach to the
    a. atrioventricular valves.
    b. atrial walls.
    c. ventricular walls.
    d. base of the aorta.
    e. Both a and c are correct.

12. Blood in the coronary sinus next enters the
    a. right ventricle.
    b. left ventricle.
    c. right atrium.
    d. left atrium.

13. Blood in the aorta has just passed a(n)
    a. atrioventricular valve.
    b. semilunar valve.
    c. spiral valve.

14. Which of the following is a correct sequence of chambers and vessels for blood passing through and around the heart?
    a. vena cava, right atrium, pulmonary artery, lungs, left atrium
    b. vena cava, left atrium, left ventricle, aorta, lungs, pulmonary veins, right atrium, right ventricle
    c. right ventricle, pulmonary artery, lungs, pulmonary veins, left atrium
    d. right ventricle, aorta, vena cava, left atrium pulmonary artery, lungs

19.4   Systemic Circulation—The Rat—Learning Outcomes Review

15. The first branch off the aorta of the rat as the aorta leaves the heart is the
    a. right subclavian artery.
    b. right common carotid artery.
    c. innominate artery.
    d. left common carotid artery.

16. Blood in the coeliac artery of the rat is flowing to the
    a. small intestine.
    b. kidneys.
    c. liver, spleen, and stomach.
    d. large intestine.

17. Blood is flowing from the origin of the aorta to the leg of a rat. This blood would pass the origin of the following vessels that branch directly off the aorta except one. Select the exception.
    a. innominate artery
    b. renal artery
    c. left common carotid artery
    d. right common carotid artery
    e. coeliac artery

18. The following veins all contribute blood to the anterior vena cavae except one. Select the exception.
    a. internal jugular vein
    b. subclavian vein
    c. common iliac vein
    d. external jugular vein

19.5   Comparative Heart Structure—Learning Outcomes Review

19. A major change in the lungfish heart as compared to the heart of other fish is
    a. a fully divided ventricle is present.
    b. right and left atria are present.
    c. the conus arteriosus is gone as a chamber, but is incorporated into the bases of major arteries.
    d. the ventricle is partially divided.
    e. Both b and d are correct.

20. A major change in the reptilian heart (other than crocodiles and birds) as compared to the lungfish heart is
    a. the ventricle is fully divided.
    b. right and left atria are present.
    c. the conus arteriosus is gone as a chamber, but is incorporated into the bases of major arteries.
    d. the ventricle is partially divided.
    e. Both b and d are correct.

21. Which of the following is a major change in the bird heart as compared to the reptile (other than crocodiles) heart
    a. A fully divided ventricle is present.
    b. Right and left atria are present.
    c. The conus arteriosus is gone as a chamber, but is incorporated into the bases of major arteries.
    d. The left systemic artery is gone.
    e. Both a and d are correct.

22. The changes in the vertebrate heart described in this section are associated with
    a. gas exchange with air via lungs rather than with water via gills.
    b. larger body size.
    c. support against gravity in terrestrial habitats.

23. What is the fate of the sinus venosus of early vertebrates in the hearts of birds and mammals?
    a. It is completely gone.
    b. It is still there as a chamber, but much smaller.
    c. It is gone as a chamber but it is retained as a patch of cells that function as the pacemaker of birds and mammals.
    d. It is incorporated into the base of the pulmonary artery.

19.6   Circulatory Physiology—The Electrocardiogram—Learning Outcomes Review

24. The initial spread of the an electrical impulse over ventricular muscle is called
    a. depolarization.
    b. repolarization.
    c. ventricular contraction.

25. The spread of the electrical impulse over ventricular muscle is detected in an ECG as the
    a. p wave
    b. QRS waves
    c. t wave

26. The ECG is a measure of events associated with the cycle of contraction and relaxation of heart muscle.
    a. true
    b. false

19.1 Formed Elements—Analytical Thinking

1. How would the blood smear from a person suffering from a chronic bacterial infection or leukemia be different from the blood smear that you observed?

19.2 Systemic Circulation—The Shark—Analytical Thinking

2. Evaluate the statement that "arteries of fish carry oxygenated blood and veins of fish carry deoxygenated blood."

19.3 The Mammalian Heart—Analytical Thinking

3. Evaluate the statement that "the mammalian heart is a double pump."

19.4 Systemic Circulation—The Rat—Analytical Thinking

4. Trace a drop of blood from the left ventricle of the rat to the right leg and back to the right atrium, naming all of the vessels involved that were studied in this laboratory.

5. Trace a drop of blood from the left ventricle of the rat to the stomach and back to the right atrium, naming all the vessels involved that were studied in this laboratory.

19.5 Comparative Heart Structure—Analytical Thinking

6. Study figure 19.5 and figure 16.5 in worksheet 16.1. Compare possible reasons for the resemblance between the bird heart and the mammalian heart versus the resemblance between the bird heart and the crocodile heart.

19.6 Circulatory Physiology—Analytical Thinking

7. The function of the AV node is to delay the impulse for 0.1 second as it passes between the atria and the ventricles. Explain why this delay is important. What is an AV block?

Exercise 20

Vertebrate Respiration

Prelaboratory Quiz

Study this week's laboratory exercise and then complete the following quiz to assess your preparation for the laboratory.

1. The portion of the alimentary tract shared by the digestive and respiratory systems is the
   a. pharyngeal region.
   b. visceral region.
   c. nasal region.
   d. trachael region.

2. The skeletal framework of the gills of a fish consists of
   a. gill filaments.
   b. gill arches.
   c. gill lamellae.
   d. gill rakers.

3. Water is in close contact with capillary beds as it passes over
   a. gill filaments.
   b. gill arches.
   c. gill lamellae.
   d. gill rakers.

4. The bony roof of the mouth of a mammal is called the
   a. soft palate.
   b. bony palate.
   c. glottis.
   d. hard palate.

5. The function of goblet cells that line respiratory passageways of mammals is to
   a. secrete mucus that traps foreign material.
   b. engulf foreign material.
   c. promote gas exchange.
   d. warm inspired air.

6. True/False   A countercurrent exchange occurs as blood and water move in opposite directions across gill rakers.

7. True/False   Major respiratory passages of the lungs are lined by simple squamous epithelium.

8. True/False   The spiracle of a shark serves as an alternate route for water entering the pharynx.

9. True/False   Most reptiles carry on cutaneous and buccal respiration.

10. True/False   Bronchioles divide into smaller branches called bronchi.

The respiratory system develops evolutionarily in conjunction with the anterior portion of the digestive tract. The **pharyngeal region** is that portion of the alimentary tract that is shared with the respiratory system. The pharynx is also the location for the evolution of the swim bladder (fishes), openings to the eustachian tubes (remnants of gill pouches), and masses of lymphoidal tissue (palatine and lingual tonsils). With respect to the respiratory system, the pharynx provides a pathway for water moving over the gills of fishes and a pathway for air entering the lung passageways of tetrapods.

In this laboratory, we will again look at the shark and the rat. The two sets of respiratory structures are not homologous; they are simply adaptations to living in different environments.

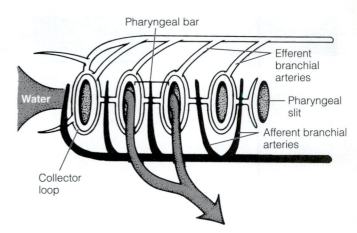

Figure 20.1 Path of blood flow to and from the gills of a fish.

20.1 GILL STRUCTURE AND FUNCTION

LEARNING OUTCOMES

1. Identify structures of the shark respiratory system.
2. Describe the path of blood and water flow through the gills of the shark.
3. Describe the countercurrent exchange of gases between blood and water in the shark gills.

Pharyngeal slits primitively served in filter feeding, but in vertebrates they are primarily respiratory.

▼ Observe a demonstration dissection showing the pharyngeal cavity of the shark. In this demonstration, the pharyngeal cavity is opened by cutting into the pharynx at the corner of the jaw and posteriorly through the pharyngeal bars. Just posterior to the pectoral fin, the cut is extended to the opposite side of the body. The mouth cavity can then be folded open (see figs. 19.6 and 19.7).

On the inner surface locate the following:

spiracle   homologous to the first pharyngeal slit, located dorsal to the corners of the jaw. Thought to serve as an alternate route for water entry to the pharynx.
internal pharyngeal slits   five pairs of openings from the pharyngeal cavity into the gill chamber.
gill chamber   the cavity between the internal and external pharyngeal slits.
gill rakers   cartilage-supported projections that line the internal pharyngeal slits. Prevent food from entering the gill chamber.

On the side of the shark that has been cut, note the following:

gill (visceral) arches   the main skeletal framework of each gill.
gill filaments   membranous branches from the gill arches. Contain blood vessels that carry blood to and from gill lamellae.

gill lamellae   thin, vascular folds of epithelium through which gas exchange takes place.

Return briefly to figure 19.6 to recall the arterial pathway: from the heart, through afferent branchials to capillary beds of the gills, to efferent branchials, and then through the dorsal aorta to the rest of the body. If the shark you are studying has the arterial system injected, efferent branchials will be filled with red latex. Efferent branchials arise as collector loops that encircle each pharyngeal slit (fig. 20.1). Collector loops receive blood after blood has passed through gill capillaries. In your specimen, locate sections through collector loop vessels on each side of one of the pharyngeal slits. Envision this collector loop circling the pharyngeal slit. Each pharyngeal bar will have two collector loop vessels: the posterior element of the anterior collector loop, and the anterior element of the posterior collector loop. Between these will be an uninjected afferent branchial that should be located at this time.

The path of blood flow through a gill is diagrammatically illustrated in figure 20.2a. In the pharyngeal bar, each afferent branchial artery gives off branches to the gill filaments. As shown in figure 20.2b, blood then moves across the gill filament through capillary beds in the gill lamellae to efferent vessels that carry blood to the efferent branchial artery. Water enters the pharynx through the mouth and spiracle. It passes through the pharyngeal slits and then between gill lamellae. Note that as water is moving between the lamellae, water and blood are moving in opposite directions. This **countercurrent exchange** provides very efficient gas exchange by maintaining a concentration gradient between the blood and the water over the length of the capillary bed. Compare the diagrams for parallel and countercurrent exchanges in figure 20.2c. ▲

20.1 LEARNING OUTCOMES REVIEW—WORKSHEET 20.1

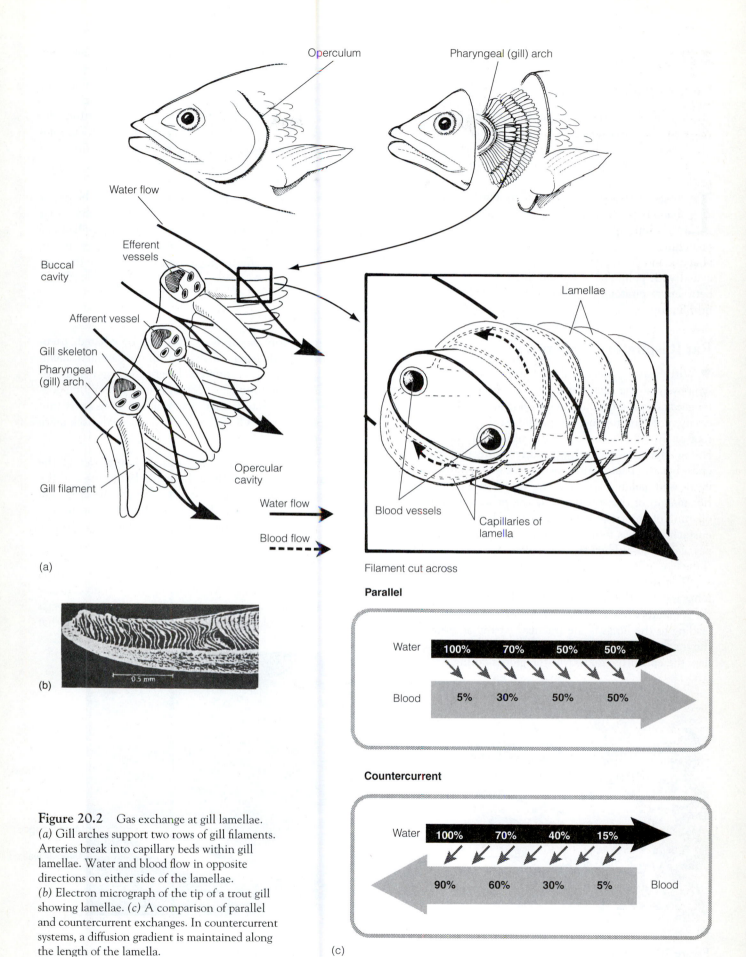

Operculum

Pharyngeal (gill) arch

Water flow

Efferent vessels

Buccal cavity

Afferent vessel

Gill skeleton

Pharyngeal (gill) arch

Gill filament

Opercular cavity

Water flow

Blood flow

Lamellae

Blood vessels

Capillaries of lamella

Filament cut across

(a)

(b)

0.5 mm

Parallel

Water   100%   70%   50%   50%

Blood   5%   30%   50%   50%

Countercurrent

Water   100%   70%   40%   15%

90%   60%   30%   5%   Blood

(c)

Figure 20.2   Gas exchange at gill lamellae. (a) Gill arches support two rows of gill filaments. Arteries break into capillary beds within gill lamellae. Water and blood flow in opposite directions on either side of the lamellae. (b) Electron micrograph of the tip of a trout gill showing lamellae. (c) A comparison of parallel and countercurrent exchanges. In countercurrent systems, a diffusion gradient is maintained along the length of the lamella.

20.2 TETRAPOD LUNGS

LEARNING OUTCOMES

1. Identify structures of the rat respiratory system.
2. Describe the histology of the mammalian lung.
3. Describe major changes that occurred during the evolution of the vertebrate lung.

Lungs are structurally and evolutionarily distinct from, but functionally similar to, gills. In most tetrapods, the breathing apparatus is similar. It consists of a series of branching tubes originating at the posterior aspect of the pharynx and ending at a moist respiratory epithelium associated with pulmonary capillary beds. Movement of gases between respiratory epithelium and the blood is always by diffusion.

Rat Respiratory Structures

▼ With heavy scissors, cut the lower jaw of the rat at the symphysis of the mandibles. Extend the cut through the tongue and into the mouth cavity and pharynx. In the mouth cavity, note the tongue, teeth, and **hard palate** (the bony roof of the mouth that separates nasal and oral cavities). Follow the hard palate posteriorly. Note that the bone ends, but the palate continues as a fleshy extension. This is the **soft palate.** The opening just dorsal to the posterior margin of the soft palate is the **posterior nares**—the opening of the nasal cavity into the pharynx. In the rat, **eustachian tubes** from the middle ear open into the nasal passageway just above the soft palate. Do not try to locate these openings.

The series of branching passageways leading into the lungs is called the **respiratory tree.** The anterior portion of the respiratory tree is supported by cartilage. The single tube leading to the lungs is the **trachea.** At the anterior end of the trachea is an opening, the **glottis.** A cartilage-supported flap, the **epiglottis,** folds over the glottis during swallowing

to prevent food from entering the trachea during swallowing. The **larynx** is the expanded, cartilaginous structure surrounding the glottis. If a vertebrate has vocal chords, they are located here in the larynx. The trachea branches into **bronchi** posteriorly, and bronchi subdivide within the lung—eventually to the level of **bronchioles.** Bronchioles end at **alveolar ducts,** which supply grapelike clusters of **alveoli.** Alveoli are the small air sacs where gas exchange occurs (fig. 20.3).

Examine a fresh cow or pig lung. Cut through the larynx and down the trachea. Follow a branch of the trachea as far as possible. Note that the lung is not just an air-filled bag but a spongelike tissue containing thousands of tiny alveoli. If possible, inflate a fresh lung using a blower. Note the elasticity of the lung. ▲

Lung Histology

Major respiratory passageways are lined by **ciliated, pseudostratified epithelium.** All cells reach the basement membrane, but cells are of different heights, giving a stratified appearance. The lining also contains mucus-secreting **goblet cells.** Mucus traps foreign material that gains entry into the respiratory tree. Cilia beat trapped materials back up to the pharynx (fig. 20.4).

▼ Examine a tracheal section under low power of the compound microscope. Note the epithelial lining and the cartilage that supports the trachea. Observe the epithelial lining under high power. The "fuzzy" border indicates the presence of cilia. Find the translucent goblet cells. Notice also that nuclei are at different levels, giving the tissue a stratified appearance.

Obtain a lung section. Note the very loose appearance. The section should be filled with hundreds of alveoli (see fig. 20.3), as well as sections through larger air passageways and pulmonary blood vessels. Under high power, note that

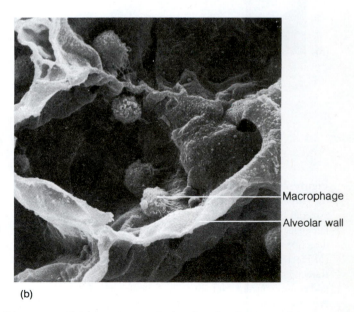

(a)   Alveoli

Alveolar wall consisting of simple squamous epithelium

(b)   Macrophage

Alveolar wall

Figure 20.3   Alveoli of the mammalian lung: (*a*) photomicrograph; (*b*) scanning electron micrograph showing alveolar macrophage.

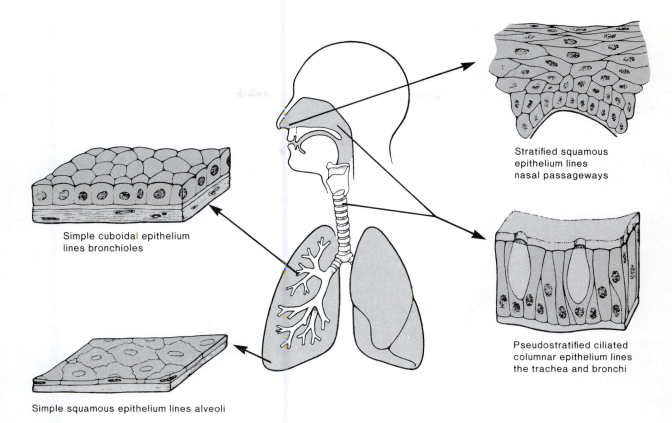

Simple cuboidal epithelium
lines bronchioles

Stratified squamous
epithelium lines
nasal passageways

Pseudostratified ciliated
columnar epithelium lines
the trachea and bronchi

Simple squamous epithelium lines alveoli

Figure 20.4 Epithelial types found within the mammalian respiratory system.

individual alveoli are lined by simple squamous cells. This single-cell boundary is the only barrier between lung gases and the capillary wall. We will not attempt to identify lung passageways further. ▲

Lung Evolution

Lungs of tetrapods and swim bladders of fishes share a common evolutionary origin as **pneumatic sacs** that arose as a ventral outgrowth of the pharynx. In evolution, the lung changed from the simple, saclike structures of lungfish and amphibians to the complex construction just observed in mammals. The saclike form seen in lungfish and amphibians presents much less surface area for the diffusion of gases, but it is adequate for the respiratory needs of these groups. It is noteworthy that respiration in amphibians is supplemented by **cutaneous** (via skin) and **buccopharyngeal** (via mouth epithelium) **respiration.** In reptiles, lung organization becomes more complex, ranging from saclike structures in the primitive lizard, *Sphenodon*, to a more complex spongelike structure in turtles and crocodiles. Birds show a remarkable modification of respiratory structures that allows a continuous, one-way passage of air across respiratory surfaces. This results in greater oxygen availability for metabolically demanding flight.

▼ If demonstrations of these lungs are available, observe them at this time. ▲

20.2 LEARNING OUTCOMES REVIEW—WORKSHEET 20.1

WORKSHEET 20.1   Vertebrate Respiration

20.1   Gill Structure and Function—Learning Outcomes Review

1. Which of the following would water encounter first in moving from the pharynx of a shark to the outside through the gills?
   a. gill chamber
   b. gill rakers
   c. gill filaments
   d. internal pharyngeal slits

2. Thin vascular folds of epithelium through which gas exchange occurs in shark gills are called
   a. gill rakers.
   b. gill filaments.
   c. gill lamellae.
   d. gill arches.

3. Blood leaving the gill lamellae next enters into the
   a. afferent branchial arteries.
   b. ventral aorta.
   c. efferent branchial arteries.
   d. dorsal aorta.

4. Blood and water move in opposite directions when moving through
   a. gill rakers.
   b. the spiracle.
   c. gill lamellae.
   d. gill arches.

5. The movement of blood and water in opposite directions promotes
   a. parallel exchange.
   b. countercurrent exchange.
   c. efficient movement of oxygen from blood to water.

20.2   Tetrapod Lungs—Learning Outcomes Review

6. The _____ is dorsal to the hard palate of a rat.
   a. oral cavity
   b. nasal cavity
   c. pharynx
   d. epiglottis

7. The opening of the nasal cavity of the rat into the pharynx is called the
   a. posterior nares.
   b. glottis.
   c. epiglottis.
   d. eustachian tube.

8. All of the following are associated with the larynx except one. Select the exception.
   a. glottis
   b. epiglottis
   c. cartilage
   d. alveoli

9. A prepared slide shows cartilage, cilia, goblet cells, and epithelial cells of differing heights. The tissue being examined is
   a. adipose surrounding the lungs.
   b. simple cuboidal epithelium from a bronchiole.
   c. simple squamous epithelium from alveoli.
   d. pseudostratified epithelium from the trachea.

10. The smallest air sacs in lung tissue are
    a. bronchi.
    b. alveolar ducts.
    c. alveoli.
    d. goblet sacs.

11. The forerunner of both lungs and swim bladders was a structure called the
    a. pharyngeal sac.
    b. pneumatic sac.
    c. bronchial sac.
    d. alveolar sac.

20.1   Gill Structure and Function—Analytical Thinking

1. Use diagrams to explain why countercurrent exchange systems are more efficient than parallel exchange systems?

20.2   Tetrapod Lungs—Analytical Thinking

2. Why is the pharynx an important body region in vertebrates?

Exercise 21

Vertebrate Digestion

Prelaboratory Quiz

Study this week's laboratory exercise and then complete the following quiz to assess your preparation for the laboratory.

1. All of the following are features of the stomach EXCEPT the
   a. gastroesophageal sphincter.
   b. pyloric region.
   c. pyloric sphincter.
   d. villi.
   e. rugae.

2. All of the following are organs of the digestive system EXCEPT the
   a. soft palate.
   b. rectum.
   c. spleen.
   d. pancreas.
   e. esophagus.

3. The inner histological layer of the gut tract is the
   a. mucosa.
   b. submucosa.
   c. muscularis.
   d. serosa, or visceral peritoneum.

4. After passing through the ileocecal valve, gut contents enter the
   a. small intestine.
   b. large intestine.
   c. stomach.
   d. rectum.

5. After passing through the pyloric sphincter, gut contents enter the
   a. small intestine.
   b. large intestine.
   c. stomach.
   d. rectum.

6. True/False   The teeth of vertebrates other than mammals are uniformly conical, something known as the homodont condition.

7. True/False   A dental formula depicts the number and kinds of teeth on one side of the upper and lower jaws of a mammal.

8. True/False   The bulk of fat digestion occurs in the stomach of vertebrates.

9. True/False   Protein digestion begins in the stomach of vertebrates.

10. True/False   In this week's exercise, you will study starch digestion using salivary amylase from your saliva.

The digestive system of chordates shows many variations that reflect specializations for various feeding habits. These variations include, but are not limited to, the use of pharyngeal slits in filter feeding (amphioxus), modifications of the small intestine to increase surface area (spiral valve of sharks), modifications of the large intestine to form a fermentation pouch (cecum of rabbits), and modifications of the esophagus for storage and grinding (crop and gizzard of birds). In spite of many variations, the structure of vertebrate digestive systems is consistent in its basic form. In this laboratory, you will study the digestive system of the rat to learn basic structures and then examine a few of the modifications just described.

21.1 THE RAT DIGESTIVE SYSTEM

LEARNING OUTCOMES

1. Identify structures of the rat digestive system.
2. Describe digestive processes occurring in major regions of the rat digestive system.

▼ The digestive system of the rat was exposed in exercises 19 and 20, so little cutting will be necessary in the following observations (see fig. 19.10). As in exercise 20, find the **tongue.** Note the numerous papillae, which contain chemical receptors for the sense of taste. Identify **teeth, hard palate,** and **soft palate.** Just below the jaw, in the neck region, identify **salivary glands.** These secrete amylase into the mouth cavity to promote carbohydrate digestion.

The **esophagus** is a tube that carries food, which has been mixed with saliva, from the mouth to the stomach. The esophagus opens anteriorly just dorsal to the larynx. As you trace the esophagus posteriorly, note that it passes through the diaphragm and liver. The liver is made up of four lobes; however, unlike many vertebrates, the rat has no gallbladder associated with the liver. Bile from the liver empties into the anterior portion of the small intestine through the bile duct and functions in emulsifying fats. (The role of emulsification in fat digestion is demonstrated later in this exercise.)

The esophagus enters the **stomach** at the **cardiac end** (fig. 21.1). A **gastroesophageal sphincter** prevents the backup of stomach contents into the esophagus. Posteriorly, the stomach narrows into the **pyloric region.** A **pyloric sphincter** regulates the passage of stomach contents out of the stomach. Open the stomach and clean out its contents. Note the ridges **(rugae)** on the inner wall of the stomach. These aid in mixing as well as increasing surface area for the secretion of enzymes and acid. Within the abdominal cavity, note a fat-filled peritoneal membrane (fig. 21.2). This is the **omentum.** Locate the spleen on the left side of the liver. The spleen is a possible site for the development of B-cells that mediate a portion of vertebrate

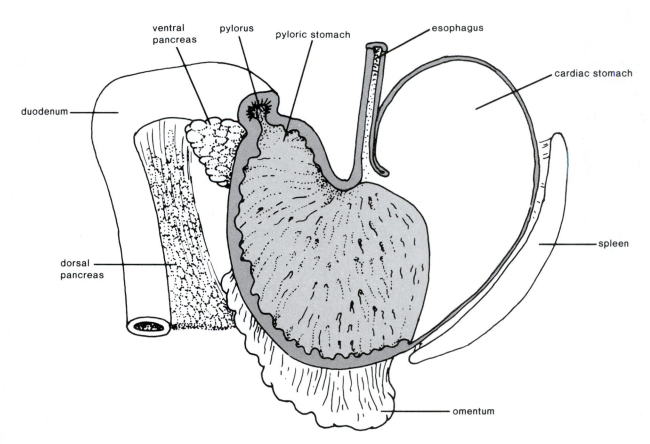

Figure 21.1 The stomach and associated structures. The wall has been cut away to expose the inner lining.

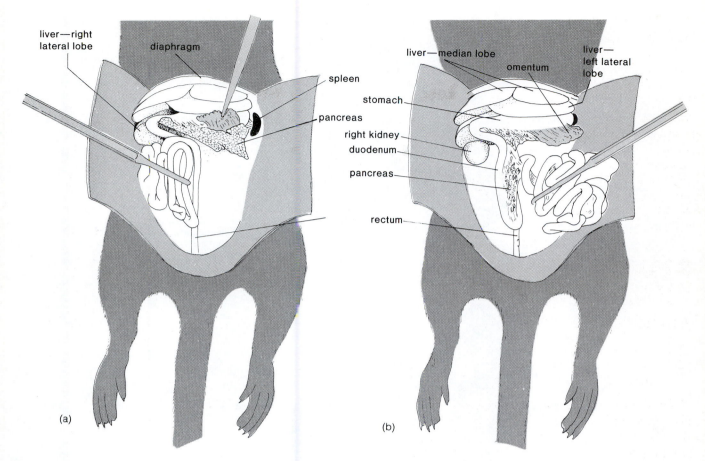

Figure 21.2 Abdominal viscera of the rat (ventral view); *(a)* intestines reflected to the right; *(b)* intestines reflected to the left.

immune responses and is the site for the destruction of old red blood cells.

The **small intestine** originates at the distal end of the stomach and is divided into three regions. (There is no clear division between these regions.) The **duodenum** is approximately the anterior one-sixth, the **jejunum** is the middle half, and the **ileum** is the remaining portion. Note that the small intestine, as well as all of the digestive tract, is suspended in the body cavity from the dorsal body wall by **mesentery.** Mesenteries are continuous with the peritoneum, the inner lining of the body wall. In the folds of the duodenum locate the **pancreas.** It is an extremely diffuse gland that empties its digestive enzymes into the small intestine near the bile duct.

At the junction of the large and small intestines is the **ileocecal valve.** It prevents backflow from the large intestine to the small intestine. The **large intestine** consists of the cecum, ascending colon, transverse colon, and descending colon. The **cecum** of the rat is a large fermentation pouch. Bacteria in the cecum aid in the digestion of plant material. Locate the sac-like cecum. Note that the ileum of the small intestine ends at the cecum and the ascending colon begins at the cecum. Follow the ascending colon anteriorly to the transverse colon and then posteriorly again to the descending colon. The colon functions in water reabsorption from, and

compaction of, digestive wastes. The colon is followed by the **rectum,** which descends through the pelvic region and ends at the anus. ▲

21.1 LEARNING OUTCOMES REVIEW—WORKSHEET 21.2

21.2 SMALL INTESTINE HISTOLOGY

LEARNING OUTCOMES

1. Describe the major histological features in a cross section of the small intestine.
2. Explain the function of the histological layers of the small intestine.

▼ Examine a cross section of the small intestine under low power of the compound microscope (figs. 21.3 and 21.4a). The cavity of the small intestine is called the lumen. Note **villi** that project into the lumen from the intestinal wall. These greatly increase the surface area for digestion and absorption. Villi are part of the **mucosal layer** of the intestine.

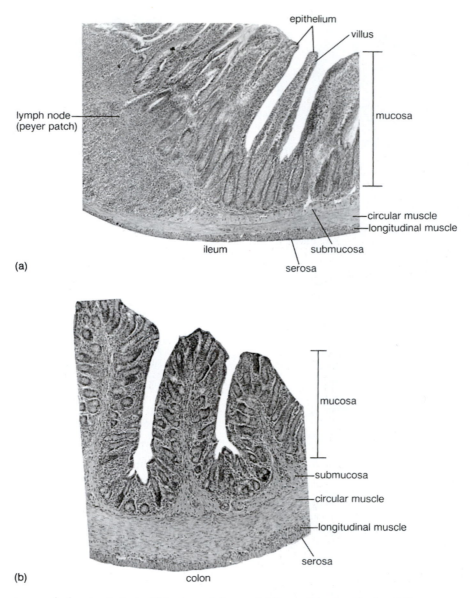

Figure 21.3 Photomicrograph showing intestinal histology of the rat: (*a*) ileum; (*b*) colon (both ×67).

The mucosa consists of simple columnar epithelium, underlying connective tissue, and some smooth muscle. Under high power note the single layer of columnar cells. Some cells are modified into mucus-secreting **goblet cells.** Mucus is secreted throughout the digestive tract to lubricate and protect the gut wall as well as facilitate the adherence of food and fecal particles. Goblet cells will appear translucent under the compound microscope. Electron micrographs reveal that the outer surface of the simple columnar cells has many **microvilli** that further increase the surface area for absorption and digestion (fig. 21.4b). Villi and the microvilli increase the surface area of the gut 600 times relative to the surface area of a simple cylinder. Next to the mucosal layer is the **submucosa.** Look for connective tissue, blood vessels, and lymphatic vessels associated with this layer. Outside the submucosa are two layers of smooth muscle. In the inner layer, smooth muscle cells are oriented circularly around the gut. These cells were sectioned longitudinally when the slide was prepared. When they contract, the gut is constricted. When waves of constriction pass along the gut, gut contents are mixed and moved through the intestine. This series of wavelike contractions is called **peristalsis.** In the outer layer, smooth muscle cells are oriented longitudinally. These cells were cut in cross section when the slide was prepared and individual cells will appear dotlike. When these cells contract, the gut shortens, which aids in mixing gut contents. The outer lining of the intestine is a layer of epithelium called the **serosa,** or **visceral peritoneum.** It is continuous with the mesentery and the peritoneum lining the inner body wall. ▲

21.2 LEARNING OUTCOMES REVIEW—WORKSHEET 21.2

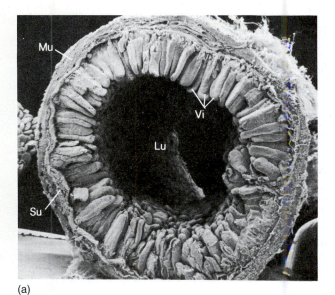

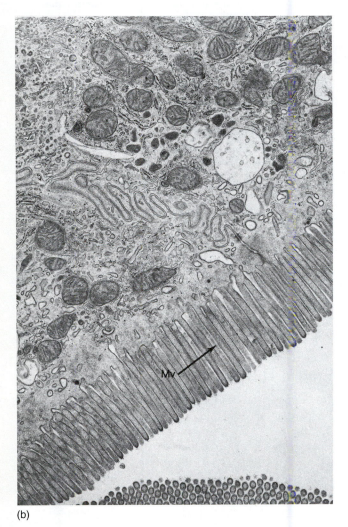

Figure 21.4 Electron micrographs of the small intestine: (a) cross section—villi (Vi), lumen (Lu), submucosa (Su), muscle layers (Mu)(×45); (b) lumenal surface of simple columnar cells lining each villus—microvilli (Mv). From Richard Kessel & Randy Kardon/Visuals Unlimited.

21.3 THE TEETH OF MAMMALS

LEARNING OUTCOMES

1. Explain the difference between homodont and heterodont tooth conditions.
2. Describe variations in the teeth of mammals.
3. Interpret a dental formula.

The structure and arrangement of teeth are important indicators of mammalian lifestyles. In vertebrates other than mammals, teeth are uniformly conical in shape. This is referred to as the **homodont** condition. In mammals, teeth along the length of the jaw are specialized for different functions. This is referred to as the **heterodont** condition.

There are four kinds of teeth in adult mammals. **Incisors** are anterior on the jaw and are chisel-like for gnawing or nipping. **Canines** are often long, stout, and conical. They are used for catching, killing, and tearing prey. **Premolars** are positioned next to the canines, have one or two roots, and have truncated surfaces for chewing. **Molars** have broad chewing surfaces and two (upper molars) or three (lower molars) roots.

A **dental formula** is an important tool used by zoologists in characterizing particular taxa. It is expressed as the number of teeth of each kind in one side of the upper jaw over the corresponding number in one side of the lower jaw. Teeth are indicated in the following order: incisors, canine, premolars, and molars.

▼ Examine a human skull that has a complete set of teeth. The human dental formula is as follows:

$$\frac{2 \cdot 1 \cdot 2 \cdot 3}{2 \cdot 1 \cdot 2 \cdot 3}$$

Notice the shape and surfaces of each tooth. Examine the skulls of the three other mammals on demonstration. Try to determine the dental formula for each mammal. **Fill in the information in table 21.1 in worksheet 21.1, and answer the accompanying questions now.** ▲

In a dog (or fox, cat) skull, notice the interaction between the fourth upper premolars and the first lower molars when the jaw is closed. This is called the **carnassial apparatus.** In a rodent, notice the large incisors. These incisors grow and wear away throughout the life of the animal. Their chisel-like shape is a result of hard enamel being present only on the outer surfaces. Gnawing causes the posterior margin of the teeth to wear more rapidly than the anterior margin, which keeps the teeth sharp. In a deer (or horse, cattle) notice that the anterior, food-procuring teeth are separated from posterior teeth by a gap, called the **diastema.** The diastema results from elongation of the snout that allows the anterior teeth to reach close to the ground or into narrow openings to procure food. Posterior teeth have high exposed surfaces and continuous growth.

21.3 LEARNING OUTCOMES REVIEW—WORKSHEET 21.2

21.4 VARIATIONS IN VERTEBRATE DIGESTIVE SYSTEMS

LEARNING OUTCOME

1. Describe variations in the digestive system of selected vertebrates.

▼ Examine demonstrations of the following vertebrates to see variations in digestive systems. Know the taxonomic group in which each occurs and its adaptive significance. ▲

Chordate Group	Modification
Cephalochordata amphioxus	Pharyngeal slits (filter feeding)
Petromyzontida lamprey	Buccal teeth and rasping tongue (ectoparasitism)
Chondrichthyes shark	Spiral valve (increased digestive and absorptive surface)
Aves	Crop (storage) Gizzard (grinding)
Mammalia rabbit	Enlarged cecum (fermentation)
human	Appendix (reduced cecum)

21.4 LEARNING OUTCOMES REVIEW—WORKSHEET 21.2

21.5 DIGESTIVE PHYSIOLOGY—ENZYME ACTION

LEARNING OUTCOMES

1. Describe conditions in the digestive tract that promote the digestion of starch.
2. Describe conditions in the digestive tract that promote the digestion of protein.
3. Describe conditions in the digestive tract that promote the digestion of fat.

In the following exercises, you will demonstrate the digestion of three classes of food by appropriate enzymes. As you do, think about your observations. Why does digestion occur under one set of conditions and not another? How do the conditions created in this exercise correspond to those actually present in the digestive tract? Where in the digestive tract would the digestion of each class of food occur?

Starch Digestion*

Starch digestion begins in the mouth, where starch is mixed with saliva (containing amylase). **Amylase** breaks starch into maltose (a disaccharide containing two glucose subunits).

Procedure

▼

1. Obtain 10 ml of saliva and dilute it with an equal volume of distilled water.
2. Number four test tubes 1 through 4.
3. Add the following to the tubes indicated:
   #1—3 ml distilled water
   #2—3 ml saliva-water mixture
   #3—3 ml saliva-water mixture + 3 drops concentrated HCl
4. Heat the remaining saliva-water mixture to boiling, then allow it to cool. Add 3 ml of boiled saliva-water mixture to tube #4.
5. Add 5 ml of cooked starch to each of the four test tubes.
6. Allow the tubes to incubate for 1 to 1.5 hours at 37° C.
7. After incubation, test the solutions for maltose by adding 5 ml of Benedict's reagent and immersing the test tubes in boiling water for 2 minutes. (Benedict's reagent changes from blue to yellow and eventually red with increasing sugar concentrations.)
8. Record and explain your results in table 21.2 in worksheet 21.1. ▲

Protein Digestion*

Protein digestion begins in the stomach. **Pepsin** breaks large polypeptides into smaller polypeptides.

Procedure

▼

1. Add the following to separate test tubes.
   #1—5 ml pepsin solution
   #2—5 ml pepsin + 1 drop concentrated HCl
   #3—5 ml pepsin + 1 drop 10N NaOH
   #4—5 ml pepsin + 1 drop concentrated HCl
2. Add a thin slice of egg white from a boiled egg to each tube. (The size of the slices is not critical; however, all four must be very thin and approximately the same size. Slices about 1 cm × 1 cm × 1 mm work well.)

3. Incubate tubes #1–3 at 37° C for 1 hour (shake periodically).
4. Incubate tube #4 at 0° C for 1 hour (shake periodically).
5. Filter each solution and visually examine each egg slice for evidence of digestion.
6. Record and explain your results in table 21.3 in worksheet 21.1. ▲

Fat Digestion*

The bulk of fat digestion occurs in the small intestine. The process is dependent on two steps: the **emulsification** of large fat droplets into smaller droplets, and the enzymatic breakdown of fat molecules into fatty acids and glycerol. Why is emulsification a necessary first step? _____

_____

In this exercise, you will monitor the breakdown of fats by observing pH changes. With digestion, an increase in free fatty acids results in a lowered pH.

Procedure

▼

1. Number three test tubes 1 through 3.
2. Add the following to the proper test tubes:
   #1—1 ml of vegetable oil + 5 ml water + a few grains of bile salts.
   #2—1 ml of vegetable oil + 5 ml of pancreatin solution + a few grains of bile salts.
   #3—1 ml of vegetable oil + 5 ml of pancreatin solution.

3. Incubate the tubes at 37° C for 1 hour. Check the pH of the tubes at the beginning of the experiment and then at 10-minute intervals.
4. Record and explain your results in table 21.4 in worksheet 21.1. ▲

21.5 LEARNING OUTCOMES REVIEW—WORKSHEET 21.2

*From Stuart Ira Fox, A *Laboratory Guide to Human Physiology,* 2nd ed. Copyright © 1980 Wm. C. Brown Publishers, Dubuque, Iowa. All Rights Reserved. Reprinted by permission.

WORKSHEET 21.1   Vertebrate Digestion

TABLE 21.1   Specializations of the Teeth of Selected Mammals

Animal	Dental Formula	Description of: Incisors Canines Premolars Molars	Food Habits
Human			
Dog			
Rodent			
Deer			

1. Briefly describe the human diet.

2. Describe specializations of human teeth that can help us understand how we process the food we eat.

3. Explain the differences between human teeth and the teeth of a dog or cat.

4. What is the usefulness of the dental formula to a zoologist who finds the skull of a mammal and wants to identify the mammal and its feeding habits?

TABLE 21.2   Starch Digestion

Tube	Results
1	
2	
3	
4	

− No digestion

+ Little to moderate digestion

++ Much digestion

TABLE 21.3   Protein Digestion

Tube	Results
1	
2	
3	
4	

− No digestion

+ Little to moderate digestion

++ Much digestion

TABLE 21.4 Fat Digestion

Time (min)	Tube 1	Tube 2	Tube 3
0			
10			
20			
30			
40			
50			
60			

5. Which tube showed the greatest starch digestion? Explain the results by comparing that tube to the other three tubes.

6. Which tube showed the greatest protein digestion? Explain the results by comparing that tube to the other three tubes.

7. Which tube showed the greatest fat digestion? Explain the results by comparing that tube to the other two tubes.

8. What is emulsification? Why is emulsification an important step in the digestion of fats?

WORKSHEET 21.2   Vertebrate Digestion

21.1   The Rat Digestive System—Learning Outcomes Review

1. After passing the gastroesophageal sphincter of the rat, gut contents would enter the
   a. stomach.
   b. small intestine.
   c. ascending colon.
   d. esophagus.
   e. transverse colon.

2. After passing the pyloric sphincter of the rat, gut contents would enter the
   a. stomach.
   b. small intestine.
   c. ascending colon.
   d. esophagus.
   e. transverse colon.

3. The ileocecal valve of the rat regulates the movement of gut contents into the
   a. stomach.
   b. small intestine.
   c. ascending colon.
   d. transverse colon.

4. After passing the ascending colon of the rat, gut contents would enter the
   a. stomach.
   b. small intestine.
   c. descending colon.
   d. esophagus.
   e. transverse colon.

5. This is a fat-filled membrane within the abdominal cavity of the rat.
   a. jejunum
   b. mesentery
   c. omentum
   d. duodenum

6. The liver of the rat is located in the
   a. thorax just anterior to the diaphragm.
   b. abdomen just posterior to the diaphragm.
   c. abdomen just posterior to the stomach.
   d. abdomen among the folds of the large intestine.

21.2   Small Intestine Histology—Learning Outcomes Review

7. This histological layer of the small intestine is comprised of simple columnar epithelium and goblet cells, and functions in secretion of enzymes and absorption of digested food.
   a. mucosal layer
   b. submucosa
   c. circular and longitudinal smooth muscle
   d. serosa

8. The histological layer on the outside of the small intestine is continuous with the peritoneum of the body wall. It is called the
   a. mucosal layer.
   b. submucosa.
   c. circular and longitudinal smooth muscle layer.
   d. serosa.

9. The wavelike contractions of the intestine that move gut contents through the intestine are called
   a. mucosal waves.
   b. peristalsis.
   c. submucosal waves.
   d. serosal waves.

21.3 Teeth of Mammals—Learning Outcomes Review

10. Toothed vertebrates other than mammals have
    a. homodont teeth.
    b. heterodont teeth.
    c. isodont teeth.

11. Teeth that are often chisel-like for gnawing or nipping are
    a. incisors.
    b. canines.
    c. premolars.
    d. molars.

12. An animal with a dental formula of 0.1.3.2/2.1.2.3 would have
    a. two premolars total in its upper and lower jaws.
    b. three molars on one-half of its lower jaw.
    c. five molars total in its upper and lower jaws.
    d. four canines total in its upper and lower jaws.
    e. Both b and d are correct.
    f. None of the above is correct.

13. A scissor-like interaction between premolars and molars in a dog is called the
    a. diastema.
    b. carnassial apparatus.
    c. molar scissors.

21.4 Variations in Vertebrate Digestive Systems—Learning Outcomes Review

14. A spiral valve increases the surface area for digestion and absorption in members of the
    a. Petromyzontida.
    b. Cephalochordata.
    c. Chondrichthyes.
    d. Aves.
    e. Mammalia.

15. The large cecum of a rabbit and a rat functions in
    a. fermentation.
    b. protein digestion.
    c. fat storage.
    d. storage of digestive enzymes.

21.5 Digestive Physiology—Enzyme Action—Learning Outcomes Review

16. Enzymatic digestion of starch
    a. begins in the mouth.
    b. halts in the stomach.
    c. is promoted by reduced pH.
    d. All of the above are true.
    e. a and b only are true.

17. The exercise on protein digestion demonstrated that
    a. amylase digestion of protein begins in the mouth.
    b. pepsin requires a low pH for its optimal activity.
    c. pepsin has both an optimal pH and an optimal temperature.
    d. protein digestion involves emulsification.
    e. Both b and c are correct.

18. The exercise on fat digestion demonstrated that
    a. fat digestion begins in the mouth.
    b. pancreatin solution requires a low pH for its optimal activity.
    c. pancreatin solution has both an optimal pH and optimal temperature.
    d. efficient fat digestion involves emulsification.
    e. Both b and c are correct.

21.1 The Rat Digestive System—Analytical Thinking

1. Unlike with the skeletal, circulatory, and nervous system exercises, little was made of the evolutionary comparisons between vertebrate digestive systems. Why do you think that might be the case?

21.2 Small Intestine Histology—Analytical Thinking

2. Describe the modifications of the small intestine that you observed in the laboratory that increase surface area for secretion and digestion.

21.3 The Teeth of Mammals—Analytical Thinking

3. Compare the dentition of an herbivore, such as a deer, to that of a carnivore, for example, a cat.

21.4 Variations in Vertebrate Digestive Systems—Analytical Thinking

4. Explain the particular digestive system modifications observed in sharks, birds, and rabbits.

21.5 Digestive Physiology—Enzyme Action—Analytical Thinking

5. Health food stores sometimes sell packaged digestive enzyme supplements to aid in digestion. Sometimes these supplements are designed to be sprinkled on food before it is eaten, and sometimes the supplements are taken orally with a meal. Based on the results of your laboratory exercise, how effective would an amylase supplement be on aiding starch digestion if the supplement were taken orally? What considerations should manufacturers keep in mind as they produce their supplement products?

Exercise 22

Vertebrate Excretion

Prelaboratory Quiz

Study this week's laboratory exercise and then complete the following quiz to assess your preparation for the laboratory.

1. The kidneys of a shark are
   a. suspended in the body cavity from the dorsal body wall.
   b. suspended in the body cavity from the ventral body wall.
   c. located adjacent to the large intestine.
   d. located along the dorsal body wall and covered by peritoneum.

2. The structure of the excretory system that runs from the metanephric kidney to the bladder is the
   a. ureter.
   b. urethra.
   c. renal pyramid.
   d. renal cortex.

3. The region of the metanephric kidney where the ureter receives urine from the tubule system of the kidney is called the
   a. medulla.
   b. cortex.
   c. renal pelvis.
   d. glomerulus.

4. The functional unit of the kidney is called the
   a. renal pyramid.
   b. nephron.
   c. calyx.
   d. glomerulus.

5. All of the following occur in the nephron of the metanephric kidney EXCEPT
   a. filtration of the blood.
   b. secretion of ions into the tubule system.
   c. reabsorption of water, ions, glucose, and amino acids from the filtrate.
   d. storage of urine.

6. True/False   In a microscope slide of a metanephric kidney, glomeruli would be found in the cortical region.

7. True/False   In a microscope slide of a metanephric kidney, glomeruli would be found in close association with collecting ducts.

8. True/False   This week's exercise involves a demonstration of active transport across the kidney tubules of a goldfish.

9. True/False   The rat has a cloaca.

10. True/False   The loop of the nephron and the collecting duct are located within the medulla of the metanephric kidney.

The excretory and reproductive systems of vertebrates are closely linked embryologically, evolutionarily, and anatomically. It is common, therefore, to study them simultaneously. They are not, however, physiologically linked. In this manual, therefore, they are placed in separate sections to reflect their physiological independence, but they will be covered one after the other to reflect their evolutionary relationship.

Excretory systems of vertebrates function in osmoregulation (water and ion regulation) and ridding the body of nitrogenous wastes. Vertebrates have adapted in various ways to osmoregulatory and excretory demands. Most of these adaptations, however, are not obvious through gross morphological observations. This laboratory, therefore, will focus on describing the anatomical characteristics of two vertebrates, the shark and the rat. These animals possess two of the more common vertebrate kidneys: **opisthonephric** and **metanephric.**

22.1 THE OPISTHONEPHRIC KIDNEY

LEARNING OUTCOME

1. Describe the excretory system of a shark.

The opisthonephric kidney of the shark is similar to the mesonephric kidney of most other adult fish, as well as that of adult amphibians, the primary difference being size and position.

▼ Examine a demonstration dissection. The kidneys are located along either side of the middorsal body wall. They are cream-colored, covered by peritoneum, and tightly pressed against the body wall in the mid and posterior region of the body cavity. Do not attempt to find the duct system, but note the **cloaca** (common opening for excretory, reproductive, and digestive waste products). A renal papilla represents the opening of the excretory system in the cloaca (see fig. 23.1). ▲

22.1 LEARNING OUTCOMES REVIEW—WORKSHEET 22.1

22.2 THE METANEPHRIC KIDNEY AND ASSOCIATED STRUCTURES

LEARNING OUTCOMES

1. Identify structures of the rat excretory system.
2. Trace a drop of urine from the renal pelvis to the outside of a rat.

▼ Examine the dorsal body wall in the anterior abdominal region of your rat for the **kidneys** (fig. 22.1, see figs. 19.12 and 19.13). They are usually embedded in fat and covered by peritoneum of the body wall rather than being suspended from the body wall by mesentery (as in the case of other abdominal organs).

Working on one side of the rat, remove fat from the kidney and surrounding area. Be careful that you do not tear the ureter in the process. Note the **adrenal gland** located just anterior to the kidney. It also is usually embedded in fat. On the medial side of the kidney is a concavity. This is the point where renal blood vessels enter and leave the kidney and the **ureter** emerges from the kidney. Follow the ureter posteriorly, cleaning fat from it as you proceed. The ureter delivers urine to the **urinary bladder.** The ureter connects to the dorsal/posterior margin of the bladder. The urinary bladder narrows posteriorly to deliver urine to the outside through the **urethra.** Follow the urethra to the pubis. In the male, the urethra receives the duct system from the testes and delivers both excretory and reproductive products to the outside.

Return to the kidneys. Remove the kidney from the body on the side opposite where you were previously working. Using a sharp scalpel, make a frontal section, dividing it into identical halves. Using a hand lens or dissecting microscope, identify the **cortex, medulla, renal pyramids, calices,** and **renal pelvis** (fig. 22.2). The latter three structures are the duct systems that deliver urine to the ureter. See any models or fresh kidneys available. ▲

22.2 LEARNING OUTCOMES REVIEW—WORKSHEET 22.1

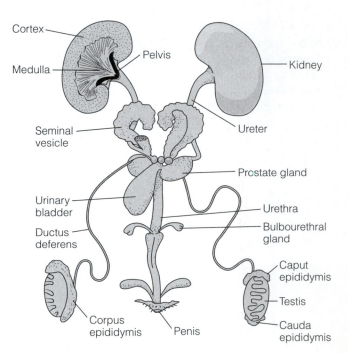

Figure 22.1   Urogenital organs of the male rat.

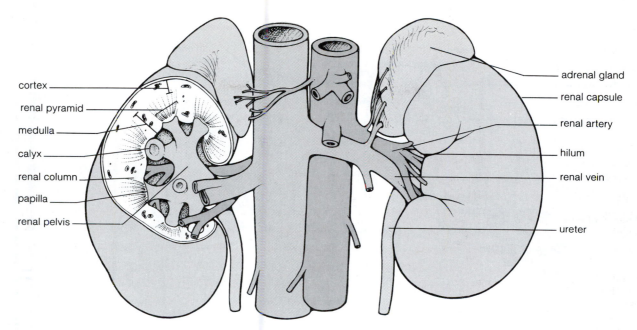

Figure 22.2 Ventral view of the mammalian kidneys. The right kidney is frontally sectioned to show internal structures. The adrenal gland of nonhuman mammals is usually positioned next to, but not attached to, the kidneys.

22.3 KIDNEY HISTOLOGY AND THE NEPHRON

LEARNING OUTCOMES

1. Identify regions of the metanephric kidney in histological preparations.
2. Identify nephron tubule systems in the kidney cortex.
3. Identify nephron tubule systems in the kidney medulla.

▼ The **nephron** is the functional unit of the kidney (fig. 22.3). Examine a slide showing a frontal section of the mammalian kidney. Under low power, distinguish between the medulla and the cortex. In the cortex, note the **glomeruli** (figs. 22.4 and 22.5). Examine a glomerulus under high power. Around it, note the **glomerular capsule.** The glomerulus is a capillary bed modified for filtration. Blood comes to the glomerulus, under pressure, through the afferent arteriole. Openings in the capillary walls of the glomerulus allow water and solutes to move from the bloodstream into the tubule system of the nephron. Blood cells and proteins are too large to pass through the pores; they move from the glomerulus into the efferent arteriole and then to the peritubular capillaries surrounding the tubules of the nephron. The filtrate enters the glomerular capsule and then the tubule system of the nephron. Around the glomerular capsule you should see sections through the tubule systems associated with the cortex. These are sections through the **proximal** and **distal convoluted tubules.** The tubules will be cut in a variety of sectional views because they are convoluted. We will not attempt to distinguish between these tubules. They represent areas where water, amino acids, glucose, and salts are reabsorbed from the tubule into the peritubular capillaries

and where ions may be secreted into the tubules. Switch back to low power and move to the medulla. There are no glomeruli in the medulla. Instead, there are linear arrays of cells descending toward the renal pelvis. These are the cells that line the **loops of the nephron** and the **collecting ducts.** These tubules allow the kidney to regulate the water content of the urine. The collecting ducts eventually lead to the renal pelvis and ureter. In the process just described, much of the urea filtered from the blood at the glomerulus is excreted, and most of the water, salts, glucose, and amino acids is reabsorbed. ▲

22.3 LEARNING OUTCOMES REVIEW—WORKSHEET 22.1

22.4 ACTIVE TRANSPORT ACROSS THE KIDNEY TUBULE

LEARNING OUTCOME

1. Observe active transport across goldfish kidney tubules.

▼ Your instructor has dissected out a goldfish kidney and placed it in saline. Cut off a small piece of the kidney, place it in a depression slide, and gently tease the tissue using dissecting pins or teasing needles. Free adhering tissue by sucking the preparation up and down in a dropper. Transfer to fresh saline, cover with a coverslip, and view under low power of a compound microscope. You should be able to see clear tubules protruding from the darker tissue mass. (You may need to repeat the process just described.) Blot saline

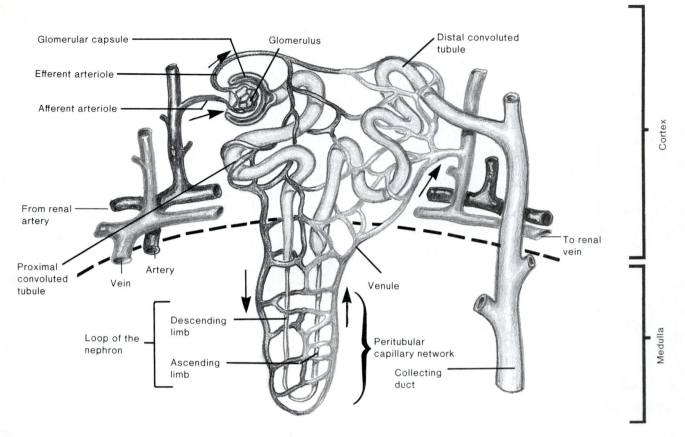

Figure 22.3   The mammalian nephron (diagrammatic).

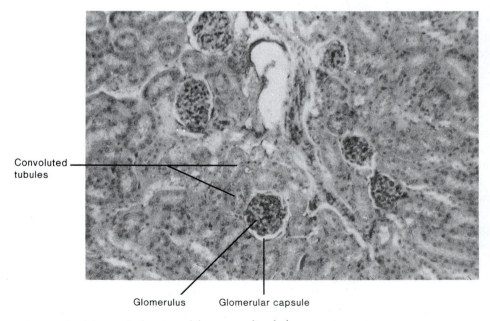

Figure 22.4   Photomicrograph of the cortical region of the mammalian kidney.

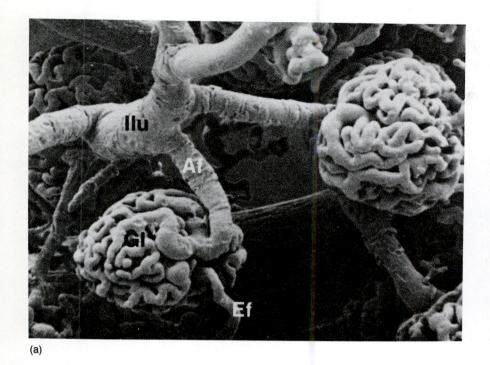

(a)

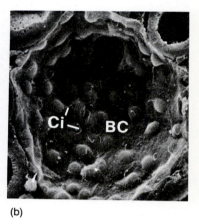

(b)

Figure 22.5 The mammalian nephron: *(a)* circulatory system of the nephron injected with plastic, kidney tissue digested away—interlobular arteries (Ilu), glomerus (Gl), afferent arteriole (Af), efferent arteriole (Ef)(×260); *(b)* the glomerular (Bowman's) capsule (BC). Note the cilia (Ci) on some cells. Cilia may be involved with moving fluids through the nephron. From R. G. Kessel and R. H. Kardon. *Tissues and Organs: A Text-Atlas of Scanning Electron Microscopy.* © 1979 by W. H. Freeman.

from the slide and add chlorophenol red (0.05 millimole per liter [mM/L] in goldfish saline). Cover again with a coverslip and observe. Watch a group of tubules (some may have been damaged in the teasing process). You should observe the interior of the tubules gradually becoming pink due to the accumulation of chlorophenol red. ▲

22.4 LEARNING OUTCOMES REVIEW—WORKSHEET 22.1

WORKSHEET 22.1 Vertebrate Excretion

22.1 The Opisthonephric Kidney—Learning Outcomes Review

1. The kidneys of the shark are located on the
   a. ventral body wall suspended in mesentery.
   b. dorsal body wall suspended in mesentery.
   c. ventral body wall between peritoneum and body-wall muscles.
   d. dorsal body wall between peritoneum and body-wall muscles.

2. The excretory system of the shark discharges
   a. at the renal papilla.
   b. into a cloaca.
   c. into a common opening with digestive and reproductive products.
   d. All of the above are true.

22.2 The Metanephric Kidney and Associated Structures—Learning Outcomes Review

3. After leaving the renal pelvis of the rat kidney, urine is in the
   a. urinary bladder.
   b. ureter.
   c. urethra.
   d. calices.

4. Before entering the urethra of the rat, urine is in the
   a. urinary bladder.
   b. ureter.
   c. renal pelvis.
   d. calices.

5. In a frontal section, the cortex of the rat kidney is
   a. outside of the medulla.
   b. inside of the renal pelvis.
   c. directly connected to calices.
   d. Both a and c are correct.

22.3 Kidney Histology and the Nephron—Learning Outcomes Review

6. Glomeruli were observed in the
   a. cortex of the kidney.
   b. medulla of the kidney.
   c. renal pelvis of the kidney.

7. Tubules observed in the cortex of the kidney were cut
   a. primarily in longitudinal sections because they are straight tubes.
   b. primarily in cross sections because they are straight tubes.
   c. in a variety of sectional views because they are convoluted.

8. The tubules observed in the medulla of the kidney were
   a. proximal and distal convoluted tubules.
   b. loops of the nephron.
   c. collecting ducts.
   d. ureters.
   e. Both b and c are correct.

9. Immediately after leaving collecting ducts, urine is in the
   a. renal pelvis.
   b. ureter.
   c. convoluted tubules.
   d. loop of the nephron.

22.4 Active Transport Across the Kidney Tubule—Learning Outcomes Review

10. In the goldfish kidney preparation you were observing the active transport of chlorophenol red
    a. out of the goldfish kidney tubules.
    b. into the goldfish kidney tubules.
    c. out of the goldfish ureter.

22.1 and 22.2 Opisthonephric and Metanephric Kidneys—Analytical Thinking

1. Compare and contrast the structure of the shark and rat excretory systems.

22.3 Kidney Histology and the Nephron—Analytical Thinking

2. The loop of the nephron of mammals is an adaptation for conserving water by producing a hypertonic urine. In general, the longer the loop, the better able the mammal is to produce urine that is strongly hyperosmotic. If you dissected a desert rodent rather than a laboratory rat, would you expect its cortex or medulla to be thicker as compared to the laboratory rat? Explain.

3. Based on your own day-to-day experiences, explain why producing hyperosmotic urine conserves water.

22.4 Active Transport Across the Kidney Tubule—Analytical Thinking

4. Explain how you know that you were observing active transport in the goldfish kidney tubules rather than passive diffusion.

Exercise 23

Vertebrate Reproduction

Prelaboratory Quiz

Study this week's laboratory exercise and then complete the following quiz to assess your preparation for the laboratory.

1. In many sharks, the development of young occurs within a uterus, but most nourishment is derived from yolk present in the egg. This kind of development is
   a. viviparous development.
   b. ovoviviparous development.
   c. oviparous development.

2. Most female mammals and some female sharks nourish young through a placenta during development. This kind of development is
   a. viviparous development.
   b. ovoviviparous development.
   c. oviparous development.

3. The tubule that carries sperm from the testes of a male shark toward the cloaca is the
   a. seminal vesicle.
   b. ductus deferens.
   c. epididymis.
   d. seminiferous tubule.

4. The ovaries of a female shark are located in the
   a. anterior region of the body cavity.
   b. posterior region of the body cavity.
   c. middle region of the body cavity.
   d. cloaca.

5. A mature follicle would contain a(n)
   a. primary oocyte.
   b. polar body.
   c. secondary oocyte.
   d. ootid.

6. True/False   Sperm-producing structures found in testes are called epididymal tubules.

7. True/False   Interstitial cells of the testes are involved in the production of the hormone testosterone.

8. True/False   A corpus luteum remains in the ovary following ovulation and produces the hormones progesterone and estrogen.

9. True/False   One meiotic cell division in female mammals produces four egg cells.

10. True/False   Seminal vesicles, the bulbourethral glands, and the prostate gland are all accessory glands that create a proper medium for sperm ejaculation and survival.

The evolutionary and developmental relationship between vertebrate excretory and reproductive systems is indicated by their common embryological origin and their common duct systems. This is especially obvious in the males of all vertebrates. The urethra of the male rat, observed in exercise 22, is a common pathway for both urine and sperm. In addition, a cloaca is found in all vertebrates through the monotremes (mammals such as the spiny ant-eater and the duck-billed platypus). The cloaca represents a common pathway for the exit of excretory and reproductive products as well as undigested food material. The trend in evolution has been to separate the two systems. This separation is essentially complete only in female mammals above the marsupials.

23.1 THE SHARK REPRODUCTIVE SYSTEM

LEARNING OUTCOMES

1. Identify structures of the reproductive system of the male shark.
2. Identify structures of the reproductive system of the female shark.

The shark will again be used to represent a basic arrangement of reproductive structures. Within the vertebrates, however, there is a considerable variation in arrangements of duct systems. As a result, one must be cautious about generalizing based on what is seen in the shark. This is especially true in regard to the male system.

▼ Examine the external structure of male and female sharks. They can be distinguished on the basis of the structure of the pelvic fin. Unlike most fishes, fertilization in the shark is internal. Sperm is transferred via modified pelvic fins, called **claspers,** that are inserted into the female cloaca. Note the dorsal sperm groove on the clasper. Internally, each clasper has a muscular walled sac, called a siphon, that aids in propelling sperm along the clasper. Sperm must then travel up the female reproductive tract to the cranial end of the oviduct, where fertilization occurs. ▲

The Male

▼ Examine a demonstration dissection of a male shark (fig. 23.1). The **testes** are located near the cranial end of the body cavity and suspended in mesentery (mesorchium). Through this mesentery, small tubules carry sperm to the **ductus deferens** (pl. deferentes). The ductus deferens is a highly coiled duct lying on the ventral surface of the kidney, which widens caudally to form the **seminal vesicle.** Posteriorly, the seminal vesicle passes dorsal to the **sperm sacs.** The latter two structures function in sperm storage. They enter into the cloaca through a urogenital papilla. A cloaca

represents a common opening for excretory, reproductive, and digestive systems ("cloaca" is derived from a Latin word for sewer). The cloaca should be cut open to see the urogenital papilla clearly. ▲

The Female

▼ Examine a demonstration dissection of a female shark (fig. 23.2). The paired **ovaries** are located in the anterior end of the body cavity and suspended by the mesentery (mesovarium). Eggs, if mature, are large, with much yolk. At ovulation, eggs are released into the body cavity and enter the oviducts at a common opening, which lies anterior and ventral to the liver. The **oviduct** runs posteriorly along the dorsal body wall. An enlargement dorsal to the ovary is the **shell gland** (nidamental gland). It secretes a proteinaceous shell around two or three fertilized eggs. A more posterior enlargement of the oviduct (posterior one-third) is referred to as the **uterus.** It is especially large in pregnant females but is almost indistinguishable from the oviduct in young females. Development occurs in the uterus, but the embryo receives its nourishment from the yolk of the egg. This is referred to as **ovoviviparous** development. Animals that have internal development and direct nourishment of the embryo through a placenta are **viviparous** (most mammals and some sharks). Animals that deposit eggs outside the body to develop from stored yolk are **oviparous** (most fishes, some primitive sharks, amphibians, reptiles, and birds). The uteri connect with the cloaca just ventral to the opening of the excretory system, the urinary papilla. The cloaca should be cut open to observe these structures. ▲

23.1 LEARNING OUTCOMES REVIEW—WORKSHEET 23.1

23.2 THE RAT REPRODUCTIVE SYSTEM

LEARNING OUTCOMES

1. Identify structures of the reproductive system of the male rat.
2. Identify structures of the reproductive system of the female rat.

In studying the rat, it should be understood that although the essence of the mammalian reproductive system is constant throughout the class, there are variations in anatomical details that cannot be covered here (e.g., variations in uterine and vaginal structures). In the following dissection, you are responsible for knowing both male and female systems. You should dissect your specimen and then exchange specimens with a classmate who has worked on the opposite sex.

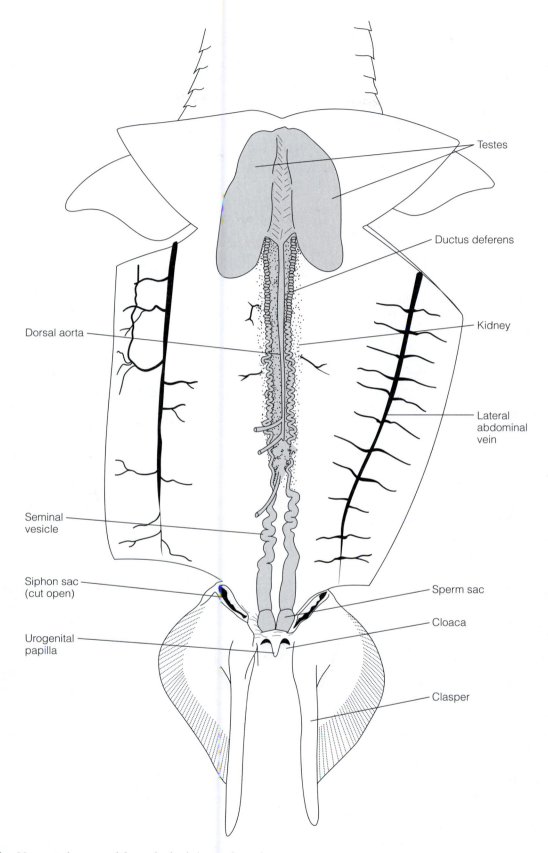

Figure 23.1 Urogenital organs of the male shark (ventral view).

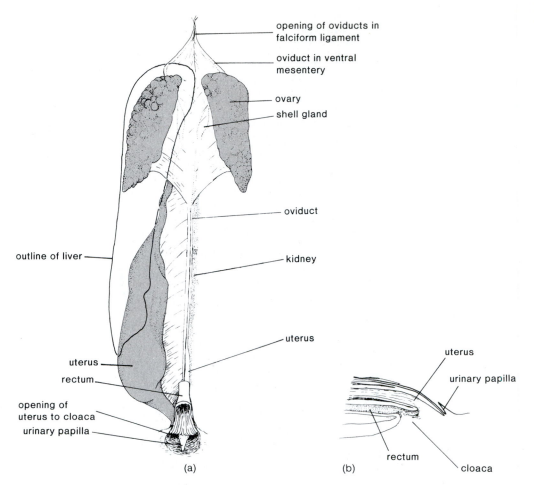

Figure 23.2 Urogenital organs of the female shark, (a) A pregnant female is illustrated in the left half of the diagram, and a nonpregnant female is illustrated in the right half of the diagram, (b) A diagrammatic sagittal view of the female cloaca.

The Male

▼ Turn the rat ventral side up. In living rats, the **testes** are normally retracted into the body cavity, leaving the **scrotum** empty. This is accomplished through the **cremaster muscle,** which pulls the testes into the abdominal cavity through the **inguinal canal.** In forms where the testes permanently descend (e.g., humans), the inguinal canal usually grows almost completely closed.

Open the scrotum and note the paired testes (fig. 23.3 and see fig. 22.1). Covering the surface of each testis is the **epididymis.** The epididymis consists of a highly coiled tubule that stores and carries sperm from the testis toward the penis. The epididymis covers the anterior, lateral, and posterior borders of the testis. Follow the epididymis to the point that it becomes a distinct, single spermatic cord. The spermatic cord consists of a connective tissue sheath surrounding a spermatic artery and vein, a nerve, and the ductus deferens. The **ductus deferens** carries mature sperm into the penis.

Follow the spermatic cord into the body cavity and toward the penis. The spermatic cord loops behind, then over the ureter before joining the urethra. Using heavy scissors, cut through the pubic bone to follow the urethra to the penis. Associated with the urethra and spermatic cords are **accessory glands** that add secretions to the sperm. These include the large **seminal vesicles,** which unite with the spermatic cord just before the spermatic cord unites with the urethra. The **bulbourethral glands** are located on either side of the proximal end of the penis. The **prostate gland** is located near the junction of the ductus deferens and the urethra. The bulbourethral glands are very small and difficult to locate. The prostate gland appears as two lobes on either side of the urinary bladder. It is actually a single gland, not a paired structure. All three accessory glands function in creating a proper medium for sperm ejaculation and survival, and they contribute to the bulk of the ejaculate.

The Female

Locate the **ovaries** just posterior and slightly lateral to the kidneys (figs. 23.4 and 23.5). The ovaries of mature females usually show **mature follicles** that represent eggs ready to be discharged. As a result, the surface of the ovary is usually

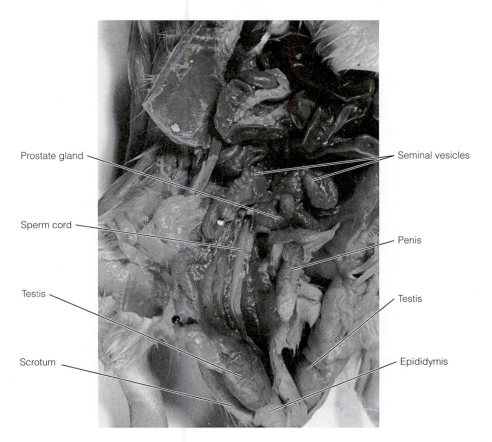

Figure 23.3 Urogenital system of a male rat.

Labels in figure:
- Prostate gland
- Sperm cord
- Testis
- Scrotum
- Seminal vesicles
- Penis
- Testis
- Epididymis

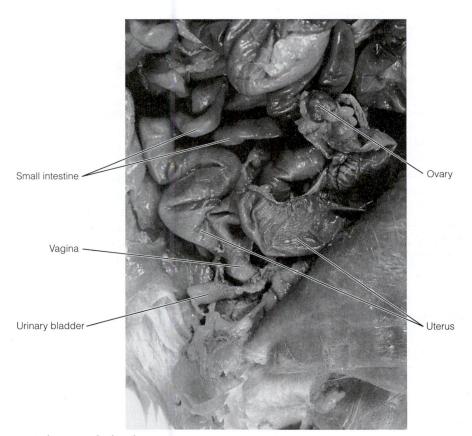

Figure 23.4 The urogenital system of a female rat.

Labels in figure:
- Small intestine
- Vagina
- Urinary bladder
- Ovary
- Uterus

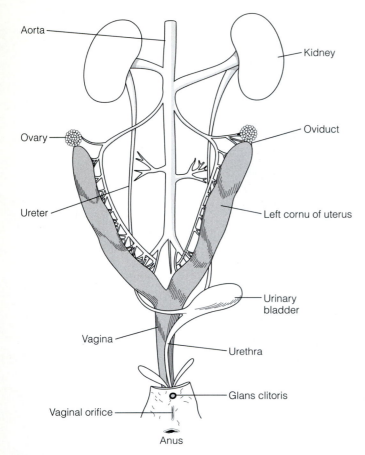

Aorta

Kidney

Ovary

Oviduct

Ureter

Left cornu of uterus

Urinary
bladder

Vagina

Urethra

Glans clitoris

Vaginal orifice

Anus

Figure 23.5 Urogenital organs of the female rat.

convoluted. The ovary of the typical mammal discharges the egg into the oviduct through a funnel-shaped opening, the **ostium.** In the rat, the ostium completely encloses the ovary. A highly coiled **oviduct** leads to the **uterus.** The uterus of the rat is divided into two "horns," or **cornua,** one associated with each ovary. The cornua unite posteriorly to form the **vagina.** The vagina opens to the exterior through the external vaginal orifice. A **clitoris (glans clitoris)** is associated with the urethral opening. It is the female homolog of the penis. ▲

23.2 LEARNING OUTCOMES REVIEW—WORKSHEET 23.1

23.3 GAMETE FORMATION

LEARNING OUTCOMES

1. Identify structures of the mammalian testes studied in histological preparations.
2. Explain the events involved with the development and maturation of sperm.

3. Identify structures of the mammalian ovary studied in histological preparations.
4. Explain the events involved with the development and maturation of an ovum.

Gametes are formed by meiotic cell division (reduction division). In meiosis, homologous pairs of chromosomes are randomly divided between daughter cells that have half the chromosome number (IN, or haploid) of the original parent cell (2N, or diploid). After a period of differentiation, the resulting gametes may be involved with fertilization. In fertilization, haploid egg and sperm nuclei fuse to restore the diploid chromosome number in the zygote. The zygote then starts its embryological development. It is assumed that students have studied meiotic cell division prior to the exercise that follows.

Spermatogenesis

Sperm cells are produced in the thousands of coiled tubules that make up the vertebrate testis. Cells involved with meiosis and sperm differentiation line these **seminiferous tubules.** Between seminiferous tubules are **interstitial cells,** which produce and secrete **testosterone,** one of the major androgens regulating sperm production and secondary sex characteristics.

▼ Obtain a prepared slide of a sectioned mammalian testis. Locate seminiferous tubules and interstitial cells (figs. 23.6 and 23.7). When observing the seminiferous tubules, note the layers of cells surrounding the central cavity of the tubule. These cells are in various stages of sperm development. The outermost cells are **spermatogonia.** These undifferentiated cells divide mitotically to produce the next layer of cells, **primary spermatocytes.** Primary spermatocytes begin meiosis. The product of the first meiotic cell division is represented by those cells within the third cell layer. These are **secondary spermatocytes** and are haploid cells. These cells undergo the second meiotic cell division and form **spermatids.** During a period of differentiation, spermatids develop a **flagellum** and a tiny enzyme-filled cap (**acrosome**). These nearly mature sperm cells can be seen lining the lumen of the seminiferous tubule. During its period of differentiation, the spermatid is associated with a **sustentacular cell,** which produces the hormone **estradiol.** Along with testosterone, estradiol helps build up the environment for spermatogenesis. Examine a smear of bull (or other) sperm. Compare your observations to figure 23.8. The flagellum is for locomotion. The caplike acrosome covering the head of a sperm cell is used in egg penetration, and the middle piece, which connects the head to the flagellum, contains mitochondria for energy conversions required for locomotion. ▲

Oogenesis

Oogenesis is similar in all animals. The meiotic processes are the same as in spermatogenesis, but the goal of oogenesis is always to produce an egg cell with a large supply of

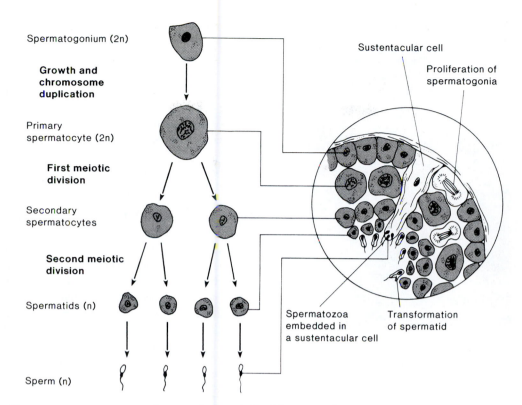

Spermatogonium (2n)

Growth and chromosome duplication

Primary spermatocyte (2n)

First meiotic division

Secondary spermatocytes

Second meiotic division

Spermatids (n)

Sperm (n)

Sustentacular cell

Proliferation of spermatogonia

Spermatozoa embedded in a sustentacular cell

Transformation of spermatid

Figure 23.6   Diagrammatic cross section of a seminiferous tubule of the mammal.

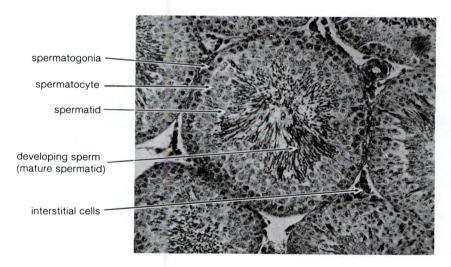

spermatogonia

spermatocyte

spermatid

developing sperm (mature spermatid)

interstitial cells

Figure 23.7   Cross section of rat seminiferous tubules (×540).

nutrients to carry the zygote at least through early embryology. Mammals do not require as much food reserve as other vertebrates because of the role of the uterus and placenta in nourishment after the first few days of development. Nevertheless, egg development in mammals is similar to that in other vertebrates. **Oogonia** divide mitotically and differentiate into **primary oocytes** that undergo meiotic cell division in the ovary. The first meiotic division results in the formation of the **secondary oocyte** (which contains the bulk of the cytoplasm) and a small **polar body.** The second meiotic division results in the formation of the **ootid** (with the bulk of the cytoplasm) and another polar body. Each meiotic cycle produces one egg with substantial food reserves in the cytoplasm.

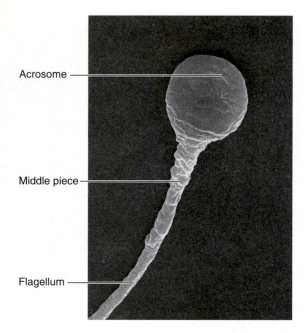

Acrosome

Middle piece

Flagellum

Figure 23.8 Scanning electron micrograph of human sperm (×8,000).

Oogenesis begins very early in female mammals. By the fourth month after conception, the human female fetus has all of her primordial germ cells (about 3 million per ovary) developed by mitotic cell division (oogonia). These oogonia begin meiotic cell division and arrest their development at the first meiotic prophase. Of these, only about 400 will develop to mature ova. At puberty, one of these primary oocytes completes its first meiotic division with each menstrual cycle. The second meiotic division occurs after penetration of the ovum by the sperm.

▼ Examine a prepared section of a mammalian ovary (figs. 23.9 and 23.10). Notice the cuboidal epithelium surrounding the ovary. Development of the primary oocyte occurs within ovarian follicles. **Primary follicles** are usually located near the ovary wall. They appear as solid spheres with the oocyte inside and are surrounded by one or two layers of cells. A **secondary follicle** has the beginning of a fluid-filled cavity (the **antrum**) surrounding the oocyte. The secondary follicle develops into a **mature follicle** in which the fully developed ovum (secondary oocyte) is held on top of a stalk of cells and is surrounded by a large central cavity

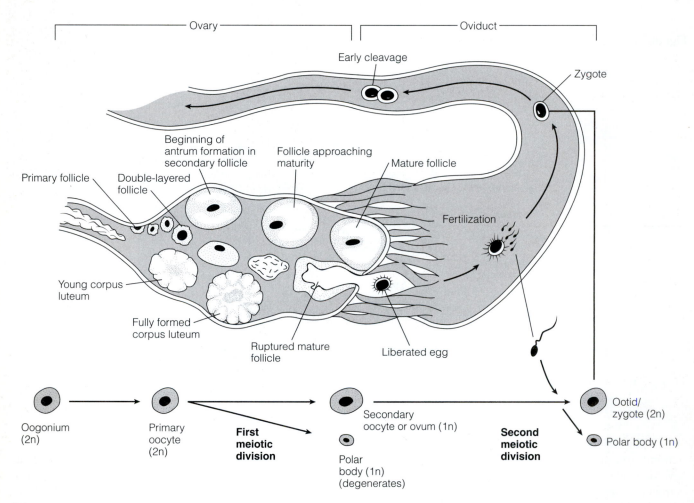

Figure 23.9 Diagrammatic section of a mammalian ovary and oviduct. Maturation of the oogonium and the first meiotic division occur during the fetal development of mammals. The development of secondary oocytes and associated follicular cells begins at puberty. When it is mature, the secondary oocyte is released from the ovary as the egg. The second meiotic division occurs, and the ootid exists very briefly, after fertilization but before the fusion of sperm and egg nuclei. Nuclear fusion results in the formation of the zygote. Polar bodies produced during meiosis degenerate.

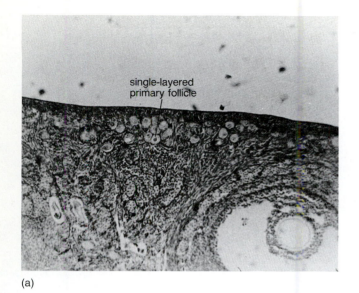

(a)

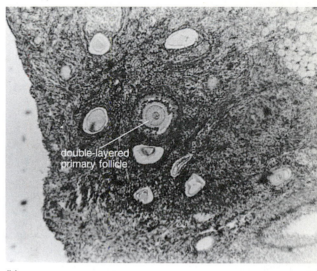

(b)

Figure 23.10 Sectioned cat ovary: *(a)* single-layered primary follicle; *(b)* double-layered primary follicle; *(c)* maturing secondary follicle.

that causes the wall of the ovary to bulge. The ovum is released into the coelom and the oviduct when the follicle wall ruptures. What remains in the ovary is transformed into the **corpus luteum,** which produces hormones (estrogen and progesterone) that maintain the uterine wall during early pregnancy. If implantation of the embryo at the uterus does not occur, the corpus luteum degenerates into a patch of scar tissue (corpus albicans). ▲

23.3 LEARNING OUTCOMES REVIEW—WORKSHEET 23.1

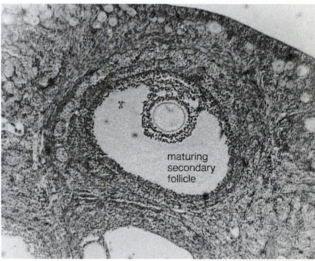

(c)

WORKSHEET 23.1 Vertebrate Reproduction

23.1 The Shark Reproductive System—Learning Outcomes Review

1. After being released from the testes of the male shark, sperm move into the (in order)
   a. ductus deferentes, seminal vesicles, and sperm sacs.
   b. sperm sacs, seminal vesicles, and ductus deferentes.
   c. seminal vesicles, ductus deferentes, and sperm sacs.
   d. ductus deferentes, sperm sacs, and seminal vesicles.

2. Fertilization in sharks is always external.
   a. true
   b. false

3. Development of young sharks is
   a. oviparous.
   b. ovoviviparous.
   c. viviparous.
   d. Either a or b is correct depending on the shark species.

4. Development of young within a female's uterus supported by nutrients stored in a yolk sack is called _____ development.
   a. oviparous
   b. ovoviviparous
   c. viviparous

23.2 The Rat Reproductive System—Learning Outcomes Review

5. After leaving the seminiferous tubules of a rat's testes sperm directly enter the
   a. ductus deferens.
   b. urethra.
   c. epididymis.
   d. seminal vesicle.

6. Which of the following is not paired in the male rat?
   a. ductus deferens
   b. seminal vesicle
   c. prostate gland
   d. epididymis

7. Accessory glands of the male reproductive system
   a. store sperm prior to ejaculation.
   b. add secretions to the sperm that create a proper medium for sperm ejaculation and survival in the female reproductive tract.
   c. contribute to the bulk of the ejaculate.
   d. All of the above are correct.
   e. b and c only are correct.

8. Which of the following sets contains terms that are most closely related functionally?
   a. prostate gland, cremaster muscle, ductus deferens
   b. cremaster muscle, scrotum, inguinal canal
   c. inguinal canal, epididymis, prostate gland
   d. seminiferous tubules, cremaster muscle, seminal vesicle

9. The uterus of the female rat
   a. consists of two cornua.
   b. unites with a single vagina.
   c. receives ova from a highly coiled oviduct between the ovary and the uterus.
   d. All of the above are correct.

23.3 Gamete Formation—Learning Outcomes Review

10. Meiotic cell division occurs within the
    a. seminiferous tubules of the testes.
    b. interstitial cells of the testes.
    c. epididymis of the testes.
    d. Both a and b are correct.

11. In a histological section of the testes
    a. seminiferous tubules comprise most of the tissue.
    b. interstitial cells are located between seminiferous tubules.
    c. sperm are seen associated with the lumens of the seminiferous tubules.
    d. All of the above are correct.

12. The cells of the testes that begin meiotic cell division are
    a. spermatids.
    b. primary spermatocytes.
    c. secondary spermatocytes.
    d. sustentacular cells.

13. All of the following cells are haploid except one. Select the exception.
    a. spermatids
    b. primary spermatocytes
    c. secondary spermatocytes
    d. mature sperm

14. Which of the following cells would not be found in the ovary of a fetal mammal?
    a. oogonium
    b. secondary oocyte
    c. polar body (at least briefly)
    d. ootid

15. A student views a histological section of an ovary in a compound microscope. She sees a circular cavity with an ovum perched on top of a stalk of follicular cells. The follicle wall is bulging from the side of the ovary. This student is viewing a(n)
    a. oogonium.
    b. primary follicle.
    c. secondary follicle.
    d. mature follicle.

16. The student described in question 15, now moves the slide slightly and observes a patch of scar tissue that she suspects is a corpus albicans, the remnant of a follicle left after ovulation. Suppose that this corpus albicans was responsible for producing an ovum that was eventually fertilized, and that it produced a young rat. During pregnancy the cells being observed would have produced
    a. additional ova.
    b. estrogen and progesterone.
    c. testosterone.
    d. chorionic gonadotropin.

23.1 The Shark Reproductive System—Analytical Thinking

1. Trace the path of sperm from the testes of a male shark until the sperm fertilize an ovum in the reproductive tract of a female shark.

23.2   The Rat Reproductive System—Analytical Thinking

2. Imagine that you could follow a single sperm cell from its point of production in the seminiferous tubules of a rat until it fertilized a rat ovum. Describe the "experiences" of that sperm cell from production to fertilization.

23.3   Gamete Formation—Analytical Thinking

3. Describe the differing "strategies" of sperm production versus ovum production in mammals.

4. Explain why human females planning on producing future offspring should be especially cautious regarding exposure of the ovaries to carcinogens and mutagens.

Glossary and Pronunciation Guide

In this pronunciation system, long vowels that stand alone or occur at the end of a syllable are unmarked. When a long vowel occurs in the middle of a syllable, it is marked with a macron (¯).

Short vowels that occur in the middle of a syllable are unmarked. When a short vowel occurs at the end of a syllable, or when it stands alone, it is marked with a breve (˘). The following are examples: cape (cāp), cell (sel), vacation (va-ka'shun), and vegetal (vej"ĕ-tal').

The vowel heard in *foot* and *book* is indicated as "foŏt" and "boŏk." The vowel heard in *loose* and *too* is unmarked and is indicated as "loos" and "too."

A

Acanthocephala (*ah-kan'thah-sef'ah-lah*)
Spiny-headed worms. A group of pseudocoelomates that are parasites of the digestive tract of fishes, turtles, swine, and other vertebrates.

acoelomate (*a-se'lŏ-māt*)
Without a body cavity.

actin (*ak'tin*)
One of the proteins involved with muscle contraction.

Actinopterygii (*ak"tin-op'te-rig-e-i*)
The class of bony fish characterized by paired fins supported by dermal rays, fin bases not expecially muscular, tail fin with approximately equal upper and lower lobes, pneumatic sacs used as swim bladders, and blind olfactory sacs. The ray-finned fishes.

active transport (*ak'tiv trans'pōrt*)
The transport of substances across a cell membrane using protein carriers and energy.

allele (*al-ēl'*)
Alternative forms of genes that occur at the same locus. Genes for white eyes and wild-type eyes of the fruit fly are alleles for the eye-color trait.

alveolus (*al-ve'o-lus*)
A small cavity. In the lungs, alveoli are the small sacs where gas exchange occurs.

amoebocyte (*ah-me'bo-sīt*)
One of the cell types in the Porifera. Amoebocytes function in spicule formation, food distribution, and reproduction.

Amphibia (*am-fib'e-ah*)
A vertebrate class whose members are characterized by skin with mucoid secretions, moist skin functioning as a respiratory organ, and, usually, metamorphosis. Frogs, toads, and salamanders.

anatomy (*ah-nat'ah-me*)
The study of the structure of living organisms.

Animalia (*an"ah-māl'ēah*)
The animal kingdom. It includes organisms that are eukaryotic and multicellular, lack cell walls, and are nourished by ingestion rather than by photosynthesis or absorption.

Annelida (*ah'nel-ĭ-dah*)
The phylum containing the segmented worms. Its members are characterized by metamerism (the serial repetition of body parts).

Anthozoa (*an'thŏ-zo-ah*)
The class of cnidarians that includes the sea anemones, corals, sea fans, and sea pens. Anthozoans lack the medusoid body form.

antibody (*an'tĭ-bod"e*)
A specific protein (immunoglobin) that is produced by B-lymphocytes to combat an antigen.

antigen (*an'tĭ-jen*)
A foreign substance that can cause an immune response in the body.

appendicular skeleton
(*ă-pen'dĭ-ku-lar skel'-ĕ-ton*)
The bones or cartilages of the paired appendages.

Arachnida (*ah-rak'nĭ-dah*)
A class of arthropods whose members are characterized as having four pairs of walking legs, one pair of chelicerae, and one pair of pedipalps. Spiders, mites, ticks, scorpions, etc.

arteriole (*ar-te're-ōl*)
A branch of the arterial system. Smooth muscle in its wall allows an arteriole to constrict or dilate to regulate blood flow to a tissue.

artery (*ar'ter-re*)
One of the vessels of the arterial system. Arteries carry blood away from the heart. Elastic connective tissue in their walls allows arteries to distend on receiving a pulse of blood and then rebound to help propel blood.

Arthropoda (*ar-throp'ah-dah*)
The phylum whose members are characterized as having a jointed exoskeleton, metamerism, and a high degree of tagmatization.

ascon (*as'kon*)
The simplest of the three sponge body forms. Asconoid sponges are vaselike, with choanocytes directly lining the spongocoel.

association neuron (*ah-so"se-a'shun nur"on*)
See interneuron.

aster (*as'ter*)
The star-shaped structure seen in a cell during the prophase of mitosis; composed of a system of microtubules arranged in astral rays around the centriole; may emanate from a centriole or from a pole of a mitotic spindle.

Asteroidea (*as"tar-oid'ēah*)
A class of echinoderms that includes the sea stars. Its members are characterized as being star shaped, with rays not distinctly set off from the central disk, having ambulacral grooves with tube feet, having suction disks on tube feet, and having pedicellariae.

atrium (*a'tre-um*)
1. In the Cephalochordata, the atrium is the chamber that water passes into after passing through the pharyngeal slits. 2. In the Vertebrata, the atrium is the chamber(s) of the heart that receives blood from major veins or the sinus venosus and delivers blood to the ventricle(s) on atrial contraction.

Aves (*a'vēs*)
A class of vertebrates whose members are characterized by feathers and other flight modifications, endothermy, and amniotic eggs. Birds.

axial skeleton (*ak'se-al skel-ĕ-ton*)
Bones on the longitudinal axis of a vertebrate. Skull, vertebral column, ribs, and sternum.

axon (*ak'son*)
The region of a neuron that propagates an all-or-none action potential (nerve impulse).

B

basal body (*ba'sel bod'e*)
A centriole that has given rise to the microtubular system of a cillium or flagellum, and is located just beneath the plasma membrane. Serves as a nucleation site for the growth of the axomene.

basement membrane (*bas'ment mem'brān*)
A noncellular layer upon which an epithelium rests. It is produced by the epithelium and the underlying connective tissue.

binomial nomenclature (*bi-no"me-al no'men-kla"cher*)
A system for naming organisms in which each kind of organism (a species) has a name of two parts, the genus and the species epithet.

biramous (*bi-ra'mus*)
Divided into two parts. The biramous appendages of crustaceans consist of a single proximal element, the protopodite, and two distal elements, the exopodite and the endopodite.

Bivalvia (*bi'valv"e-ah*)
The class of molluscs having a shell of two hinged valves. Clams, mussels, etc.

blastula (*blas'tu-lah*)
An early stage in the development of an embryo. It consists of a sphere of cells enclosing a fluid-filled cavity (blastocoel).

Brownian movement (*brow'ne-an moov'ment*)
The movement of particles in water that results from water molecules colliding with the particles in question.

brush border (*brush bor'der*)
Microvilli lining the luminal surface of some epithelial tissues that increases the surface area for cellular transport processes.

bulk transport (*bulk trans-pōrt'*)
The transport of large compounds across a cell membrane by invaginating or evaginating large areas of the cell membrane.

C

cancellous bone (*kan'sel-us bōn*)
Spongy bone. Bone made up of a network of loosely organized bony fibers called

trabeculae. As in the epiphyses of long bones.

capillary (*kap'ĭ-ler"e*)
Vessel within the circulatory system having the smallest diameter. Capillaries have walls consisting of a single cell layer, thus allowing exchanges of gases and nutrients between blood and surrounding tissues.

cardiac muscle (*kar'de-ak mus"el*)
Muscle of the heart. Muscle characterized by branching fibers, striations, and cells separated at their ends by intercalated disks.

cartilage (*kar'tĭ-lij*)
A connective tissue that gives support to structures such as the nose and earlobes, and often the intervening material between bones. Cartilage consists of cells (chondrocytes) and fibers embedded in a semisolid matrix.

cell cycle (*sel si'kel*)
The regular sequence of events during which a cell grows, prepares for division, duplicates its contents, and divides to form two daughter cells.

cell membrane (*sel mem'brān*)
See plasma membrane.

cellular respiration (*sel'u-lar res'pĭ-ra"shun*)
The series of cellular metabolic reactions by which energy in the chemical bonds of glucose is converted into a form of energy directly usable by the cell (adenosine triphosphate).

central nervous system (*sen'tral ner'vus sis'tem*)
The brain and the spinal cord. The region of the nervous system where integration takes place.

centriole (*sen'tre-ōl'*)
A cellular organelle located just outside the nucleus of the nondividing animal cell. Centrioles may play a role in the organization of spindle fibers during mitosis and meiosis.

centromere (*sen'tro-mer*)
The constricted region of a mitotic chromosome that holds sister chromatids together; also the site on the DNA where the kinetochore forms and then captures microtubules from the mitotic spindle.

centrosome (*sen'tro-sōm*)
The two centrioles found outside the nucleus of a nondividing cell.

cephalization (*sef"al-īz-a'shun*)
The development of a head, with nervous tissue accumulating to form a brain.

Cephalochordata (*sef"al-o-kor-da'tă*)
A chordate subphylum whose members are characterized as having a laterally compressed, transparent, fishlike body. All four chordate characteristics persist throughout life.

Cephalopoda (*sef"ah-lah-po'dah*)
A class of Mollusca whose members are characterized as having the foot modified

into a circle of tentacles and a siphon, the shell reduced or absent, and the head in line with the visceral mass.

Cestoidea (*ses-toid'e-ah*)
A class of the phylum Platyhelminthes. Tapeworms.

Chelicerata (*ke-lis"er-ah'tah*)
The subphylum of arthropods characterized by a body divided into prosoma and opisthosoma; first pair of appendages piercing or pincerlike (chelicerae) and used for feeding. Horseshoe crabs, spiders, scorpions.

Chilopoda (*ki-lah'pōd-ah*)
A class of Arthropoda. Chilopods have one pair of appendages per segment and a body that is oval in cross section. Centipedes.

Chondrichthyes (*kon-drik'thi-ez*)
A class of the phylum Chordata that includes the sharks, skates, and rays. Its members are characterized by jaws, paired appendages, a cartilaginous skeleton, no swim bladder, and no operculum.

Chordata (*kōr-da'tah*)
The phylum characterized by a dorsal tubular nerve cord, pharyngeal pouches, a notochord, and a postanal tail.

chromatid (*kro'mah-tid*)
One of the two (normally identical) coils of DNA that make up a chromosome.

chromatin (*kro'mah-tin*)
Nuclear material that gives rise to chromosomes during mitosis; complex of DNA, histones, and nonhistone proteins.

chromosome (*kro"mŏ-sōm*)
DNA and protein (chromatin) condensed or wound into discrete threadlike structures seen during cell division.

cilia (*sil'e-ah*)
Microscopic, hairlike processes on the exposed surfaces of certain eukaryotic cells. Cilia contain a core bundle of microtubules and are capable of performing repeated beating movements. They are also responsible for the swimming motion of many single-celled organisms.

cisternae (*sis'tern-a*)
1. Membranous vesicles that stack to form the Golgi apparatus. 2. The vesicular portion of sarcoplasmic reticulum (endoplasmic reticulum) of skeletal muscle.

cladogram (*klad'o-gram*)
Cladograms are a result of one approach to systematics called phylogenetic systematics (cladistics). Cladograms depict a sequence in the origin of taxonomic characteristics. The entire cladogram can be considered a hypothesis regarding a monophyletic lineage.

Cnidaria (*ni-dār'ēah*)
The phylum whose members are characterized by nematocysts, polymorphism, and diploblastic tissue-level organization.

Jellyfish, sea anemones, corals, sea fans, etc. Also Coelenterata.

coelom (*se'lōm*)
The fluid-filled body cavity of many animals. The term is also used to designate the eucoelom, a body cavity lined by mesoderm.

compact bone (*kom'pakt bōn*)
The hard, dense bone making up the shaft of a long bone such as the femur.

concentration gradient (*kon-sen"tra'shen gra'de-ent*)
A difference in the concentration of a substance between two points of reference. A substance diffuses along its concentration gradient from an area of higher concentration to an area of lower concentration.

connective tissue (*kŏ-nek'tiv tish'u*)
A tissue that binds and/or supports body parts. It consists of scattered cells and fibers embedded in a gel-like matrix.

control group (*kon'trol groop*)
In a scientific experiment, a control group is treated the same as the experimental group except that the variable being tested is omitted. A control group serves as a basis for comparing the data derived in an experiment.

controlled variables (*kon-trol'ĕd var'e-ă-bels*)
Controlled variables are factors or conditions of an experiment that are held constant so that they do not affect the outcome.

cortex (*kŏr'teks*)
The outer region of an organ. For example, the cortex of the kidney, adrenal gland, or cerebrum.

Craniata (*kra"ne-ah'tah*)
The chordate subphylum whose members possess a skull, a three-part brain, and neural crest tissue.

Crinoidea (*krīn-oid'ēah*)
A class of Echinoderms that includes sea lilies and feather stars. Crinoids are frequently attached to the substrate through an aboral stalk of ossicles.

cristae (*kris'tae*)
Shelf-like folds of the inner membrane of a mitochondrion. Enzymes located on cristae catalyze cellular energy conversions.

Crustacea (*krus-tās'ēah*)
A subphylum class of Arthropoda including crabs, crayfish, lobsters, etc. Crustaceans possess biramous appendages and two pairs of antennae.

cytokinesis (*si'ta-kin-e'sis*)
The division of the cytoplasm of a cell into two parts, as distinct from the division of the nucleus (which is mitosis).

cytoplasm (*si'ta-plazm*)
The contents of a cell surrounding the nucleus; consists of a semifluid medium and organelles.

D

dendrite (*den'drīt*)
The part of a neuron that initiates a graded action potential.

dependent variable (*dĭ-pen'dent var'e-ă-bel*)
The dependent variable is the parameter that the scientist is measuring. It is what will be affected during an experiment. In many experiments, scientists must monitor more than one dependent variable.

diencephalon (*di"en-sef'ah-lon*)
The portion of the embryonic vertebrate brain that develops to form the thalamus, epithalamus, and hypothalamus.

diffusion (*di-fu'zhun*)
The movement of a substance from an area of higher concentration to an area of lower concentration.

dioecious (*di-ēs'ēus*)
Having separate sexes. Derived from Greek root meaning "two houses."

diploblastic (*dip"lo-blast-ik*)
Having a body wall derived from two embryonic layers, ectoderm and endoderm (e.g., the Cnidaria).

diploid (*dip'loid*)
A cell that has two members of each set of homologous chromosomes. Somatic cells are usually diploid. 2N.

Diplopoda (*dip"lo-pōd'ah*)
The class of arthropods that contains the millipedes. Its members are characterized by two appendages per apparent segment and a body that is round in cross section.

domain (*do-mān'*)
The broadest taxonomic grouping. Recent evidence from molecular biology indicates that there are three domains: Archaea, Eubacteria, and Eukarya.

dominant (*dom'ah-nent*)
A gene that masks one or more of its alleles. In order for a dominant trait to be expressed, at least one member of the gene pair must be the dominant allele. *See* recessive.

E

Echinodermata (*ek"ĭn-o-derm'ah-tah*)
The phylum that includes the sea stars, sea urchins, sea cucumbers, and sea lilies. Its members are characterized as having a water-vascular system, pentaradial symmetry, and an internal calcareous skeleton.

Echinoidea (*ek"i-noi'de--ah*)
The class of echinoderms that includes the sea urchins and sand dollars. Members are characterized by a globular or disk-shaped test, no rays, movable spines, and a skeleton (test) of closely fitting plates.

ecosystem (*ek"o-sis'tem*)
All of the populations of organisms living in a certain place plus their physical environment.

ectoderm (*ek"to-derm'*)
The outer embryological tissue layer. It gives rise to skin epidermis, nervous tissue, glands, hair, nails, and other structures.

effector organ (*ĕ-fek'tōr ōr'gan*)
A muscle or gland that responds to an impulse from the central nervous system.

endocytosis (*en"do-si-to'sis*)
Physiological process by which substances may move through a plasma membrane into the cell via membranous vesicles or vacuoles.

endoderm (*en'do-derm"*)
The innermost embryological tissue layer. It gives rise to the inner lining of the gut tract.

endoplasmic reticulum (ER) (*en"do-plaz'mik rĭ-tik"ku-lum*)
A membrane system spreading through the cytoplasm of a cell. Its function is transporting proteins within the cell. Rough ER (granular ER) has ribosomes associated with it. Smooth ER (agranular ER) lacks ribosomes.

endostyle (*en'do-stil'*)
A ciliated tract within the pharynx of some chordates that is used in forming mucus for filter feeding. Homologous to the thyroid gland of non-filter-feeding chordates. One of five unique chordate characteristics.

epithelium (*ep"ĭ-the'le-um*)
A sheet of tissue that covers a surface or lines a cavity of an organism. Epithelium can be protective or provide surfaces for diffusion and other forms of transport.

erythrocyte (*ĕ-rith'ro-sīt*)
A mature red blood cell. It contains the respiratory pigment hemoglobin and carries oxygen and some carbon dioxide in the vertebrate circulatory system.

eukaryotic (*u"kār-e-ot'ik*)
Cells that have a membrane-bound nucleus as well as other internal membranous organelles.

evolution (*ev"ol-u'shun*)
Biological evolution is a series of changes in the genetic composition of a population over time.

evolutionary adaptation (*ev"ah-loo'she-nere ad"ap-ta'shen*)
Structures or processes that increase an organism's potential to successfully reproduce in a specified environment.

exocytosis (*eks'o-si-to"sis*)
The process by which substances are moved out of a cell; the substances are transported in the cytoplasmic vesicles, the surrounding membrane of which merges with the plasma membrane in such a way that the substances are dumped outside.

F

flagellum (*flah-jel'um*)
A long, hairlike process. Beating flagella move some protistans and sperm cells through their media, and they move fluids and particles across the surface of tissues. Flagellar organization is based on a 9+2 microtubule arrangement. Flagella are similar to cilia, though longer and fewer in number.

fluid mosaic model (*floo'id mo-za'ik mod'el*)
A model that describes the plasma membrane as a bimolecular lipid layer in which hydrophilic ends of lipid molecules are oriented to the outside of the membrane, while hydrophobic ends are oriented to the inside of the membrane. Proteins are associated with the surface of the membrane or embedded within the membrane.

food web (*food web*)
A sequence of organisms through which energy is transferred in an ecosystem. Rather than being a linear series, a food web has highly branched energy pathways.

formed element (*fōrmd el'ĕ-ment*)
Blood cells or fragments of cells. White blood cells (leukocytes), red blood cells (erythrocytes), and platelets.

Fungi (*fun'ji*)
One of the five kingdoms. Its members are characterized as being eukaryotic, multicellular, sedentary, and saprophytic.

furrowing (*fur'o-ing*)
The process by which division of the cytoplasm (cytokinesis) occurs during cell division. Microfilaments at the equator of a cell contract to constrict the plasma membrane of the cell.

G

gametes (*gam'ets*)
Mature haploid cells (sperm or eggs) that function in sexual reproduction.

gap 1 (*gap 1*)
A portion of interphase of the cell cycle characterized by rapid RNA and protein synthesis. It immediately follows mitosis.

gap 2 (*gap 2*)
A portion of interphase of the cell cycle characterized by protein synthesis prior to mitosis.

Gastropoda (*gas"tro-po'dah*)
The class of molluscs that includes the snails, slugs, and limpets. Its members are characterized by a coiled shell (when a shell is present) and distorted symmetry. Both monoecious and dioecious taxa are present in this class.

Gastrotricha (*gas'trŏ-trik-ah*)
A phylum of microscopic, freshwater, pseudocoelomate animals that have a ventrally flattened and bristly or scaly body.

gastrovascular cavity (*gas'tro-vas'ku-lar kav'ĭ-te*)
The large, central cavity of cnidarians that serves to receive and digest food.

genetics (*jĕ-net'iks*)
The study of the mechanisms of transmission of genes from parents to offspring.

glomerulus (*glo-mer'u-lus*)
The bundle of capillary-like vessels within the glomerular capsule of the nephron. It is the place where filtration of blood occurs.

goblet cells (*gob'let selz*)
Simple columnar cells in the digestive tract modified for the production and secretion of mucus.

Golgi apparatus (*gōl"je ap"ah-ra'tus*)
A cellular organelle that packages secretory products.

gonad (*go'nad*)
An organ that produces gametes by meiotic cell division (the ovary or testis).

H

haploid (*hap'loid*)
A cell having one member of each pair of homologous chromosomes. Haploid cells are usually the product of meiosis. Gametes are haploid.

head-foot (*hed-fŏŏt*)
The body region of a mollusc that contains the head and is responsible for locomotion as well as retracting the visceral mass into the shell.

heterozygous (*het"er-o-zīg'us*)
The condition of an individual or a cell having different alleles at the same locus on a pair of homologous chromosomes.

Hexapoda (*hex'să-pōd'ah*)
The subphylum of mandibulate arthropods whose members have a body consisting of head, thorax, and abdomen; five pairs of head appendages; and three pairs of uniramous appendages on the thorax. Insects and their relatives.

Hirudinea (*hīr"oo-din'ēah*)
The leeches. The class of Annelida whose members are characterized as having 34 segments with many annuli and anterior and posterior suckers. Most leeches are predators on other invertebrates. A few leeches feed on vertebrate blood.

histology (*his"tol'o-je*)
The study of tissue structure.

Holothuroidea (*hōl"o-thur-oid'ēah*)
The class of echinoderms whose members are characterized as being elongate along the oral-aboral axis, having microscopic ossicles embedded in a muscular body wall, having circumoral tentacles, and lacking rays. The sea cucumbers.

homologous (*ho-mol'ŏ-gus*)
1. Structures that have a common evolutionary history. The wing of a bat and the arm of a human are homologous; each can be traced back to a common ancestral appendage. 2. Chromosomes with genes that code for the same traits, but not necessarily the same expression of those traits.

homozygous (*ho"mo-zīg'us*)
The condition of having identical alleles for a particular gene pair.

Hydrozoa (*hi"drŏ-zo'ah*)
A class of cnidarians whose members are characterized as usually having well-developed polypoid and medusoid stages. Medusa with velum.

Hyperotreti (*hi"per-ot're-te*)
The craniate infraphylum that is fishlike, skull consisting of cartilaginous bars, jawless, no paired appendages, mouth with four pairs of tentacles, 5 to 15 pairs of pharyngeal slits, and ventrolateral slime glands. Hagfishes.

hypertonic (*hi"per-ton'ik*)
A solution having a greater number of solute particles than another solution to which it is compared.

hypothesis (*hi-poth'es-is*)
A tentative explanation of a question; an explanation based on careful observations.

hypotonic (*hi"po-ton'ik*)
A solution having a lesser number of solute particles than another solution to which it is compared.

I

independent variable (*in-dĭ-pen'dent var'e-ă-bel*)
The parameter being tested is the independent variable. It is what the scientist varies. In any experiment, there must be only one independent variable. When there is only one independent variable, any changes in the dependent variable(s) must be the result of the scientist's manipulation of the independent variable.

Insecta (*in-sekt'ah*)
The class of arthropods whose members are characterized as having three pairs of legs, one pair of antennae, and often wings (usually two pairs). The insects.

integration (*in"tĕ-gra'shun*)
Combining parts into an integral whole. The role of the central nervous system is to combine sensory input from a variety of sources and initiate a coordinated response to that input.

interneuron (*in'ter-nu'ron*)
A nerve cell entirely confined to the central nervous system. Interneurons carry information from one part of the central nervous system to another. Also association neuron.

interphase (*in'ter-faz*)

Period between two cell divisions when a cell is carrying on its normal functions. Long period of the cell cycle between one mitosis and the next. Includes G_1 phase, S phase, and G_2 phase. The replication of DNA occurs during interphase.

isotonic (*i"so-ton'ik*)

In a comparison of two solutions, both have equal concentrations of solutes.

K

kingdom (*king'dom*)

The level of classification above phylum; the traditional classification system includes five kingdoms: Monera, Protista, Fungi, Plantae, and Animalia. Recent evidence from molecular biology indicates that these five kingdoms may not be monophyletic lineages.

Kinorhyncha (*kin'o-ring'kah*)

A phylum of pseudocoelomate marine worms. Its members are characterized by a body of 13 segments, spines, and a retractile head and proboscis.

L

leucon (*lu'kon*)

The sponge body form that has an extensively branched canal system. The canals lead to flagellated chambers lined by choanocytes. Flagellated chambers discharge water into excurrent canals that eventually lead to an osculum.

leukocyte (*lu'ko-sīt"*)

A white blood cell.

lungs (*lungz*)

1. The highly vascular tissues at the distal end of the respiratory tree that allow gas exchange between air and the blood of lungfish and tetrapods. 2. The highly vascular modification of the mantle of some gastropod molluscs that allows gas exchange between the air and blood.

lymphatic system (*lim-fat'ik sis'tem*)

The system of vessels, glands, and nodes that is responsible for returning excess tissue fluids and proteins to the circulatory system and for developing and maintaining acquired immunity.

lysosome (*li'so-sōm*)

Vesicular intracellular structures that contain enzymes capable of digesting all biologically important macromolecules. They digest ingested particles as well as old, nonfunctional organelles.

M

Malacostraca (*mal-ah-kos'trah-kah*)

The class of crustaceans whose members have appendages modified for crawling along the substrate or the abdomen and body appendages modified for swimming. Lobsters, crabs, shrimp.

Mammalia (*mam-āl'e-ah*)

The vertebrate class whose members are characterized by hair, mammary glands, amniotic eggs, and endothermy (Chordata).

mantle (*man'tel*)

The outer, fleshy tissue of molluscs that secretes the shell.

matrix (*ma'trix*)

The gelatinous interior of a mitochondrion containing DNA, ribosomes, and enzymes. Enzymes located in the matrix catalyze cellular energy conversions.

meiosis (*mi-o'sis*)

The type of nuclear division in which homologous chromosomes are separated into different cells, thus reducing chromosome number by one half. Meiosis usually results in the formation of haploid gametes.

meninges (*mĕ-nin'jēs*)

The three connective tissue membranes that cover and protect the brain and spinal cord of vertebrates.

Merostomata (*mer'o-sto'mah-tah*)

The class of arthropods whose members are characterized as having the first pair of appendages modified to form chelicerae, having one pair of pedipalps and four pairs of legs, having a body divided into a cephalothorax and abdomen, and lacking antennae. Eurypterids (extinct) and horseshoe crabs.

mesencephalon (*mez"en-sef'el-on*)

The region of the embryonic brain that gives rise to the adult midbrain.

mesentery (*mez'en-ter-e*)

A sheet of peritoneum (mesodermal in origin) that suspends most abdominal viscera from the body wall. It is continuous with the parietal and visceral peritoneum.

mesoderm (*mez"o-durm*)

The middle of the three embryological germ layers. Mesoderm gives rise to muscles, blood, bone, cartilage, etc. Mesoderm is found only in triploblastic animals.

metamerism (*met-am'er-izm*)

The serial repetition of body parts. The two major metameric phyla are Annelida and Arthropoda.

metamorphosis (*met-ah-morf'ĕ-sis*)

A transformation from one form to another. As used here, metamorphosis generally refers to a transformation from the larval to the adult body form.

metencephalon (*met"en-sef'al-on*)

One of the five embryonic regions of the brain of vertebrates. It develops into the pons and cerebellum of the adult.

microfilament (*mi'kro-fil'ah-ment*)

Intracellular proteins arranged in linear arrays or networks. Microfilaments are involved with movement within cells and with maintaining or changing cell shape.

microtubule (*mi"kro-tu'bŭl*)

Cylindrical arrangements of proteins that, like microfilaments, are the basis of cellular locomotion and structure.

microvillus (*mi"kro-vil'us*)

An extension of the cell membrane of some epithelial cells that increases surface area for secretion, digestion, and/or absorption.

mitochondrion (*mi"to-kon'dre-on*)

The organelle that is responsible for aerobic energy conversions within a cell.

mitosis (*mi-to'sis*)

Nuclear division resulting in two cells that are genetically identical to a parent cell.

mitotic spindle (*mi-to'tic spin'del*)

In a cell undergoing mitosis, the centrioles and the microtubules radiating between them form the mitotic spindle.

Mollusca (*mol-usk'ah*)

A phylum of triploblastic coelomate animals whose members are characterized by a body of three regions: the visceral mass, the head-foot, and the mantle. Molluscs include the snails, bivalves, chitons, squid, octopuses, and others.

monoecious (*mon-ēs'e-es*)

A species in which both male and female sex organs occur in the same individual. Derived from Greek root meaning "one house."

Monoplacophora (*mon"o-plah-kof'o-rah*)

The molluscan class whose members are characterized by a single arched shell, a broad and flat foot, and serial repetition of certain structures.

motor unit (*mo'tor unit*)

A nerve cell plus the muscle cells associated with it.

mucosa (*mu-ko'sah*)

The inner lining of the digestive tract or other tube or cavity that opens to the outside of the body. Mucous membrane.

muscular tissue (*mus'kŭ-lar tish'u*)

A tissue characterized by the ability to shorten and conduct action potentials.

myelencephalon (*mi"ĕ-len-sef'al-lon*)

One of the five regions of the embryonic vertebrate brain. It develops into the medulla oblongata of the adult.

myosin (*mi'o-sin*)

One of the proteins involved with muscle contraction.

Myriopoda (*mir"e-a-pod'ah*)

The subphylum of arthropods whose members have a body consisting of two tagmata and uniramous appendages. Cenitpedes and millipedes.

Myxini (*mik'si-ne*)

The sole class within the craniate infraphylum Hyperotreti. Hagfishes. *See* Hyperotreti.

N

nematocyst (*nĭ-mat′ah-sist*)
An organelle characteristic of the Cnidaria that is used in defense, food gathering, and as a holdfast structure. Nematocysts develop in and are discharged from cnidoblasts.

Nematoda (*nem′ah-tōd″ah*)
The pseudocoelomate phylum whose members are characterized as being worm-like and round in cross section; having a flexible, nonliving cuticle; lacking motile cilia and flagella; and having only longitudinal muscles. The roundworms.

Nematomorpha (*ne″mat-o-morf′ah*)
A phylum of pseudocoelomate worms commonly called horsehair worms. Adults are elongate and free living in streams and ponds. Larvae are parasites of arthropods.

neoteny (*ne-ot′en-e*)
The retention of the larval body form in a reproductive individual.

nephron (*nef′ron*)
The functional unit of the kidney of vertebrates. It filters excretory products from the blood, reabsorbs water and nutrients from the filtrate, and actively transports ions and other substances from the blood into the filtrate.

nerve (*nerv*)
A bundle of nerve fibers (axons or dendrites) outside of the central nervous system.

nervous tissue (*ner′vus tish′u*)
The type of tissue composed of individual cells called neurons and supporting neurological cells.

neuron (*nur′on*)
A nerve cell. The basic cellular unit of the nervous system.

neurotransmitter (*nu″ro-trans-mit′er*)
A chemical released from synaptic vesicles at the presynaptic membrane. It binds to receptors at the postsynaptic membrane to either cause or inhibit depolarization of the postsynaptic membrane.

notochord (*no′to-kord*)
A connective tissue sheath that is filled with vacuolated cells and that runs along the dorsal longitudinal axis of chordates. This supportive, yet flexible, structure is present in all embryonic and many adult chordates. One of five unique chordate characteristics.

nuclear envelope (*nu′kle-ar en-vel′op*)
The double membrane forming the surface boundary of a eukaryotic cell nucleus.

nucleolus (*nu-kle′o-lus*)
A darkly stained region within the nucleus where ribosomal RNA is synthesized.

nucleus (*nu′kle-us*)
The genetic control center of the cell. It contains DNA and protein in a highly dispersed state called chromatin.

O

objective lens system (*ob-jek′tiv lenz sis′tem*)
The lens system of a compound microscope closest to the specimen.

ocular lens system (*ok′ya-ler lenz sis′tem*)
The lens system of a compound microscope closest to the eye of the observer.

ocular micrometer (*ok′ya-ler mi′kro-me-ter*)
A thin circle of glass or plastic etched with a nonunit scale. When placed in the ocular lens of a compound microscope, the scale image is superimposed over the specimen and is used to measure the size of microscopic objects.

Oligochaeta (*ol″ig-o-ke′tah*)
A class of annelids whose members are characterized by few setae, no parapodia, no distinct head, direct development, and monoecious individuals. The earthworms and related organisms.

Ophiuroidea (*ōf″e-ūr-oid′ēah*)
The class of echinoderms whose members are characterized by arms that are sharply set off from the central disk, a body cavity limited to the central disk, and tube feet without suckers. Brittle stars.

organ (*or′gan*)
An assemblage of tissues cooperating to perform a specialized function.

organ system (*or′gan sis′tem*)
An assemblage of organs cooperating to perform specialized functions.

osteon system (*os-te′on sis′tem*)
The organizational unit of compact bone. It consists of concentric rings of bone called lamellae, organized around a central canal. Bone cells are located in lacunae between adjacent lamellae.

osmosis (*oz-mō′sis*)
The movement of water from an area of high concentration to an area of low concentration through a selectively permeable membrane.

osmotic pressure (*oz-mot′ik presh′er*)
A measure of the pressure that would be needed to prevent the net diffusion of water from one solution to another.

ovary (*o′var-e*)
The primary female reproductive organ. An egg-producing organ.

oviduct (*o′vĭ-dukt*)
A tube that carries eggs from an ovary to an area where eggs develop or are stored. In animals with internal fertilization, the oviduct is frequently where eggs are fertilized.

oviparous (*ōv-ip′er-us*)
Organisms that lay eggs that develop outside of the body of the female.

ovovivparous (*ōv″o-vīv-ip′er-us*)
Organisms with eggs that develop within the reproductive tract of the female. Embryos are nourished by yolk stored within the egg.

P

parfocal (*par-fo′kel*)
The property of a compound light microscope that allows one to switch from one objective lens to another without refocusing. An object that is in focus using one objective lens will remain in focus when another objective lens is moved into place.

parthenogenesis (*par″then-o-jen′es-is*)
The development of an egg without its first being fertilized.

passive transport (*pas′iv trans′pōrt*)
The movement of substances into and out of cells without the use of energy.

peripheral nervous system (*pĕ-rif′er-al ner′vus sis′tem*)
The portion of the nervous system that includes the nerves going to somatic and visceral structures outside of the brain and spinal cord.

Petromyzontida (*pet′ro-mi-zon″tid-ah*)
The vertebrate class whose members have a sucking mouth with teeth and rasping tongue, seven pairs of pharyngeal slits, and blind olfactory sacs. Lampreys.

phagocytosis (*fag′o-si-to′sis*)
Process by which a cell engulfs bacteria, foreign proteins, macromolecules, and other cells and digests their substances; cellular eating.

pharyngeal slits (pouches) (*far-in′je-al slitz [pouches]*)
A series of openings between the digestive tract, in the pharyngeal region, and the outside of the body. One of the five unique characteristics of the chordates.

pharynx (*far′ingks*)
The region of the alimentary tract shared by the digestive and respiratory systems.

physiology (*fiz″e-ol′o-je*)
The study of the function of cells, tissues, organs, or organ systems.

pinocytosis (*pin′o-si-to″sis*)
Cell-drinking; the engulfment into the cell of liquid and of dissolved solutes by way of small membranous vesicles.

Plantae (*plant′a*)
One of the five kingdoms of life. Its members are characterized as being eukaryotic and multicellular and as having rigid cell walls and chloroplasts.

plasma membrane (*plas′ma mem′brān*)
The selectively permeable outer boundary of a cell. It is composed of proteins and lipids. Also cell membrane and plasmalemma.

Platyhelminthes (*plat″e-hel-min′thēz*)
The animal phylum whose members are characterized as being triploblastic and acoelomate and as having dorsoventrally flattened bodies, tissues organized into organs, and bilateral symmetry with cephalization.

Polychaeta (*pol″e-kēt′ah*)
The class of annelids whose members are characterized by parapodia bearing numerous setae, and a head with eyes and tentacles. Mostly marine.

polymorphism (*pol″e-morf′izm*)
The condition in which a species has more than one body form (as in the Cnidaria).

Polyplacophora (*pol″e-plak-of′er-ah*)
The class of molluscs whose members are characterized by an elongate, dorsoventrally flattened body; a reduced head; and a shell of eight dorsal plates. The chitons.

Porifera (*por-if′er-ah*)
The animal phylum whose members consist of loosely organized cells, including choanocytes, amoebocytes, and pinacocytes.

postanal tail (*post-a′nal tāl*)
A projection that extends posterior to the anus of chordates and that is supported by either the notochord or the vertebral column. It is one of the five unique chordate characteristics.

prediction (*pri-dik′shen*)
A forecast of the outcome of a scientific experiment that is based upon the hypothesis—usually in the form of an if/then statement.

prokaryotic (*pro″kar-e-ot′ik*)
A cell without a membrane-bound nucleus or other internal membranous compartments.

pseudocoelom (*sŭd″o-se′lōm*)
A body cavity that is not lined by a mesodermal, epidermal sheet.

Pycnogonida (*pik″no-gon′id-ah*)
A class of marine arthropods called sea spiders.

R

radula (*raj′oo-lah*)
The rasping tonguelike structure of most molluscs. It is composed of minute chitinous teeth that move over the cartilaginous odontophore. The radula is used in scraping food.

receptor (*rĭ-sep′tor*)
Something that receives. A sensory receptor receives an environmental stimulus and converts the stimulus into a nerve impulse.

recessive (*ri-ses′iv*)
A gene whose expression is masked when it is present in combination with a dominant

gene. In order for a recessive gene to be expressed, both members of the gene pair must be the recessive form of the gene.

replicate (*rep′le-kit*)
The duplication of an experiment to ensure that the results are consistent and the experiment is well conceived.

Reptilia (*rep-til′e-ah*)
A class of vertebrates whose members are ectothermic, lay amniotic eggs, have dry skin, and have epidermal scales.

respiration (*res″pah-ra′shun*)
1. The energy-converting reactions of the cell. The energy in organic compounds is converted to that in adenosine triphosphate. 2. The act of breathing.

ribosomes (*ri-bo-soms*)
Cytoplasmic organelles that consist of protein and RNA, and function in protein synthesis.

Rotifera (*ro′tĭ-fer-ah*)
A pseudocoelomate phylum whose members have a body consisting of head, trunk, and foot and who have a ciliated corona that functions in gathering food. Rotifers.

S

sarcomere (*sar′ko-mēr″*)
A contractile unit of skeletal muscle.

Sarcopterygii (*sar-kop-te-rij′e-i*)
The class of vertebrates whose members possess paired fins with muscular lobes, pneumatic sacs that function as lungs, and atria and ventricles that are at least partially divided. Lungfishes.

Scaphopoda (*skaf″o-po′dah*)
A class of marine molluscs whose members have a tubular shell, tentacles, and no head. Tooth shells or elephant-tusk shells.

science (*si′ens*)
Knowledge of the natural world gained by observation.

scientific method (*si′en-tif″ik meth′od*)
A series of steps that reflects a logical approach to solving problems in our physical world. It usually involves an experimental approach involving questions based upon careful observation, the formulation of a hypothesis, a test of the hypothesis, and a conclusion based upon the results of the test.

scientific theories (*si″en-tif″ik ther′es*)
Generalized statements of predictive value that are supported by many individual experiments, each testing one small aspect of a larger concept.

Scyphozoa (*sīf″ŏ-zo′ah*)
The cnidarian class in which the polyp is reduced or absent and the medusa is without a velum.

sensory neuron (*sen′so-re nu′ron*)
A neuron that carries nerve impulses from a sensory receptor toward the central nervous system.

serosa (*si-ro′sah*)
The peritoneal lining on the outside of a visceral organ. The serosa is a serous membrane that is sometimes called the visceral peritoneum.

sinus venosus (*si′nus ven-o′sus*)
A chamber of the vertebrate heart (exception, the mammals). It receives blood from the venous system and initiates the heartbeat. Evolution of the vertebrates has resulted in a gradual reduction in the size of the sinus venosus. The pacemaker function, however, remains unchanged.

skeletal muscle (*skel′ĕ-tal mus′el*)
Muscle in which a regular arrangement of contractile filaments gives a banded appearance when viewed through the compound microscope. Skeletal muscle is voluntary and usually associated with the skeleton.

smooth muscle (*smooth mus′el*)
Muscle that lacks a regular arrangement of contractile filaments, that is involuntary, and that usually is associated with visceral structures.

Solenogastres (*sole′no-gas′trez*)
Solenogastres. A class of worm-like molluscs whose members lack a shell, mantle, and foot.

species (*spe′shez*)
The fundamental unit of classification of living organisms.

spindle fibers (*spin′del fi′berz*)
Microtubles that originate at the centriole and attach at the centromere of individual chromosomes.

submucosa (*sub″mu-ko′sah*)
A layer of connective tissue that lies below the mucosa of an organ.

sycon (*si′kon*)
A sponge body form characterized as having choanocytes lining radial canals (Porifera).

Symphyla (*sim-fil′ah*)
A class of centipedelike arthropods whose members occupy soil and leaf mold.

symplesiomorphy (*sim-ples′e-o-mor′fe*)
Characters common to all members of the lineage. Symplesiomorphies indicate that all members of a study group are related. Because they are present in ancestors of all members of the study group, they cannot be used to distinguish between members of the study group.

synapomorphy (*sin-ap′o-mor′fe*)
A character that has arisen within a group since it arose from a common ancestor. It is also called a derived character. A synapomorphy can be used to indicate relatedness within a group.

synapse (*sin'aps*)
The region where an impulse is conducted from an axon of one neuron to a dendrite of a second neuron, or from an axon of a neuron to a muscle cell.

synthesis (*sin'the-sis*)
The portion of interphase of the cell cycle during which DNA is replicated so that each chromosome will consist of two identical chromatids.

systematics (*sis-tem-at'iks*)
The study of the classification and phylogeny of animals.

T

tagmatization (*tag"mah-ti-za'shun*)
The specialization of body regions of a metameric animal for specific functions, usually, feeding and sensory functions (head tagma); locomotion (thoracic tagma); and visceral functions (abdominal tagma).

telencephalon (*tēl"en-sef'al-on*)
The portion of the embryonic vertebrate brain that develops into cerebral hemispheres.

testis (*tes'tis*)
The primary male reproductive organ. A sperm-producing organ.

tissue (*tish'u*)
A group of similar cells that cooperate in a common function.

Trematoda (*trem'ah-tōd-"ah*)
The class of Platyhelminthes whose members are characterized as being leaflike or cylindrical, having oral and ventral suckers, and having a nonciliated, syncytial tegument. Adults are parasites of vertebrates. The flukes.

Trilobitamorpha (*tri"lo-bit'a-mor'fah*)
A subphylum of extinct arthropods that is believed to be ancestral to many (if not all) arthropods. Trilobites had a body divided into three longitudinal lobes as well as head, thoracic, and abdominal tagmata; compound eyes; and biramous appendages.

triploblastic (*trip'lo-blast"ik*)
Having a body derived from three embryonic layers: ectoderm, mesoderm, and endoderm. All animals above the Cnidaria are triploblastic.

trochophore (*trok'o-for"*)
A name given to the ciliated, free-swimming larval stages of some annelids and some molluscs.

Turbellaria (*tur"bel-ar'e-ah*)
The class of Platyhelminthes whose members are free living, have a ciliated epidermis containing rhabdites, and have a ventral mouth opening.

U

Urochordata (*ūr"o-kord-a'tah*)
A subphylum of chordates in which the adults are sessile or occasionally planktonic and are enclosed in a cellulose tunic. A notochord, a nerve cord, and a postanal tail are present only in free-swimming larvae. The tunicates.

V

vacuole (*vak'u-ōl*)
A membrane-bound vesicle within the cytoplasm of a cell. It is used for storage and transport.

vein (*vān*)
A blood vessel that carries blood toward the heart.

veliger (*vēl'ĭ-jer*)
The second free-swimming larval stage of many molluscs. It develops from the trochophore and develops rudiments of the shell, visceral mass, and head-foot before settling to the substrate and undergoing metamorphosis.

ventricle (*ven'tri-kl*)
A fluid-filled cavity. The ventricles of the brain are filled with cerebrospinal fluid. The ventricles of the heart are filled with blood. This term is also used to refer to the muscle that pumps blood from the heart to the systemic circulation of an animal.

venule (*ven'ūl*)
A blood vessel that carries blood from a capillary bed to a vein.

Vertebrata (*ver"te-brah'tah*)
The chordate infraphylum whose members have vertebrae that provide primary axial support and protect the nerve cord. The skeleton is modified anteriorly into a skull for protection of the brain. Also Craniata, fishes, amphibians, reptiles, birds, and mammals.

visceral mass (*vis'er-al mas*)
The region of a mollusc's body that contains visceral organs.

viviparous (*viv-ip'er-us*)
An animal in which young develop within the reproductive tract of the female and are nourished directly by the female.

Z

zygote (*zi'gōt*)
The cell that results from the fusion of gametes of opposite mating types.

Credits

Photographs

Chapter 1

1.1b: Courtesy of Carl Zeiss, Inc., Thornwood, NY; **1.2a&b:** Courtesy of Reichert Scientific Instruments; **1.5:** Courtesy of Bausch and Lomb, Inc.

Chapter 2

2.2b: © Gordon Leedale/Biophoto Associates/Photo Researchers; **2.3a&b:** © D. Fawcett/Visuals Unlimited; **2.4 (top):** © Gordon Leedale/ Biophoto Associates/Photo Researchers; **2.5:** Courtesy of Dr. Paul Heidgar; **2.6:** © Professors Pietro M. Motta & Tomonori Naguro/Photo Researchers; **2.7 (left):** Courtesy of Dr. Keith Porter; **2.8:** Courtesy of James Burbach, University of South Dakota, School of Medicine; **2.9a&b:** © L.E. Roth, Y. Shigenaka, and D.J. Pihlaja/Biological Photo Service; **2.10a:** Courtesy of Jan Lofberg, Animal Development and Genetics, Uppsala University, Sweden; **2.10b:** © 2012 William L. Dentler/Biological Photo Service; **2.11:** © Gordon Leedale/Biophoto Associates/Photo Researchers; **2.12:** Courtesy of Joseph Viles, Iowa State University; **2.13a:** © Dwight Kuhn; **2.13b:** © Ed Reschke; **2.13c:** © B.F. King/Biological Photo Service; **2.14, 2.15, 2.16 and 2.17 (bottom):** © Ed Reschke; **2.18:** © Victor P. Eroschenko

Chapter 3

3.2: © Steve Miller; **3.7a&b:** Carolina Biological Supply Company/ Phototake

Chapter 4

4.2a&b: © Steve Miller

Chapter 5

5.2: © Design Pics/David Chapman/Getty RF; **5.3a:** © R. Calentine/ Visuals Unlimited; **5.3b:** © Nature's Images/Photo Researchers; **5.4:** © T.E. Adams/Visuals Unlimited; **5.5:** © G. Meszaros/Visuals Unlimited; **5.6b & 5.7:** © William H. Amos; **5.8:** © Stephen Dalton/Photo Researchers; **5.9:** Courtesy of Douglas A. Craig, Department of Biological Sciences, University of Alberta, Edmonton; **5.10:** © Lawrence Pringle/ Photo Researchers

Chapter 7

7.1b: © Paul W. Johnson/Biological Photo Service; **7.2a, 7.3a & 7.4a:** Carolina Biological Supply Company/Phototake; **7.5b:** © M.I. Walker/ Photo Researchers; **7.8:** Courtesy Barbara Grimes; **7.9:** © Bruce Russell/ BioMedia Associates; **7.10:** Carolina Biological Supply Company/ Phototake

Chapter 8

8.4 & 8.5: Courtesy of Dr. Louis De Vos; **8.6:** © Carolina Biological Supply Company/Visuals Unlimited

Chapter 10

10.7: © Steve Gschmeissner/Getty; **10.8:** From J.E. Ubelaker, et al., *Journal of Parasitology*, 59:667–671, American Society of Parasitologists. Reprinted by permission of Dr. John Ubelaker

Chapter 11

11.2a: © Dr. William Weber/Visuals Unlimited; **11.4, 11.5, 11.7:** © Steve Miller; **11.9c:** © Ed Reschke; **11.10:** Carolina Biological Supply Company/ Phototake

Chapter 12

12.1a: © The McGraw-Hill Companies; **12.1b:** © Malcolm Storey, www.bioimages.org.uk; **12.5 & 12.6:** © Steve Miller; **12.7:** © John D. Cunningham/Visuals Unlimited; **12.8, 12.10a, 13.5:** Carolina Biological Supply Company/Phototake

Chapter 13

13.7: © Steve Miller

Chapter 14

14.1: © Michael DiSpezio; **14.5:** © Steve Miller

Chapter 15

15.1: Carolina Biological Supply Company/Phototake; **15.2:** © Steve Miller; **15.6:** © John Cunningham/Visuals Unlimited; **15.7:** Carolina Biological Supply Company/Phototake

Chapter 16

16.3a: © Runk/Schoenberger/Grant Heilman; **16.4a&b:** Carolina Biological Supply Company/Phototake

Chapter 17

17.1b: From Richard Kessel & Randy Kardon/Visuals Unlimited; **17.2, 17.3 & 17.6b:** Carolina Biological Supply Company/Phototake; **17.7:** © Dr. Michael Reedy, Duke University Medical Center; **17.9b:** Carolina Biological Supply Company/Phototake; **17.10b:** © Stuart Fox

Chapter 18

18.3: Carolina Biological Supply Company/Phototake; **18.4b:** Courtesy of J.D. Robertson; **18.5:** © John Cunningham/Visuals Unlimited; **18.6:** © EM Research Services, Newcastle University; **18.9, 18.10 & 18.11:** © Steve Miller

Chapter 19

19.1: © Susumu Nishinaga/Photo Researchers; **19.2:** D.M. Phillips/ Visuals Unlimited; **19.4, 19.7, 19.10, 19.11 & 19.13:** © Steve Miller

Chapter 20

20.2b: © Dr. G.M. Hughes; **20.3a&b:** Carolina Biological Supply Company/Phototake

Chapter 21

21.3a&b: From R.B. Chiasson, *Laboratory Anatomy of the White Rat*, 4th edition, 1980, McGraw-Hill Co., Inc.; **21.4a&b:** From Richard Kessel & Randy Kardon/Visuals Unlimited

Chapter 22

22.4: Carolina Biological Supply Company/Phototake; **22.5a&b:** From Richard Kessel & Randy Kardon/Visuals Unlimited

Chapter 23

23.3 & 23.4: © Steve Miller; **23.7:** From R.B. Chiasson, *Laboratory Anatomy of the White Rat*, 4th edition, 1980, McGraw-Hill Co., Inc.; **23.8:** © Steve Gschmeissner/SPL/Getty RF; **23.10a–c:** © Steve Miller